Data Privacy, Data Property, and Data Sharing

An Interdisciplinary Perspective for Post-Pandemic Transnational Scientific Research

Editors

Mariavittoria Catanzariti
University of Padua
Padua, Italy

Francesca Incardona
EuResist Network GEIE, Rome, Italy
InformaPRO s.r.l., Rome, Italy

Giorgio Resta
University of Roma Tre
Rome, Italy

Anders Sönnerborg
Karolinska Institutet
Stockholm, Sweden
Karolinska University Hospital
Stockholm, Sweden

CRC Press
Taylor & Francis Group
Boca Raton London New York

CRC Press is an imprint of the
Taylor & Francis Group, an **informa** business
A SCIENCE PUBLISHERS BOOK

Cover credit: Reworked from Henri Matisse, La danse (1909–1910), image in the public domain available at https://it.m.wikipedia.org/wiki/File:La_Danse_II,_par_Henri_Matisse.jpg

First edition published 2025
by CRC Press
2385 NW Executive Center Drive, Suite 320, Boca Raton FL 33431

and by CRC Press
4 Park Square, Milton Park, Abingdon, Oxon, OX14 4RN

CRC Press is an imprint of Taylor & Francis Group, LLC

Library of Congress Cataloging-in-Publication Data (applied for)

ISBN: 978-1-032-72756-1 (hbk)
ISBN: 978-1-032-75854-1 (pbk)
ISBN: 978-1-003-47592-7 (ebk)

DOI: 10.1201/9781003475927

Typeset in Times New Roman
by Prime Publishing Services

Preface

In the midst of the COVID-19 pandemic, the European Union (EU) issued an emergency call to scientists to urgently collect and analyze real-world data and derive information that could inform evidence-based decisions to fight the disease.

Major multinational research projects were launched in 2020 and 2021, including the EuCARE project (European Cohorts of Patients and Schools to Advance Response to Epidemics), an international effort led by the EuResist Network, which focuses on the acute and long-term phases of the disease, the immunological aspects of virus-host-vaccine interaction, and COVID-19 in schools. EuCARE launched three cohort studies and one randomized controlled trial, covering eleven countries in four continents.

From the outset, the ethical and legal issues related to the processing of sensitive personal data emerged as a critical aspect of the research. The project has a support infrastructure for ethics and data management, nevertheless, a wide and diverse range of unforeseen issues came to light, some of which hampered the research.

While there is no doubt on the need to protect confidentiality and personal control over data - in terms of deciding what happens to one's own data - and, more generally, the rights of people, particularly vulnerable people, it has become clear to us, especially after the COVID-19 pandemic, that there are still many concrete issues related to regulation that should be addressed to enable research practice while respecting both the individual and the public interest.

This urgent need led first to an interdisciplinary workshop on the subject, which took place in June 2023, and then to this book. The chapters are based precisely on the fundamental issues of data sharing, such as data protection, ownership and reuse, and the ethical approach based on altruism or public interest. They feature contributions from various experts and stakeholders.

One of the main aims of this book, is to promote an interdisciplinary dialogue between people involved in healthcare and biomedical research, on the one hand, and people involved in the regulation of such activities, on the other. Scientists, doctors and lawyers do not share the same perspectives and do not use the same language. This leads to potential misunderstandings, especially when decisions are taken autonomously and without prior consultation. This was evident during the EuCARE project: on several occasions, the mismatch between scientific jargon and legal categories came to the fore, leading to intense debate among the participants.

Let us take two examples. First, the nature of biological samples: doctors and researchers tend to consider them as materials that can be owned and freely transferred (between institutions and between countries); lawyers, on the other hand, take a different view, considering them as a source of personal data and therefore subject to the General Data Protection Regulation (GDPR) regime, which is very demanding both in terms of compliance obligations and data transfer restrictions.

The second example is the ambiguity of the notion of data ownership. While scientists tend to take a flexible and non-technical view of what can be owned, resulting in the concept of ownership being applied to a wide range of objects relevant to biomedical research (raw data, databases, biological materials), lawyers - at least those belonging to the civil law tradition - are much more selective, restricting the same concept to a narrow range of entities (having a physical character or falling within the traditional forms of intellectual property) and, in particular, rejecting the assumption that raw data can be the subject of exclusive rights that can be protected and transferred with erga omnes effects.

Apart from the difficult interplay between different fields of knowledge, each with its own perspectives and languages, the research activities carried out under the EuCARE project have highlighted objective weaknesses in the European regulatory framework for biomedical research. In Chapter 1, Incardona presents the "cahiers de doléances" resulting from the scientific community's efforts to share and reuse medical data for the benefit of the public within this large multinational project. Problems include: (i) overly strict consent requirements for processing of personal data for biomedical research purposes; (ii) insufficient harmonization of national laws on the processing of personal data for scientific research; (iii) obstacles to international data sharing for research purposes; (iv) lack of clarity on ownership of raw data.

Issues looked more serious when the studies involved minors. This aspect is relevant, as it may lead to a reduction in the number of studies involving minors and thus, certainly contrary to the will of the regulators, to a reduced or slower contribution of science to the benefit of children.

Compliance with an unclear or overly rigid regulatory framework can lead to friction or deadlocks in the conduct of research. Consequently, as widely recognized, research that could potentially benefit society as a whole may be hindered or discouraged if the above issues are not adequately addressed at national and transnational level. As argued in detail by Bincoletto in Chapter 19, this overriding concern is reinforced by the fact that the regulatory framework, including legal and ethical solutions, should be tailored to facilitate and support biomedical research, rather than slow it down. On this specific issue, Chapter 3, by Catanzariti highlights the most controversial and critical tensions between the public interest in scientific research and the individual control over health data, mostly related to the different meanings evoked by the right to health, and the stakes of governance and decision-making.

In fact, biomedical research in Europe has historically been regulated mainly at local and national level. From this point of view, the EU has played a marginal role, leaving the initiative mainly to the Council of Europe (who had little success, apart from the soft law dimension) and the Member States. Apart from the regulation of clinical trials and the production of pharmaceuticals, the field of biomedical research remains highly fragmented. The GDPR plays an important operational role. It is a remarkable piece of legislation, because, as Scorza points out in Chapter 4, it reinforces the fundamental right to the protection of personal data (Article 8 of the European Charter of Fundamental Rights) and, simultaneously, allows for a pan-European flow of data. Moreover, as explained by Kofod and Xiang in Chapter 12, the GDPR introduces specific exemptions for data processing aimed at scientific research. Indeed, the authors argue that the GDPR and the Chinese Personal Information Protection Law are based on different assumptions about the value and objectives of data protection legislation, with the former based on fundamental rights and the latter on the logic of control and maintaining social harmony and national security.

However, if the focus is not only on the black letter rule, but also on the actual implementation of the GDPR in national legal systems, two main shortcomings of the regulation become apparent: firstly, the GDPR focuses on the important but conceptually narrow perspective of data protection, leaving aside other equally important societal interests such as innovation, data solidarity and scientific progress. In practice, these interests tend to be sacrificed in favour of the value of data protection. Secondly, despite its form, the Regulation retains the veiled character of a directive in certain areas, which prevents the regulatory framework in Europe from being properly harmonised. The secondary use of data for scientific research is a perfect example. Article 9 (4) grants Member States the right to introduce further restrictions on the processing of genetic, biometric and health data, and various States, such as Italy, have made use of this derogation, thus strengthening the role of consent as a necessary condition for data processing. Article 89 (2), on the other hand, introduces the possibility of limiting the rights of the data subject in the case of data processing for scientific purposes. Member States have followed different paths. While northern countries, such as the Netherlands and Norway, have restricted the right to object, many other countries have not. As Resta explains in Chapter 18, this results in a fragmented legal framework, which increases transaction costs, given that the most important research projects are transnational and work with patient cohorts that are rarely confined to a single country.

Overall, this legal patchwork creates uncertainty and increases the risk of non-compliance. Moreover, funds and time that could be devoted to scientific research are diverted to less productive tasks, such as interacting with lawyers and ethics committees in several countries to find practical solutions that could withstand legal objections.

A striking example, although a positive one, that highlights the differences that exist within Europe, is Sweden. Whereas most European countries use informed consent as the legal basis to process personal data, Sweden can use the public interest of scientific research as a principle for lawfulness. This fact allows for rapid and extensive inclusion of study subjects from Sweden in research projects. Furthermore, Sweden boasts several nationwide or countywide data sources that are widely used by Swedish researchers and provide important opportunities. Study populations can be identified from the general population register using the unique national registration number and can be linked to other registers, such as the "patient register", which contains data on health care episodes in inpatient and outpatient specialist care. There are also several disease-specific registers. By linking these databases, scientific collaborations can generate a wealth of new knowledge and inform evidence-based policies.

In 2003 a register for HIV infection was established which currently includes > 99% of patients living in Sweden. This database has substantially contributed to international collaborations despite the very low prevalence of HIV in Sweden. Drawing from this experience, researchers at Karolinska Institute rapidly adapted the already-established infrastructure and network to efficiently collect COVID-19 data from the area of Stockholm already in Spring 2020. The Region Stockholm COVID-19 cohort follows all residents (2.4 million) through the epidemic. Sönnerborg highlights in Chapter 2 the enormous opportunities for research that such data structure provides. The InfCareHIV and EuCARE research projects demonstrate the impact of database-based research on clinical practice and health policy. By demonstrating innovative approaches to data linkage, integrating diverse data sources and translating findings into clinical practice, they show the immense potential of data-driven research to address complex public health challenges, for the ultimate good of the patient.

And we must not forget that the patient, as underlined by Norén (Chapter 5) is not only producing "data", but can provide a different and useful perspective. It is thus important to integrate patients' insights into medical practice and research. In the current knowledge gap, for instance, on the pathogenesis and treatment of the newly discovered long COVID, the patients offer invaluable real-time expertise, dedicated focus and robust support networks that can fill this gap.

Scientific research in health today is highly data-driven. The growth rate of electronic health data in healthcare is unprecedented. This is due to five major phenomena: the digitization of imaging; the digitization of medical records and health records; the explosion of the Internet of Things; the proliferation of e-health applications, biobanks and the development of sequencing technologies; and the collection and processing of data by public and private health institutions for public health. In this changing scenario, scientists often feel left out of the rules governing scientific research. This book reconsiders health data-related issues in the context of a multidisciplinary task that can be realized through cooperation between scientists and lawyers. In so far as scientists clarify to lawyers their needs in relation to the use of health data and the legal obstacles that hamper their daily activities, lawyers can better reflect on the applicability of legal frameworks that are more responsive to societal needs and can engage in dialogue with scientists through a process of joint reflection.

In addition, the increasing use of Artificial Intelligence (AI) plays an important role in research with big data, but also poses risks to privacy. Pfeifer, in Chapter 15, points out that privacy-preserving AI-based solutions are preferable options for research. In particular, he highlights how AI approaches require large data sets, but keeping health data in a central location or sharing data between multiple locations both pose privacy risks. Researchers have begun to build systems that can query data and train machine learning models on data stored in different locations. Different platforms, frameworks and cryptographic approaches are discussed in this chapter.

One of the side effects of the increasing digitization of healthcare and its research is the growing economic value of health data and health data collections.

The importance of this aspect is widely recognized by researchers, who seek private funding and count on institutional datasets as one of the key elements attracting the interest of the pharmaceutical industry and other sources of capital. The central question is whether data, and health data in particular, can be owned and, if so, by whom.

This book deals with different dimensions of data ownership, understood differently by lawyers and scientists.

This was for instance the concern of Chatzimichail, a member of the European students' organization OBESSU, who, in Chapter 11, discusses the importance of students' ownership of their data, and of this ownership to be conscious, in order to promote students' empowerment. He highlights the paramount importance of ethical considerations in ensuring that data is used to support students' learning and well-being.

The legal aspect of data ownership is addressed in three different chapters.

Chapter 9, by Xiong, deals with the regulation of health data under Chinese law. In particular, he highlights China's highly innovative approach to the allocation of data rights. While the EU Commission ultimately rejected the proposal to recognize exclusive property rights over raw data, China has recently established a coherent system based on the idea of data property. Xiong analyzes the implications of such an approach in the field of health data, discussing four main issues: the protection of first-order legal rights, the allocation of rights to jointly contributed data, the protection of data ownership in data processing services, and the exploitation of public data.

Chapter 7, written by Calzolaio, is symmetrical to that of Xiong: it also explores the fundamental issue of data ownership, but it focuses on the Western legal tradition, namely EU law. While China openly opted for the recognition of exclusive rights to data, the situation in Europe is less clear. Formally rejected by the Commission, the property logic has infiltrated data law. As shown by Calzolaio, the provision on non-personal data introduces a property-oriented right, thus challenging the non-proprietary stance mentioned above. When it comes to health data, however, the proprietary logic tends to recede, not only because the personal nature of such information brings the discourse of fundamental rights into play, but also because the societal interest in open science shifts the regulatory focus to the opposite dimension of data sharing.

Riccio, in Chapter 8, reflects on the distinction between data control under data protection law and data ownership. He clarifies that, under current European law, individual personal data cannot be the subject of property rights. However, it is possible to obtain exclusive rights over a collection of personal or non-personal data (including health data).

The logic of ownership can be in conflict with the logic of sharing, which should be one of the key features of publicly funded research. The legal, ethical and technical dimensions of data sharing are extensively explored in this book.

Rinck describes in Chapter 14 the history and solutions available for data sharing in Europe, providing a comprehensive and useful guide for researchers and consortia. She covers the online resources that support the scientific community to achieve FAIR data (Findable, Accessible, Interoperable, and Reusable), the available repositories for sharing different types of medical data with different solutions according to the GDPR, and data harmonization guidance for efficient data use.

In Chapter 21 Thorogood emphasizes the importance of international data sharing as an essential component of cross-border collaborations in health research and critically reflects on the various forms (mainly normative) of transfer restrictions that make such collaborations difficult. He argues that the protection of individual privacy must be balanced against the global interest in advancing society's understanding of human health and disease. Thorogood concludes that, in the interests of open science, the restrictions created by legal regimes such as the GDPR or the US Executive Order of 2024 should be removed, and the health research community should proactively pursue novel collaborative strategies at the transnational level.

Canziani et al. describe in Chapter 17 how the Cohort Coordination Board (CCB) was conceived and how it has evolved into a formal coordination mechanism funded by the European Commission, within a broader framework that supports a coordinated approach to research as a cornerstone of pandemic preparedness. The activities of the CCB in the area of data sharing and data harmonization are presented, including reflections on the merits of this type of network as a means of support to the sharing of health data in the context of pandemic preparedness.

The participatory dimension of biomedical research, which balances individuals′ rights to control the flow of health data with their participation in decisions about its use for scientific research, also seems particularly important.

This aspect is discussed transversally in four chapters with regard to specific examples of vulnerability, the interplay between law and ethics, the issue of transparency in health policy and, finally, a reconceptualization of health data governance.

In Chapter 16 Cascione examines how the relationship between older people and e-health can create legal vulnerabilities with regard to the processing of their health data. She notices that one of the consequences of the digital transition in public health is the creation of huge collections of health data, within which it is difficult to orient oneself; this may result in misinformation for elderly people. Starting from this consideration she proposes an elderly-oriented interpretation of the GDPR.

Mantelero explores in Chapter 6 the importance of a human-centred approach in scientific research, including the interplay between AI technologies and research ethics. By looking at the main technological components that underpin data-driven scientific research (data generation, data access and AI) the author focuses on some possible approaches for establishing expert committees, inspired by ethics committees, to advise AI providers and users in their research and development activities.

Curtin and Della Corte, in Chapter 20, investigate the issue of transparency in public decision-making during the pandemic, with a particular focus on the implementation of the European Commission's Vaccine Strategy. They examine the features of this strategy: on the one hand, the process of gradual disclosure that the European Medicines Agency undertook in approving the COVID-19 vaccines and, on the other, the lack of transparency in the Advanced Purchase Agreements and the Health Emergency Preparedness and Response Authority.

Karavas proposes in Chapter 10 an innovative reconceptualization of governance mechanisms for data flows that transcend the paradigms of privacy and property. He highlights how both these paradigms, which relate data to individual sources, do not address the disruptive effect of data aggregation. Against this background, the author reworks the notions of data solidarity and data trust within a broader dimension of democratization of the science system.

In Chapter 13, Dove examines data-related ethical issues in EU-funded research projects. Starting from the case study of the Framework Programmes for Research and Technological Development, the author describes the ethics review process in Horizon Europe and focuses on data-related ethical issues as well as the impact of AI on research. He demonstrates the close intertwining of the work of research funders and ethics experts with regard to the use of data for scientific research.

The perspectives opened up by the European Health Data Space Regulation are touched upon in several contributions, and namely in the chapters by Catanzariti (Chapter 3), Riccio (Chapter 8), Resta (Chapter 18), Bincoletto (Chapter 19), and Thorogood (Chapter 21). They all emphasize that, under the new regulation, the sharing and re-use of health data will be better supported by the European normative framework, perhaps leading to a more balanced relationship between the private value and the collective dimension of data.

Acknowledgements and Funding

The editors would like to extend their sincere gratitude to Maria Pia Fabiani, Chiara Mommo, Sara Roccu, Sara Serrao and Davide Sili for their invaluable support through the editing process.

This book was co-financed by the EuCARE project, a research project funded by the European Union's Horizon Europe Research and Innovation Programme under Grant Agreement No 10104601.

Finally, we extend our deepest appreciation to the Foundation Brocher, where the seminal EuCARE workshop on "Data privacy, data ownership and data sharing: an interdisciplinary dialogue for post-pandemic transnational research" was held in June 2023, as a source of inspiration for most of the contributions included in this volume.

Contents

SECTION 1

Scientific Research at the Crossroads of the Law and Ethics of Health Data

DOI: 10.1201/9781003475927-1

1

Emerging Issues in Transnational Biomedical Research

The Value of Patient Cohorts and Databases

Francesca Incardona

The initiative to bring medical and medical-data scientists together with law scientists in a workshop, and consequently in this book, was conceived within the EuCARE project, a research project on the SARS-COV-2 pandemic. The idea was born from a group of medical scientists and lawyers who collaborated on the project to address the numerous legal and ethical issues it raised. We, at EuCARE, thought that the in-field collaboration occurring in the project among scientists from the biomedical and the legal worlds was a great chance for mutual understanding that should be further pursued. We also thought that a rich and complex health research project like EuCARE was a kind of test bed for the European legal framework for the use of personal data in healthcare, a good starting point for discussion.

So, I will give a brief overview of the EuCARE project and then I will summarize some of the main legal and ethical issues that emerged during the project that required, and still require, the collaboration among medical researchers and lawyers.

1.1 The EuCARE Project

European Cohorts of Patients and Schools to Advance Response to Epidemics, in short EuCARE (www.eucareresearch.eu), is a research project, funded by the European Union's Horizon Europe Programme. It involves almost 30 partners from all over the world: not only partners in Europe and the UK, but also in Africa, Central and South America. It started in late 2021 and will last five years with a maximum funding of almost 10 million euro.

EuCARE is a cohort project encompassing patients, healthcare workers, and school cohorts. It also foresees horizontal activities and work packages designed to maximize the value of these cohorts (Fig. 1).

EuResist Network GEIE, Rome, Italy; InformaPRO s.r.l., Rome, Italy. Email: f.incardona@informa.pro

DOI: 10.1201/9781003475927-2

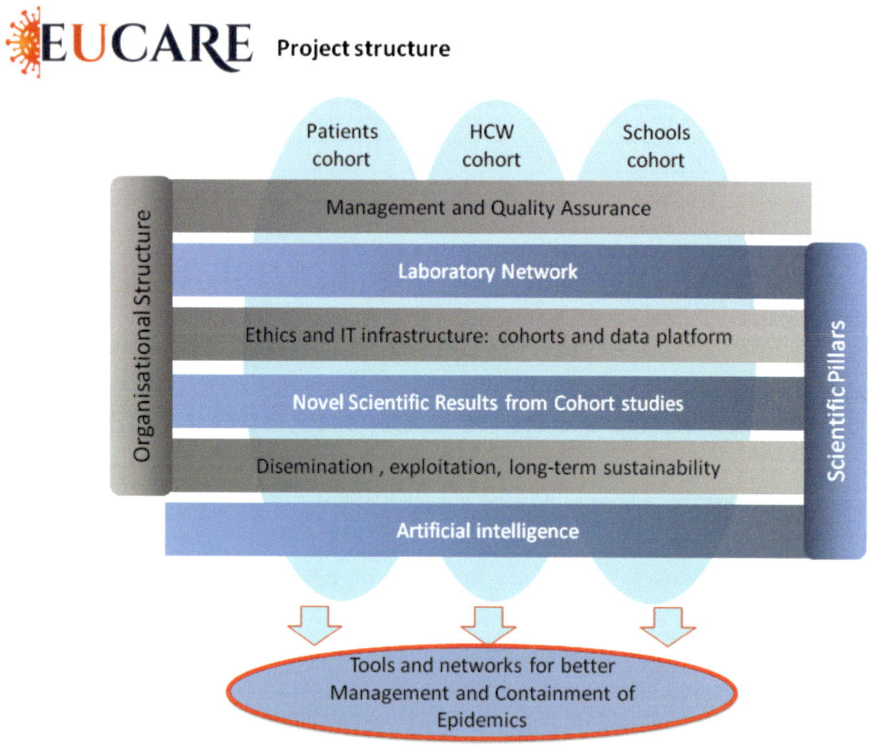

Figure 1: EuCARE project structure.

The project activities include a laboratory network, an artificial intelligence work package and organizational activities. All of these efforts contribute to the development of tools, knowledge and, hopefully, long-lasting networks for improved epidemic management.

The project's four main cohorts allow for multiple studies.

Healthcare workers are a population of healthy individuals undergoing periodic medical visits and vaccinations. As such, they are a crucial for the population to understand vaccine escape by the virus. Vaccine hesitancy studies have also been conducted within this group.

Hospitalized patients permitted us to study the severity of the disease induced by different SARS-COV-2 variants (Hedberg et al. 2023, Hedberg et al. 2024) and to look for determinants of severe COVID-19. Studies in specific fragile populations (ongoing) are of great relevance also after the acute, pandemic phase of the infection.

Patients suffering from long term Post COVID-19 Conditions (PCC or Long-COVID or PASC) who are enrolled at Long-COVID hospital centres allowed us to study the determinants, physiology and pathogenesis of this serious, widespread and still unknown condition (Varisco et al. 2023, Bai et al. 2024).

Finally, the SCHOOLS cohort. We included schools in the project because they represented a crucial point of vulnerability during the COVID-19 pandemic. Our goal is to support schools in their mission and to identify better response strategies for the future. To this end, we evaluated a pooled saliva screening strategy known as the "Lolli Methode" (Dewald et al. 2022) against the other COVID-19 containment measures adopted in schools.

We designed an international cluster Randomized Controlled Trial (RCT) and an observational study to compare regular screening using the Lolli Methode to the standard of care. These studies also aim to determine the prevalence and pathways of SARS-CoV-2 infections in schools, as well as to compare the effectiveness of different preventive measures and to investigate the psychological issues and learning loss associated with the containment measures (Raimondi et al. 2023).

These cohorts all contribute samples for advanced immunological studies conducted by the laboratory network. They also offered data to, or required help from, the Artificial Intelligence (AI) work package in multiple forms.

To give an idea of the complexity of the project, we have so far developed six study protocols that have been approved by multiple centres in ten different countries, and more studies are coming. We also established new collaborations: with the Italian National Learning Evaluation Agency (INVALSI), the University of Milan, the PTP lab Milan, CHLN Portugal, the Calamandrei Foundation, the ORCHESTRA cohort project, and the Cohort Coordination Board (CCB), an entity coordinating all the cohort projects funded by the European Commission. In some of these collaborations we established joint studies on the collected data, either based on data exchange or on AI solutions such as federated learning.

The IT and ethics infrastructure

As per the Information Technology (IT) infrastructure that we purposely built for the project (Fig. 2), data are collected from the centres in a pseudonymized fashion and are managed and stored centrally in Europe by the EuResist Network, in accordance with the GDPR. EuResist Network is a joint controller of the data together with the data providers.

In order to manage such a structure dealing with sensitive personal data, we set up an ethical and legal infrastructure coordinated by Prof. Giorgio Resta, our Regulatory and Ethics Manager, together with Prof. Mariavittoria Catanzariti. An external Ethics and Advisory Board periodically reviews and advises on the project activities; its members are Prof. Alessandro Mantelero and Prof. Claudia Morgana, for legal advice, Mr Petr Franc and Mrs Nadine Toye representing the students, and Dr Lisa Norén representing the patients. The project has a Data Protection Officer (DPO), Eng PhD Antonia Bezenchek, and a Data Management Board, and of course it interacts with the local Ethics Committees and DPOs at the contributing centres. It set up a number of agreements with the European Union, amongst the partners, as well as bilateral data transfer and material transfer agreements.

1.2 The Importance of Cohort Studies and of Retrospective Data

As I said, EuCARE is a cohort project, but what is a cohort?

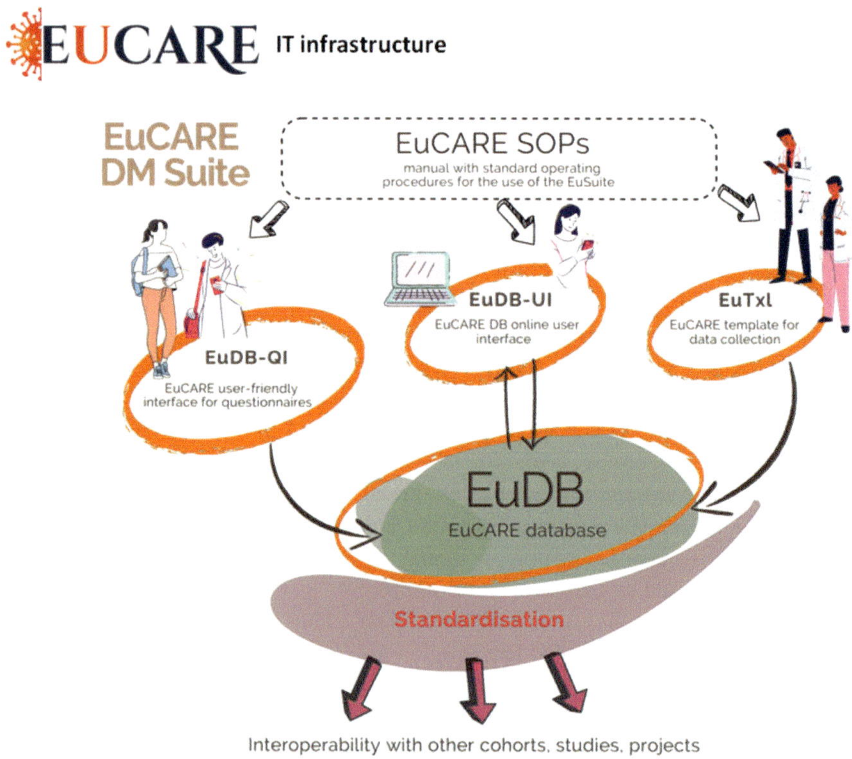

Figure 2: EuCARE IT infrastructure.

A cohort is a group of people who share some characteristics, which are the entry (and exit) characteristics of the cohort, and who are followed over time with periodic observations, thus providing longitudinal data and/or sample collections.

Cohorts are really useful for collecting scientific evidence.

In fact, the gold standard of evidence in medical research is represented by (double blinded) randomized controlled trials: studies in which the subjects are randomly assigned either to the group that undertakes the intervention or to the "control" group. In this way, if the numerosity of participants is big enough, the background differences among the subjects are evenly distributed among the groups and any difference in the results can be safely related to the sole intervention.

But, just below the randomized controlled trials in terms of evidence strength (see Fig. 3) we have the cohort studies. They are close to the top of the pyramid of evidence because they provide longitudinal patient-level information.

Particularly during a pandemic, cohort studies can be very precious and it is for this reason that the European Commission launched, during the COVID-19 pandemic, several emergency research calls devoted to cohort studies.

Why are cohorts so important? Cohort studies are excellent for observing how diseases develop and progress over time and are particularly valuable in conditions where interventions might be unfeasible or unethical. In our case, for instance, it would have been unethical to assign hospitalized patients to mechanical ventilation

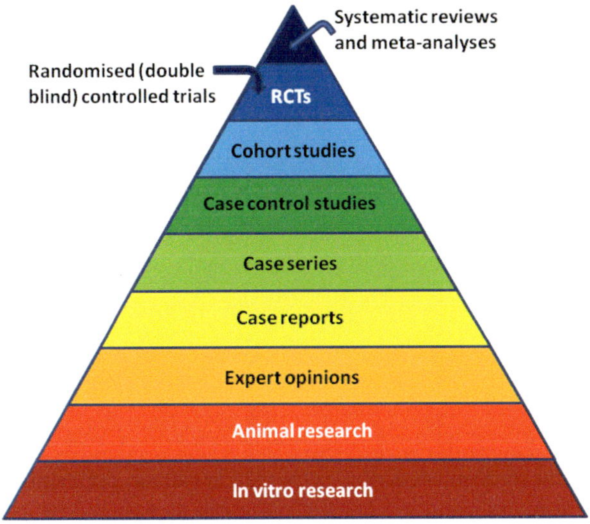

Figure 3: The pyramid of evidence-based medicine.

or not, just for study purposes. Cohort studies allow researchers to observe the effects of such exposures without actively manipulating them.

The conditions in cohort studies are often closer to those of the real world, compared to RCTs. For example, they may involve more diverse populations and settings, providing insights into how interventions or exposures work in real-world scenarios.

Also, unlike RCTs that typically focus on a single intervention, cohort studies can examine the impact of various factors on a range of health outcomes. This makes them useful for exploring complex relationships and identifying potential risk factors.

They are more agile and flexible: while data collection is ongoing, researchers can decide to gather information on new exposures or outcomes.

Also, they can be used for various research designs, including the classical longitudinal observational cohort studies, nested case-control studies, or as the basis to set-up RCTs, like in our SCHOOLS cohort, or non-randomized controlled trials, like in our Long-COVID cohort.

Data can be analyzed at any time to get a hint of the situation, without the fixed time constraint of an RCT.

Finally, one of the crucial advantages of cohort studies is that, with respect to RCTs, they are usually cheaper and easier to set-up, especially when leveraging existing infrastructure and data, allowing retrospective or retrospective/prospective studies.

Retrospective data

One example is the EuResist cohort (Rossetti et al. 2023). This is a long-standing project that began in 2006, collecting data on treatment resistance of people living with HIV and is now a very large dataset with more than 115,000 subjects, the largest of its kind in the world. These data have been used to train artificial intelligence models to predict the best treatment and to understand the landscape of resistance. he EuResist cohort is of particular relevance to study the emergence of resistance related to new treatments when they are introduced in healthcare.

In EuCARE, although we created our cohorts from scratch, we also merged data from ongoing cohorts and we have a strong contribution from retrospective data collections, which is incredibly valuable and constitutes one of the strengths of the project. However, these data would be of no value if there were no way to use them. If we had to go back and collect consent from 30,000 people, the data would be unusable.

1.3 Ethics Issues in EuCARE

Data reuse

The possibility to reuse, or further process, sensitive personal data originally collected for a different objective has thus emerged as one of the key issues that we encountered.

Not only was data reuse essential to build our cohorts and pursue our primary objectives, but once done this, it is also fundamental in order to use the same data to answer new, emerging questions.

To have the possibility of reusing data is always important but it is particularly relevant in the framework of a rapidly changing situation, such as the one we had to face with COVID-19.

In accordance with the GDPR, personal medical data are usually collected and processed based on consent for its use for a specific purpose. The GDPR also provides public interest as a legal basis for data processing and this possibility was used during the pandemic. However, already at the time of our project (study protocols submitted to the ethics committees in 2022) it was generally not considered viable anymore.

For further processing, the GDPR requires the new purposes to be compatible with the initial ones for which the data have been collected and specifies that scientific research purposes should, in general, not be considered incompatible.

In our experience, the interpretation of these principles varies greatly between European countries, with Sweden and Italy at opposite ends of the spectrum. In Sweden, reference to public interest and scientific research purposes are widely used (still respecting rigorous conditions) as a legal basis for data processing and further processing. In Italy, on the other hand, informed consent of the subject is, in practice, the only accepted basis for medical research. Also, the informed consent itself is highly scrutinized and there is a growing demand for its purposes to be more and more specific. This makes data reuse increasingly problematic: if we ask the subject for their consent to use their data to identify determinants of Long-COVID, can we also use it to identify determinants of long post-infection syndrome in general?

Furthermore, there are concerns that the informed consent given to a doctor may not be truly free. Thus we may find ourselves without a legal basis at all.

Respect for the free and autonomous will of individuals is indisputable, inalienable.

But what do patients think?

In our project, the association of Long-COVID patients states: "We want doctors to use our data because we want doctors to find treatments for us".

Why then not allow a general, revocable, authorization to use data for scientific research purposes?

It is also hardly understandable for me that a plethora of companies can claim legitimate interest as a legal basis for the use of personal data and, on this basis, intrude into our lives at any time, calling us by name on our personal cell phones against our helpless will, while hospitals can be in practice not entitled to legitimate interest for carrying out medical research.

Lack of regulation uniformity across Europe

As anticipated above, we have come up against the existence of different national interpretations of the GDPR and of different national regulations.

For interventional studies involving pharmaceuticals, the legislation in Europe has been recently unified and a centralized platform has been set-up for studies' ethics evaluations (the Clinical Trials Information System—CTIS). But, for non-pharmaceutical studies and for observational studies, the approval rules and pathways can be quite different from country to country and even from centre to centre in the same country.

EuCARE is, by nature, an international project involving several countries in southern, central and northern Europe as well as non-European countries from three continents (see Fig. 4). This is necessary for a project that aims to study a widespread disease, and not only for its epidemic/pandemic potential. Large numbers are necessary for statistical significance and good geographic distribution is essential to take into account variability of ethnicity, clinical protocols, vaccines and treatments available and, last but not least, of viral variants.

The fragmentation of regulation has led to several problems.

For instance, in Italy and Germany multicentre observational studies require a study coordination site, while this is not the case in Sweden. Our hospitalized patients' cohort was coordinated by the Karolinska Institutet, in Sweden, and it was not easy to firstly understand the issue and then to find a common version of the study protocol.

Requests from the ethical committees and their times for approval were different between centres, and the problem was greater for studies on minors (which we will see in more detail later). As a result, some studies started with significant delays in some centres or were even rejected in one centre and approved in others, putting the entire study at risk.

A recent editorial on The Lancet dedicated to the Italian health system (The Lancet RHE, 2025) reports the issue, along with Italy's restrictive ethical legislation (also described above) and the fragmentation of the health data management system, as a reason for extra spending, waste of resources and rise in inequalities, ultimately representing an obstacle to progress.

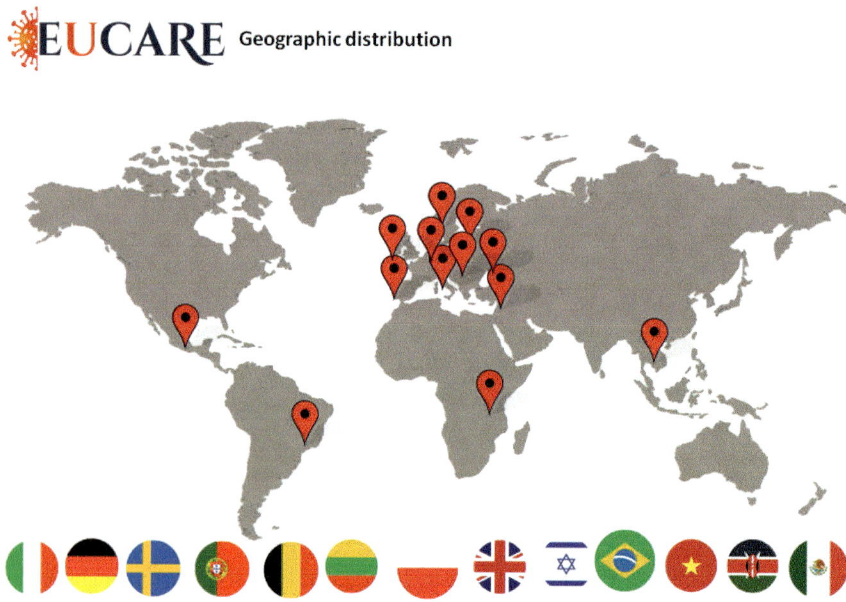

Figure 4: The geographic coverage of the EuCARE project.

International data sharing

The international nature of the EuCARE project also brings up the issue of international data sharing. We had to face this problem for instance with IBM, a partner in Israel, who is involved in modelling the data from the Long-COVID and the SCHOOLS cohorts with the support of AI techniques.

In fact, the European Commission issued an adequacy decision for Israel in 2011, stating that Israeli data protection laws offer a level of protection essentially equivalent to that of the EU, allowing personal data to flow freely between the EU and Israel. But this decision was made under the previous Data Protection Directive (95/46/EC). With the entry into application of the GDPR in May 2018, such adequacy decision remained in force but was subject to review. The European Commission concluded positively its review in January 2024, finding that Israel continues to provide an adequate level of protection for personal data transferred from the EU.

When our project started, the review was not completed and, to be on the safe side, IBM required to receive completely anonymized data from the EuCARE's database.

Complete anonymization is an asymptotic concept and a field of study. There are several proposed methodologies, some based on the introduction of noise in the data. We adopted data aggregation based on the analysis of their granularity.

In any case, data anonymization inherently brings loss of information and the result is that more effort is required to obtain lower (or similar) quality outcomes.

In order to cope with this issue and, more in general, with the data sharing issue, distributed learning techniques are being developed and refined to allow data processing on separated datasets, without data merging, as described in more detail elsewhere in this volume.

While we write, a study on Long-COVID adopting a distributed learning method is ongoing among EuCARE and the ORCHESTRA project, another European cohort project. It is an exciting and very promising study but we should not conceal that it requires a substantial effort for data pre-processing.

The EuCARE project also collects data from non-EU partners in Kenya, Mexico, Brazil. No European data are transferred to these countries, thus the data transfer is not subject to the GDPR. However, our External Ethics Board advised that, with reference to approvals of the ethics committees in these countries, minimum standards have to be met in order to comply with the European principles, especially in relation to patient enrolment.

Although this requirement prompted the question of whether it runs against the general principle of territoriality of the law, we decided to adopt minimum standards, based on Article 3 of Charter of Fundamental Rights of the European Union and on Article 29 of the Additional Protocol to the Oviedo Convention on Human Rights and Biomedicine.

Data sharing with international organizations

Further to international data sharing, there is the issue of data sharing with international organizations.

Since its beginning, EuCARE had an agreement with the World Health Organization (WHO) to share its hospitalized cohort data with them. If this data transfer had to find its legal basis in Article 46 of the GDPR (appropriate safeguards), it would have been a nightmare, requiring considerable efforts not only from the project but also from the WHO and probably causing delays that would have made the entire operation useless. Luckily, after considering that option, the EuCARE External Ethics Board advised to adopt the public interest, Article 49 (d), as the legal basis, and the transfer was made without further delays.

Studies involving minors

As briefly described at the beginning, EuCARE includes an observational study and a randomized controlled trial in schools that involve pupils from preschool to secondary schools.

The studies were conceived to sustain the school mission and to protect the children's rights.

In fact, during the COVID-19 pandemic, school closure was adopted as a widespread Non-Pharmacological Intervention (NPI), together with social distancing, wearing of masks, lessons in hybrid format with classes split online and at home.

In some European countries and for certain grades, schools were kept closed to in-person teaching for as long as two years.

This policy was adopted despite the fact that the COVID-19 disease spread and virulence by age soon showed less frequent infection and milder manifestations in children compared to adults (Viner et al. 2020, Bhopal et al. 2021, Zhu et al. 2020).

Furthermore, already in the early stages of the pandemic, several studies have shown that in Western countries (where appropriate mitigation measures are implemented) the prevalence of positive cases in schools is lower and the number and size of clusters is smaller than in the general population (Macartney et al. 2020, Otte im Kampe et al. 2020, Ismail et al. 2020, Gandini et al. 2021).

Importantly, school closure, supported or not by remote teaching, causes harm to the health of children (UNESCO 2020), psychological consequences (Ashikkali 2020, Loades 2020) and learning loss (Engzell 2021) with consequent long-term earning losses (Dorn 2020, Hanusheck-OECD 2020), that are higher for the socio-economically disadvantaged (Agostinelli 2020).

Thus, it appears that the school closure was even violating the third article of the Convention on the Rights of the Child: "In all actions concerning children, undertaken by public or private social welfare institutions, courts of law, administrative authorities or legislative bodies, the best interests of the child shall be a primary consideration".

In order to protect the health and the rights of the child, there was a great need for studies and solutions to support a safe opening of schools during the pandemic based on scientific evidence of harms and benefits, and this was the aim of the EuCARE SCHOOLS studies (see Fig. 5).

The involvement of minors in research studies is rightly subjected to rigorous ethical assessment and the EuCARE SCHOOL studies have received the most attention from the entire ethical infrastructure of the project.

The study protocol involved centres in Italy, Portugal, Mexico and Germany.

In Italy and Mexico the study was approved without problems in 2022 (one but not the other of the two Italian centres required a data transfer agreement with the sponsor). In Portugal, the approval required a double step, with conclusion in 2023 so the study had to be performed in the following school year compared to Italy. In Germany the study was rejected in early 2023 by the ethics committee of the University of Cologne.

Other minor and manageable discrepancies among the ethics committees were related to informed consent: whether the signature of both parents and the minor was required starting at a certain age, and at what age.

The delays in Portugal and the halt to the study in Germany had the paradoxical effect that the utmost (but fragmented) ethical attention to children's rights jeopardized a study designed to protect and enforce their very rights.

1.4 Acknowledgements and Disclaimer

The EuCARE project is funded by the European Union's Horizon Europe Research and Innovation Programme under Grant Agreement No 101046016.

Views and opinions expressed here are however those of the author only and do not necessarily reflect those of the European Union. Neither the European Union nor the granting authority can be held responsible for them.

Figure 5: The EuCARE SCHOOLS study in an Italian school.

References

Agostinelli, F., M. Doepke, G. Sorrenti and F. Zilibotti. 2020. When the Great Equalizer Shuts Down: Schools, Peers, and Parents in Pandemic Times. IZA – Institute of Labor Economics, Dic 2020.

Ashikkali, L., W. Carroll and C. Johnson. 2020. The indirect impact of COVID-19 on child health. Paediatrics and Child Health Journal, Sept 2020.

Bai, F., A. Santoro, P. Hedberg, A. Tavelli, S. De Benedittis, J. de Morais Caporali et al. 2024. EuCARE Project. The Omicron Variant Is Associated with a Reduced Risk of the Post COVID-19 Condition and Its Main Phenotypes Compared to the Wild-Type Virus: Results from the EuCARE-POSTCOVID-19 Study. Viruses. 2024 Sep 23; 16(9): 1500. doi: 10.3390/v16091500.

Bhopal, S.S., J. Bagariac, B. Olabib and R. Bhopal. 2021. Children and young people remain at low risk of COVID-19 mortality. The Lancet Child & Adolescent Health, Volume 5, Issue 5, Published March 10, 2021 - DOI: 10.1016/S2352-4642(21)00066-3.

Dewald, F., I. Suárez, R. Johnen, J. Grossbach, R. Moran-Tovar, G. Steger et al. 2022. Effective high-throughput RT-qPCR screening for SARS-CoV-2 infections in children. Nat Commun. 2022 Jun 25; 13(1): 3640. doi: 10.1038/s41467-022-30664-2.

Dorn, E., B. Hancock, J. Sarakatsannis and E. Viruleg. 2020. COVID-19 and student learning in the United States: The hurt could last a lifetime. McKinsey, Jun 2020.

Engzell, P., A. Frey, MD Verhagen. 2021. Learning loss due to school closures during the COVID-19 pandemic. Proc Natl Acad Sci U S A. Apr 27;118(17):e2022376118. doi: 10.1073/pnas.2022376118.

Gandini, S., M. Rainisio, M.L. Iannuzzo, F. Bellerba, F. Cecconi and L. Scorrano. 2021. A cross-sectional and prospective cohort study of the role of schools in the SARS-CoV-2 second wave in Italy. Lancet Reg. Health Eur. 5: 100092. doi: 10.1016/j.lanepe.2021.100092.

Hanushek, E.A. and L. Woessmann. 2020. Economic impact of learning losses. OECD/OCSE Sett 2020.

Hedberg, P., M. Parczewski, K. Serwin, G.C. Marchetti, F. Bai, B. Jensen et al. 2024. Twenty-Eight-Day In-Hospital Mortality During the Wild-type, Alpha, Delta, and Omicron SARS-CoV-2 Waves: A Multinational Cohort Study in the EuCARE project. Lancet Reg Health Eur. 2024 Feb. 2; 38: 100855. doi: 10.1016/j.lanepe.2024.100855.

Hedberg, P., B. Varisco, F. Bai, A. Sönnerborg, P. Naucler, N. Pfeifer et al. 2023. EuCARE-hospitalised study protocol: a cohort study of patients hospitalised with COVID-19 in the EuCARE project. BMC Infect Dis. 2023 Oct. 16; 23(1): 690. doi: 10.1186/s12879-023-08658-2.

Ismail, S.A., V. Saliba, J.L. Bernal, M.E. Ramsay and S.N. Ladhani. 2021. SARS-CoV-2 infection and transmission in educational settings: a prospective, cross-sectional analysis of infection clusters and outbreaks in England. The Lancet Infectious Diseases Volume 21, Issue 3p344-353March 2021. DOI: 10.1016/S1473-3099(20)30882-3. Published December 8, 2020

Loades, M.E., E. Chatburn, N. Higson-Sweeney, S. Reynolds et al. 2020. Rapid Systematic Review: The Impact of Social Isolation and Loneliness on the Mental Health of Children and Adolescents in the Context of COVID-19. J. Am. Acad. Child Adolesc Psychiatry. 2020 Nov; 59(11): 1218–1239.e3. doi: 10.1016/j.jaac.2020.05.009. Epub 2020 Jun 3.

Macartney, K., H.E. Quinn, A.J. Pillsbury, A. Koirala, L. Deng, N. Winkler et al. 2020. Transmission of SARS-CoV-2 in Australian educational settings: a prospective cohort study. The Lancet Child & Adolescent Health, Volume 4, Issue 11, 807 - 816, DOI: 10.1016/S2352-4642(20)30251-0. Published August 3, 2020.

Otte im Kampe, E., A.S. Lehfeld, S. Buda, U. Buchholz and W. Haas. 2020. Surveillance of COVID-19 school outbreaks, Germany, March to August 2020. Euro Surveill. 2020; 25(38): pii=2001645. https://doi.org/10.2807/1560-7917.ES.2020.25.38.2001645. Accepted: Sept 23, 2020.

Raimondi, S., S. Gandini, G.H. Rubio Quintanares, A. Abecasis, P.L. Lopalco, O. D'Ecclesiis et al. 2023. European Cohorts of patients and schools to Advance Response to Epidemics (EuCARE): a cluster randomised interventional and observational study protocol to investigate the relationship between schools and SARS-CoV-2 infection. BMC Infect Dis. 2023 Jan 3;23(1):1. doi: 10.1186/s12879-022-07947-6. Erratum in: BMC Infect Dis. 2023 Feb 16;23(1):96. doi: 10.1186/s12879-023-08065-7. PMID: 36597074; PMCID: PMC9808677.

Rossetti, B., F. Incardona, G. Di Teodoro, C. Mommo, F. Saladini, R. Kaiser et al. 2023. EuResist Network. Cohort Profile: A European Multidisciplinary Network for the Fight against HIV Drug Resistance (EuResist Network). Trop Med Infect Dis. 2023 Apr 23; 8(5): 243. doi: 10.3390/tropicalmed8050243. PMID: 37235291; PMCID: PMC10222321.

The Lancet RHE Editorial, The Italian health data system is broken. Lancet Reg. Health Eur. 2025;48;101206, January 2025, doi: 10.1016/j.lanepe.2024.101206.

UNESCO's COVID-19 Education Response - Adverse consequences of school closures https://en.unesco.org/covid19/educationresponse/consequences.

Varisco, B., F. Bai, S. De Benedittis, A. Tavelli, A. Cozzi-Lepri, M. Sala et al. 2023. EuCARE-POSTCOVID Study: a multicentre cohort study on long-term post-COVID-19 manifestations. BMC Infect Dis. 2023 Oct 13; 23(1):684. doi: 10.1186/s12879-023-08595-0. PMID: 37833640; PMCID: PMC10576381.

Viner, R.M., O.T. Mytton, C. Bonell, G.J. Melendez-Torres, J. Ward, L. Hudson et al. 2021. Susceptibility to SARS-CoV-2 Infection Among Children and Adolescents Compared With Adults—A Systematic Review and Meta-analysis. JAMA Pediatr. 2021; 175(2): 143–156. doi:10.1001/jamapediatrics.2020.4573 September 25, 2020.

Zhu, Y., C.J. Bloxham, K.D. Hulme, J.E. Sinclair, Z.W.M. Tong, L.E. Steele et al. 2021. A Meta-analysis on the Role of Children in Severe Acute Respiratory Syndrome Coronavirus 2 in Household Transmission Clusters. Clinical Infectious Diseases, Volume 72, Issue 12, 15 June 2021, Pages e1146–e1153, https://doi.org/10.1093/cid/ciaa1825 Published: December 06, 2020.

2

Opportunities for Research with Data

Anders Sönnerborg

2.1 Background

Infections, such as the Human Immunodeficiency Virus (HIV), which causes AIDS, and SARS-CoV-2, which causes COVID-19, do not know any borders. These viruses have rapidly spread around the globe resulting in devastating pandemics. National and international research collaborations using big data is the key to successfully combating such pandemics. At the national level, Sweden has several data resources, both at disease-specific level and at the generalized level. These include databases on HIV and COVID-19, which can be adapted and used to efficiently contribute to the counteracting of pandemics through international collaborations.

2.2 Swedish Database Resources

Several different Swedish data sources are available, creating important opportunities for research.[1,2] Study populations can be identified from the Total Population Register using the unique national registration number received at the time of birth or immigration into Sweden, thereby linking individuals' data across different registers. For specific diseases, the health registers are nationwide. This is the case, for example, for the National HIV register[3] (Carlander et al. 2023), launched in 2003, and for the National Cancer register, launched in 1958. Swedish law dictates that all new HIV diagnoses and all cancer cases are to be reported, which occurs in virtually all cases. For COVID-19, the National COVID-19 register[4] was launched in 2022

Karolinska Institutet, Stockholm, Sweden; Karolinska University Hospital, Stockholm, Sweden.
Email: anders.sonnerborg@ki.se

[1] https://snd.se
[2] https://www.registerforskning.se
[3] https://www.infcarehiv.se
[4] https://covid19register.se

DOI: 10.1201/9781003475927-3

and, as of October 2024, covered around one third of diagnosed patients in Sweden, as well as all diagnosed patients in the region of Stockholm. Another register which can be used is the Patient Register, maintained by the Swedish National Board of Health and Welfare, which comprises data on health care episodes in inpatient (hospital) and outpatient specialist care. The Swedish National Board of Health and Welfare also maintains the Cause of Death Register, which is a virtually complete, high quality register of all deaths in Sweden since 1952. The Longitudinal Integration Database for Health Insurance and Labour Market Studies (LISA) is maintained by Statistics Sweden and includes all people from 16 years of age onwards, who were registered in Sweden on the 31st of December each year. Data is available from 1990 onwards and the completeness is high, for example, 95% of data on occupation and > 98% of data on education are complete. Data can be extracted based on the level of income, civil status, main source of income, total income, and days of sick leave. The Prescribed Drug Register contains data on prescribed medications, allowing to identify and assess medication-based comorbidities and treatments.

By linking these databases, interprofessional and intersectoral collaborations can generate new knowledge that bridges fields such as epidemiology, clinical medicine, virology, infectious diseases, molecular biology and public health, ultimately informing evidence-based policies.

2.3 Opportunities for HIV Research: The Swedish InfCareHIV Database

It is estimated that there are about 40 million persons living with HIV (PWH) in the world. In Sweden, there were 8,593 diagnosed living patients, as of December 2024. The total number of patients in the database were 13,743, which includes patients who migrated or who are deceased, corresponding to around 80% of all patients ever diagnosed in Sweden. This means that Sweden has one of the lowest national prevalences of HIV in the world, reaching only 0.08%. The question to ask is: how were Swedish researchers so successful in contributing to the global international research efforts against HIV with such a small number of patients at the national level?

The answer lies in the establishment of a health IT software, InfCareHIV,[5] which is a clinical decision support system used at each PWH visit, employed at all Swedish HIV clinics. The system collects HIV data generated in the clinic, which is automatically transferred to a pseudonymized research database, once a day. InfCareHIV is equipped with a long-distance electronic consultation tool to be used by consultants from university clinics, allowing them to discuss pseudonymized patients with less experienced physicians at "smaller" clinics. This further increases the value and impact on national HIV healthcare.

As the system is based on clinical usefulness, physicians at all HIV clinics accepted to be part of the National Swedish HIV cohort and its research database. Thereby, Swedish clinical/translational researchers have contributed substantially to a vast number of European HIV research projects, including the EuResist project,[6]

[5] https://www.infcarehiv.se

[6] https://www.euresist.org

and to several global research projects, despite the very low prevalence of HIV in the country. On top of clinically generated data, the InfCareHIV database also includes laboratory data generated in routine diagnostics and data generated through laboratory research. This interactivity is further facilitated by the connection of a biobank to the research database.

The InfCareHIV database includes all categories of patients who are otherwise underrepresented in research, especially in randomized clinical trials. By focusing on high-risk populations and stratifying the analyses by factors such as country of birth, age and gender identity, this register-based research can be used to develop targeted public health interventions, and to promote health equity. This data is also accessible online, making it easier for doctors, who do not always have much time to dedicate to research, to participate in research activities.

As described above, in addition to the specific InfCareHIV data, Swedish researchers have access to several other healthcare registers, which are run by the Board of Health and Welfare in Sweden and Statistics Sweden. There are also databases for health insurance and labour market studies. This Swedish health data space allows many possibilities to use healthcare digital information for HIV research (and other clinical research) in Sweden with or without combining retrospectively and prospectively collected clinical information, as well as laboratory information obtained in routine care or research.

2.4 Challenges and Results for HIV Research

When InfCareHIV was established in 2003, several challenges had to be overcome. Physicians and healthcare personnel were reluctant, fearing increased workload and worrying of the risks for personal integrity. However, after a short time period using the system in clinical practice, the reluctance waned. Another great hurdle was the initial diffidence of the hospital IT-departments' staff who saw the introduction of InfCareHIV as an intrusion in their area of interest. An important step was to involve the patient-organization HIV-Sweden at an early stage. A patient-representative from HIV-Sweden became an adjunct member of the InfCareHIV steering committee. The InfCareHIV system is based on an opt-out approach and participation is discussed with every patient once a year. Each research project using the InfCareHIV database is approved by the steering committee of the national InfCareHIV cohort and the Swedish Ethical Review Authority.

The development, implementation, and use of the InfCareHIV system resulted in several important achievements in clinical care and research. When initiating the implementation of InfCareHIV, pronounced health inequities existed in Sweden, depending on whether the patients were followed by experienced HIV physicians at larger clinics or not. The use of the InfCareHIV system and its data in HIV clinical care and in integrated translational research resulted in great improvements, eventually eliminating these large health inequities.

In 2014, Sweden was the first country to reach the UNAIDS/WHO 90/90/90 goal of HIV care targets (Gisslén et al. 2016), which stipulated that 90% of all people living with HIV in a country should be identified, of which 90% should be connected to clinical care, and of which 90% should have reached the treatment goal of undetectable plasma HIV-viremia. In 2021, UNAIDS/WHO changed their goal to 95/95/95. Sweden reached this new demanding goal in 2017 and published the results in 2024, being the first country to attain this, through a solid documentation (Lundgren et al. 2024). Another important consequence of this success is that reaching undetectable viremia renders the patients non-contagious during sexual intercourse.

By using InfCareHIV, Swedish translational/clinical researchers have been involved in, and have substantially contributed to many international research projects, despite the very low prevalence of HIV in the country, including participation in almost all large European HIV cohorts financed by EU, and in several collaborations with low to middle income countries. The high-quality data and the rapid evaluation of requests for participation in clinical studies have also been attractive for the pharmaceutical industry.

2.5 Opportunities for COVID-19 Research—the Region Stockholm COVID-19 Cohort

In Sweden, the first case of SARS-CoV-2 infection, causing COVID-19, was diagnosed on the 31st of January 2020. Shortly thereafter, clinical/translational researchers at Karolinska University Hospital/Karolinska Institutet, Stockholm, Sweden, implemented an already-established structure to efficiently collect clinical health and laboratory COVID-19 data from the Stockholm area. The key to this rapid action was the existing network that was established before the COVID-19 pandemic. The delayed responses to the H1N1 swine flu pandemic, occurring in 2009, warned the Karolinska researchers of the need of such networks to handle future epidemics, as well as for pandemic competitiveness.

The Region Stockholm COVID-19 Cohort follows all residents (2.4 million inhabitants) through the epidemic (225,399 COVID-19 cases, as of the 25th of April 2021). In this population-based study, residents are characterized by health administrative databases including all primary care, high resolution clinical data for hospitalized COVID-19 patients, microbiological data from hospital electronic health record systems and laboratories, intensive care unit data from the Swedish Intensive Care Register and data from the Swedish Vaccination Register at Public Health Sweden.

2.6 The Multinational EuCARE Collaboration

When the COVID-19 outbreak broke out in early 2020, there was an urgent need for multinational collaborations, and the EuCARE consortium was established in 2021. At the initial phases of this collaboration, the Karolinska researchers contributed with around 20,000 hospitalized patients, increasing to around 35,000 patients by 2023 (Hedberg et al. 2024) due to the preexisting network at Region Stockholm and the established infrastructure for collecting data.

The importance of having a multinational collaboration is obvious: there are large international differences in terms of outcomes and characteristics of the COVID-19 pandemic, including number and type of deaths according to different socioeconomic or income groupings across the world. This highlights the need to consider data from multiple sites, ethnic groups, socioeconomic groups, etc., as done in the EuCARE project.

An important opportunity of the EuCARE project is that the multinational collaboration increases the external validity, that is to say the generalizability to other countries than those represented in the consortium, and also includes sites that are generally underrepresented. Furthermore, having the same inclusion and exclusion criteria for all patients considered in the study, provides the possibility to reach a higher internal validity in terms of stronger statistical power and possibility to rule out systematic errors within the study. Importantly, the collaborative efforts give researchers a chance to introduce different competencies and different expertise into the project, a key for the EuCARE building foundations and networks of great potential also beyond the actual research project.

In such a large multipartner project including countries from three continents, several challenges exist. Firstly, extensive data collection is needed in many of the global COVID-19 projects. For example, the COVID-19 data portal from the WHO includes a quite extensive number of variables, and the challenge referred to as data collection fatigue is obvious. At Karolinska, there is the opportunity to have digital health record systems that facilitate the collection of data, but this is not the case in many settings where data needs to be collected manually. At Karolinska, it would have been an impossible task to go through 20,000 patients' health records and curate 50 different variables.

2.7 Interdisciplinary Research in EuCARE

The research environment of EuCARE brings together experts from diverse fields, such as infectious diseases, epidemiology, virology, molecular biology, artificial intelligence, and social science. This interdisciplinary collaboration enables the EuCARE consortium to tackle the complex challenge of COVID research in a way that would be impossible for individual researchers working separately. This synergy allows to combine large-scale epidemiological data with clinical insights and molecular analyses and leverage the international collaboration within EuCARE to enhance the scope and impact of the findings of individual partners. For example, the ability to link registered clinical and epidemiological data on COVID-19 and then correlate them with molecular biomarkers, such as the SARS-CoV-2 variants and their genetic sequences, is only possible through combined expertise. The research environment in EuCARE addresses several key scientific questions that require the integration of multiple national registers, clinical data, and advanced laboratory analyses. By linking diverse data sources, the power of database-based research is demonstrated to address complex health challenges that span multiple medical disciplines and health system levels. This approach shows how database-based research can be enriched and validated through direct clinical observations and advanced laboratory analyses.

2.8 Differences in Ethical Principles and Rules

The large EuCARE project on COVID-19 involved extensive use of sensitive personal and medical data from multiple cohorts in several countries from three continents. There are significant differences in ethical principles and rules between the participating countries and sometimes even between different universities of the same country. Informed consent is a great challenge. In Sweden, the researchers may not need to ask for informed consent for large data-base studies, if the Swedish Ethical Review authority has given such a permission for a given project. Patients are informed about the register and have the possibility to opt-out. They are also informed that data may be used for research given that ethical permission has been granted. As regards to registers that are held by Statistics Sweden and the National Board of Social Affairs and Health, the use of these registers for research is governed by the Patient Data Act and the Public Access to Information and Secrecy Act which allow for the use of health data for research purposes without individual consent under certain conditions and, again, given that ethical permission has been granted. Obtaining individual consent from each person would be impractical and could potentially introduce bias into the research if only a subset of individuals have consented.

The differences in ethical principles and rules across different settings and countries, which leads to quite long lead times, does not harmonize well with something so ever-changing as the COVID-19 pandemic. Also, differences in data sources, data availability, languages, units used in the clinical measurements, and how the data are harmonized and translated into one common language are another challenge. The changing pandemic is always present and has a major impact on the understanding of risk factors including a potential need to update established protocols early on during the pandemic, where the principles and rules between countries again may differ.

Despite these challenges related to ethical principles and rules, heterogeneous data sources and the rapid phases of research, the collaborations across countries and continents must and will continue to be integral to understanding emerging infectious diseases, increasing external and internal validity and offering lessons to be learned from for future, similar situations.

2.9 Translating Register-Based Findings into Clinical Practice

The InfCareHIV and EuCARE translational projects showcase the practical impact of database-centred research on clinical practice and health policy. These projects not only utilize database-centred research methods, but also significantly contribute to their advancement. By demonstrating innovative approaches to data linkage, integration of diverse data sources, and translation of findings into clinical practice, the immense potential of data-centred research is showcased to address complex public health challenges. The InfCareHIV and EuCARE projects serve as a model for how such research can evolve to meet the demands of modern medical science and public health practice.

2.10 Acknowledgements and Disclaimer

The EuCARE project is funded by the European Union's Horizon Europe Research and Innovation Programme under Grant Agreement No 101046016.

Views and opinions expressed here are however those of the author only and do not necessarily reflect those of the European Union. Neither the European Union nor the granting authority can be held responsible for them.

References

Carlander, C., J. Brännström, F. Månsson, O. Elvstam, P. Albinsson, S. Blom et al. 2023. Cohort profile: InfCareHIV, a prospective registry-based cohort study of people with diagnosed HIV in Sweden. BMJ Open. 2023 Mar 17; 13(3):e069688. doi: 10.1136/bmjopen-2022-069688.

Gisslén, M., V. Svedhem, L. Lindborg, L. Flamholc, H. Norrgren, S. Wendahl et al. 2016. The UNAIDS/WHO 90-90-90 continuum of HIV care targets achieved in Sweden. HIV Medicine, 2016 Aug 18. doi: 10.1111/hiv.12431. [Epub ahead of print].

Hedberg. P., M. Parczewski, K. Serwin, G.C. Marchetti, F. Bai, B. Jensenet et al. 2024. Twenty-Eight-Day In-Hospital Mortality During the Wild-type, Alpha, Delta, and Omicron SARS-CoV-2 Waves: A Multinational Cohort Study in the EuCARE Project. The Lancet Regional Health – Europe. Lancet Regional Health – Europe.

Lundgren, E., M. Locke, E. Romero-Severson, M. Dimitrijevic, M. Axelsson, E. Andersson et al. 2024. Sweden surpasses the UNAIDS 95-95-95 Target: Estimating HIV-1 incidence between 2003-2022. Euro Surveill. 2024 Oct; 29(42): 2400058. doi: 10.2807/1560-7917.ES.2024.29.42.2400058.

3

What Does Health Data Stand for Scientists and Lawyers?

Mariavittoria Catanzariti

3.1 Introduction

Big data applications in biomedical sectors such as bioinformatics, clinical informatics, digital imaging and predictive public health are increasingly fostering scientific research and predictive capabilities for management of diseases. Data storage and retrieval as well as data analysis represent some of the most used informatics applications, while infectious disease surveillance and population health management are at the core of public policies. In addition, the use of raw health data implies their conversion into a more readable format, such as a graph, report or chart, either manually or through an automated tool. All these tasks are based on data sharing and involve the action of different actors with different roles. Healthcare is nowadays entirely dependent on data collection and processing by public and private health institutions. This chapter will investigate some controversial issues related to scientific freedom of data sharing and to the right to science that lie at the crossroads of law and ethics.

3.2 Health Data Sharing For Scientific Research

Data sharing has various implications for healthcare and scientific research, either if we consider data extracted from clinical trials or patients' data. As reported by Recital 19 of the Regulation EU 2025/327 on the European Health Data Space (hereinafter, EHDS),[1] "timely and full access of health professionals to the medical records of patients is fundamental for ensuring continuity of care, avoiding duplications and errors, and reducing costs [...] Providing health professionals access services is a task in the public interest". The relation between patients and health professionals

---

University of Padua. Email: Mariavittoria.catanzariti@unipd.it

[1] Regulation (EU) 2025/327 of the European Parliament and of the Council of 11 February 2025 on the European Health Data Space and amending Directive 2011/24/EU and Regulation (EU) 2024/2847.

DOI: 10.1201/9781003475927-4

is generally related to the primary use of data and it is legally subject also to ethical principles.[2] However, in the logic of health data sharing, primary use normally deployed for healthcare and secondary use related to the access of researchers and innovators to health data need to be connected, in light with the general aims of the EHDS.

This chapter tries to explore the way by which scientific research is nowadays stuck between law and ethics, and how this can affect data usage. The underlying argument of the chapter is that the aim of legislation in this field should be the regulation of the production of knowledge, whereas the role of ethics committees that authorize or deny a certain pathway for a clinical trial, for instance, should only be limited to identify unethical scientific practices not undermining the scientific results. The fact that both may have effects on scientific results may create confusion in terms of clarity and blurring of purposes (Aurucci 2023). A more confusing factor is also the fact that ethical requirements in most cases overlap with the basic requirements of data protection law. Ethics is indeed increasingly overlapping with data protection law, but there are certainly other relevant factors that have an ethical relevance, such as the empowerment of minors in giving their consent to actively participate in the production of scientific outcomes. Another relevant issue is that in the design of clinical trials one must take care how to convince the patients to participate. In Europe, in fact, one is not allowed to be subject to trials, but in other countries this is possible. The rationale of the ban is that individuals cannot be paid because this of course would imply that only poor people would participate in dangerous research. These examples provide a number of situations that are highly controversial for scientists to be assessed and managed.

From the perspective of the law, the collection, analysis, and timely sharing of health data is a key component of transnational scientific research that should be as far as possible universal in the sense of non-local. For lawyers, it is intuitive that legal requirements are different from ethical precepts. However, scientists are often misled in taking care of scientific research while trying to be compliant with ethical and legal rules (McNabb 2019). For scientists, legal and ethical rules are ultimately rules to be respected, without much distinction (Viberg Johansson et al. 2022). The legislative formulation is sometimes not clear enough, for example if we look at the Regulation on clinical trials, Article 4 sets out that clinical trials shall be subject to scientific and ethical review and shall be authorized in accordance with it. Such ambiguity informs the way by which rules are perceived by scientists. Furthermore, the regulation of health data is split into different sectoral silos that scientists should confront with to be able to conduct their research and use results, without knowing exactly what the difference between legal and ethical precepts is, except for the awareness of written legal provisions. However, for scientists, standards of compliance involve both ethical and legal obligations, but the blurring of requirements may largely puzzle scientific freedom.

[2] For example, the compliance with the European ethical principles for digital health adopted by the eHealth Network on 26 January 2022.

If we look at the European framework of research on health-data we can realize that the substantive piece of legislation to face ethical and legal issues in the collection of health data for scientific purposes is the General Data Protection Regulation (GDPR), which addresses health data as a special category of personal data. We can add to the picture the European Union (EU) pharmaceutical legislation known as the consolidated version Clinical Trials Regulation (Reg EU 536/2014) that entered into application on 31 January 2022 repealing the Clinical Trials Directive (EC) No. 2001/20/EC. It aims to ensure the EU offers an attractive and favourable environment for carrying out clinical research on a large scale, with high standards of public transparency and safety for participants to clinical trials. The Clinical Trial Regulation has institutionalized the Clinical Trials Information System (CTIS), a single-entry point for submission of data and information relating to clinical trials via the EU Portal.

At the European level, we do not have anything else in terms of binding legislation, except for the EHDS, and this creates some issues of reference for scientists. What does a scientist need to have in mind to access and use health data? What is the interest at stake? Is scientific research instrumental to the right to health? The way by which scientists look at this is basically affecting maybe two sets of problems. First of all, the scientific freedom of producing knowledge, secondly, the conditions of data usage in so far as the collection and finally sharing of health data represents a key component of the unprecedented global research environment.

There are three main problems due to the lack of clarity:

(i) The need to unpack the idea of health data ownership. Most researchers, pharmaceutical corporations, innovators, public institutions tend to regard the data that are collected and processed for research (as well as the research results) as objects of property. The EHDS fosters data sharing between doctors and digital health authorities for public interest without the consent of these patients. This would severely compromise trust between patients and their doctor as well as individual participation in public decision making. Different actors dealing with health data – public bodies, doctors, researchers, individuals - unreflectively use the lexicon of "ownership" and structure their legal relationships accordingly. This may be misleading, because ownership in many legal systems is not applicable to intangibles such as data, and data cannot be the object of exclusive rights, except in the hypotheses recognized by the law (such as database protection, copyright or patent). Moreover, extending beyond reasonable limits property rights in data, risks creating substantial obstacles to data sharing for scientific research and data analytics for AI applications as well as concerns in terms of surveillance and securitization. This is why it is urgent to clarify the main differences between 'data ownership' and 'data control', identifying the specific hypotheses in which samples and data may truly be regarded as object of rights, and when, by contrast, they should be regarded either as commons or as goods held in trust.

(ii) The need to identify the conditions for a proper balance between data protection and data sharing in the field of transnational biomedical research and public management. The current EU regulatory framework suffers from two main

of Fundamental Rights, besides the general prescription on the right to respect for physical and mental integrity, states important limitations that inform the logic of sectoral regulation of health data sharing, being also rationales of the EHDS: the free and informed consent of the person concerned, according to the procedures laid down by the law; the prohibition on making the human body and its parts as such a source of financial gain, the prohibition of the reproductive cloning of human beings. It is, however, important to acknowledge that the blurring of issues involved in the public management and regulatory framework of health requires the re-conceptualization of the right to health in the context of a multidimensional perspective. Not only individuals, but also communities are increasingly involved in the management of huge amounts of data. The group perspective as well as the societal impact of health data flows – as strategic assets for states, relevant information for social safety, examples of common trends in epidemiology – often shows the inadequacy of a model centred on the individual consent.

If this may appear obvious for people who perceive health-related matters as a public service, it is not completely adherent to a universal model of rights protection that fosters the evolution of humanity.

The General Comment n. 25 of the Committee on Social, Economic and Cultural Rights on science and economic, social and cultural rights has clearly identified science as part of cultural life that encompasses the right of every person to the development of a critical mind (so called "citizen science"). It provides a definition of science as a public participatory good where the collective and social dimension reflects the idea of common responsibility: "the enterprise whereby humankind, acting individually or in small or large groups, makes an organized attempt, by means of the objective study of observed phenomena and its validation through sharing of findings and data and through peer review, to discover and master the chain of casualties, relations or interactions; brings together in a coordinated form subsystems of knowledge by means of systematic reflection and conceptualization; and thereby furnishes itself with the opportunity of using, to its own advantage, understanding of the processes and phenomena occurring in nature and society".

3.4 Health Data as Private Life or Public Interest?

We do not have time in this chapter to examine the reasons and the implications of the evolution of the right to science, to what extent it can be explained and developed also in light of data sharing, but the short reference to the international protection of such a right sheds light on the ambiguity and complexity that the right to health may imply.

It is not sufficiently clear, in fact, if we can look at health data sharing as a branch of the right to health or of the right to science, or if we should shrink its meaning to the individual control of health data flows and the potential detrimental use by third parties. If this is true from the legal perspective that works with differentiated concepts with very precise meanings and purposes, it is a bit puzzling for non-lawyers. Confronting legal concepts, we can see how health data protection can only be regarded as a procedural step with respect to more relevant values, the right to health (the right to access better treatment through the use of data) and the right to science (the right to take part in the scientific development and be beneficiary of scientific results).

This means that the individual protection of health data as a safeguard of personal data protection should not be a dogma that disadvantages other human rights or that hinders the public interest in the effectiveness of science but should be necessarily read in light of the above.

For the purposes of this chapter, the question is precisely if and how health data usage may be configured as a part of the right to health, or if instead the only possibility for individuals to enjoy legal safeguards is related to the control of health data flows. This aspect crisscrossed another issue, why health data are identified as the most intimate part of ourselves and whether the right to personal data may exhaust all the possibilities that health data usage imply for humankind as a whole, in its contextual, relational and overlapping dimensions of individuals, scientists, scientific, communities and creators of the public opinion.

Health data sharing appears, under safe conditions, to be the way by which the right to health, the right to science and the right to personal data protection can be reconciled and be interdependent, not excluding.

To fill various gaps, not only between law and ethics, e.g., legal provisions and ethical committees' authorizations, but also between the perception of scientists of their scientific freedom and individual control on personal data flows, the European Union has developed a regulatory model, which is based on different normative spaces for data sharing. These normative spaces of data sharing will be captured by sectoral silos, where data can be used by trustworthy users and under safe conditions of fair sharing and transparency, delegating the responsibility for granting access to electronic health data for secondary use to public bodies designated by Member States (health data access bodies). Secondary use of electronic health data means the processing of electronic health data for purposes of scientific research, public interest, health statistics. The data used may include personal electronic health data initially collected in the context of primary use, which means but also electronic health data collected for the purpose of the secondary use itself.

One of these normative spaces is the EHDS which will become a legal basis for the secondary use of data as an additional opportunity for scientists. However, quite some skepticism emerges from the legal architecture surrounding the secondary use of data for public interest within which one of the key purposes is obviously scientific research. This piece of legislation is important as scientists situate themselves precisely as secondary users of a special category of personal data which is health data and that as a matter of principle should not be processed unless a legal basis (provided by a taxonomy) is provided.

The approach that has been followed by the EU is the so called "intervention with medium intensity", that aims at strengthening the rights of natural persons to control their data flow and provide an EU framework for the secondary use of electronic health data. National authorities will be responsible to implement policies and establish national bodies for primary and secondary use of electronic health data while at the EU level two digital infrastructures will be used to support cross border sharing and secondary use of electronic health data. Implementation would be supported by a mandatory certification for EHR systems and a voluntary label for wellness applications, thus ensuring transparency for authorities, procurers and users. A mandatory certification for Electronic Health Records will be required while a voluntary label for wellness applications will be encouraged in order to have a cascading effect on medical devices to be made interoperable with Electronic Health Records.

It is useful to shortly compare the two legal regimes.

Under the GDPR a triple legal authorization is required for international data sharing: a legal basis for processing personal data under Article 6, a legitimation of processing special categories of data (e.g., health and genetic data) under Article 9, and eventually a basis for data transfers outside the EEA under Articles 44-49. Even after the pandemic, research still requires a legal basis for the processing of sensitive data, additional justification for its processing, and a basis for any transfer of data outside Europe. The possibility of transfers outside the EU is only theoretical where important grounds of public interest, for reasons of public health are met (Recitals 111 and 112 GDPR).

Health and genetic data are considered to be "special categories" of data. Article 9(1) of the GDPR prohibits processing of special categories of data by default. Research institutions therefore not only need a legal basis but also an additional legitimation under Article 9(2) to process health and genetic data that can be of four different types: necessary processing for reasons of substantial public interest, on the basis of the Union or Member State law in accordance with fundamental rights protection; necessary processing for the purposes of preventive or occupational medicine, for the assessment of the working capacity of the employee, medical diagnosis, the provision of health or social care or treatment or the management of health or social care systems and services on the basis of the Union or Member State law or pursuant to contract with a health professional; necessary processing for reasons of public interest in the area of public health, such as protecting against serious cross-border threats to health or ensuring high standards of quality and safety of health care and of medicinal products or medical devices, on the basis of the Union or Member State law with respect to individual rights and liberties, including professional secrecy; necessary processing for archiving purposes in the public interest, scientific or historical research purposes or statistical purposes based on Union or Member State law in accordance with fundamental rights protection and the conditions required by Article 89 GDPR implying specific safeguards consisting in technical and organizational measures in order to ensure respect for the principle of data minimization, derogations to Articles 15 (right to access), 16 (right to rectification), 18 (right to restriction of processing) and 21 (right to object) The right to erasure shall not apply in the case of processing for reasons of public interest in the area of public health and archiving for scientific research.

According to Recital 159 GDPR processing of personal data for scientific research purposes should be interpreted in a broad manner including for example technological development and demonstration, fundamental research, applied research and privately funded research, including studies conducted in the public interest in the area of public health.

Under the EHDS the secondary use of health data is instead limited to the EU and should be compliant to the GDPR.

There are purposes for which the EHDS permits a secondary use that are compatible with the GDPR and somehow overlapping: activities for reasons of public interest in the area of public and occupational health, such as protection against serious cross-border threats to health, public health surveillance or ensuring high levels of quality and safety of healthcare and of medicinal products or medical devices; scientific research related to health or care sectors; AI systems and digital health applications used for public health or social security.

In this complex scenario, the GDPR still represents the bastion of individual safeguards, or an elephant in the room: it depends on the perspectives. What is not clear to scientists is to what extent GDPR could be an opportunity for research rather than a limitation. What is instead crystal clear that GDPR exerts a sort of traction towards all the issues: but GDPR is not a panacea.

In practical terms, if Recital 33 GDPR states that "...data subjects should be allowed to give their consent to certain areas of scientific research when in keeping with recognized ethical standards for scientific research", we know that legally speaking consent can be withdrawn anytime, and this may inhibit research. The problem is that the relevance of individual consent for scientists and lawyers is a different matter beyond its rationale. We all agree that infectious disease or public health laws should provide the necessary legal basis as a task in the public interest. Where infectious disease legislation does not authorize a public research institution or a university to process personal data for pandemic research, they may still be able to rely on their research mission given to them by the law. Some countries, such as Finland with the Act on the Secondary Use of Health and Social Data and Norway with the Personal Data Act, have specified in their national legislation that the public interest legal basis may be relied upon where the processing is necessary for scientific purposes.

3.5 Concluding Remarks

When we draw pictures of data sharing, we are talking about something very technical from the perspective of lawyers, and nowadays the Data Governance Act (DGA)[3] defines data sharing as the provision by a data holder of data to a data user for the purpose of joint or individual use of the shared data, based on voluntary agreements, directly or through an intermediary. The envisaged possibility to choose the public interest, instead of legitimate interests or consent, is really at the core of scientists' expectations, but still not of individuals' expectations. This recent legislative piece introduces the logic of data re-use as applicable to data sharing, defined as the use by natural or legal persons of data held by public sector bodies, for commercial or non-commercial purposes other than the initial purpose within the public task for which the data were produced, except for the exchange of data between public sector bodies purely in pursuit of their public tasks. The concept of data altruism is also introduced with reference to "the voluntary sharing of data on the basis of the consent of data

[3] Regulation (EU) 2022/868 OF THE EUROPEAN PARLIAMENT AND OF THE COUNCIL of 30 May 2022 on European data governance and amending Regulation (EU) 2018/1724 (Data Governance Act).

subjects to process personal data pertaining to them, or permissions of data holders to allow the use of their non-personal data without seeking or receiving a reward that goes beyond compensation related to the costs that they incur where they make their data available for objectives of general interest as provided for in national law, where applicable, such as healthcare" (Article 2(16) DGA) and upon conferral by recognized data altruism organizations.

There are of course different choices in this sense, and I think that the inspiration provided by data sharing could help in the sense of the obligation to make data available based on exceptional needs after the experience of pandemics. This was probably something that has emerged as a need. In fact, upon request, the Data Act[4] includes the provision pursuant to which a data holder shall make the data available to a public sector body or an agency or body demonstrating the exceptional needs, that can be assessed in response to a public emergency or to prevent a public emergency or to assist the recovery from a public emergency.

The demonstration of the exceptional needs could be the response to a public emergency.

All these issues compel us to reflect upon the sense of bona fide research, intended as discovering new knowledge for the general interest in health and to be made publicly accessible without undue delay, in the sense that of course what scientists should do after taking care of their task is not only a matter of compliance with the legal frameworks that are of course binding, but it is also a sort of qualification in terms of responsiveness towards future generations and what is intended to be the current and prospective meaning of public interest. While in order to apply the legal basis of public interest for health data processing, we need to identify under the GDPR an applicable national or EU law, in order to apply the legal basis of the legitimate interest we do not need to identify an applicable national or EU law.

Finally, in the Data Act we have an odd provision enshrined in Article 27 on international transfers stating that providers of data processing services shall take all reasonable technical, legal and organizational measures including contractual agreements in order to prevent international transfers or governmental access to non-personal data. We can explain this trend in light of a mimesis towards the restrictive trends about personal data transfers. And this makes me connect also with a trend of an interesting discourse today that started, of course, from the market of knowledge and lies in the hands of scientists. Scientists should be granted scientific freedom in conducting scientific research and they must comply with legal and ethical provisions, so that they create some knowledge. Such a knowledge should be, in a sense, publicly accessible under the conditions for which some disclosures could be made or not. At the end, if we have different layers of knowledge – the scientific, the legal, the ethical layer – it is hard to distinguish at different institutional and political layers and with multiple stakeholders. The scientific community first of all uses data in order to be compliant with a legal framework that ensures them to produce scientific research and should communicate scientific results in a sort of

---

[4] Regulation (EU) 2023/2854 of the European Parliament and of the Council of 13 December 2023 on harmonised rules on fair access to and use of data and amending Regulation (EU) 2017/2394 and Directive (EU) 2020/1828 (Data Act).

political oriented manner. This makes us wonder how the market of knowledge can be modified or also altered by the use of the law and also by other interdisciplinary reflections.

The creation of health data then needs to be considered as a creation of value which is much stronger as it is based on legal safeguards in order to preserve science and its authority as well as scientific freedom. However, what is avoidable is the application of proprietary paradigms to health data under the veil of data protection. Individuals should be aware of what scientists do with their data but cannot risk overusing their legitimate right to control data flows hampering scientific research as well as their own participatory right to scientific results. This, of course, has nothing to do with the "primary use" of data, that is "the processing of personal electronic health data for the provision of health services to assess, maintain or restore the state of health of the natural person to whom that data relates, including the prescription, dispensation and provision of medicinal products and medical devices, as well as for relevant social security, administrative or reimbursement services" (EHDS, Article 2, 2 (d) Definitions). The potential obstacle for data sharing is in fact a psychological extension of the logic surrounding the primary use of data to the secondary use of data, which is completely different. On the contrary, a strong protection of the primary use can represent the most suitable protection for a safe secondary use.

In this respect, this chapter highlighted how the potential of a legal and ethical model of data sharing for purposes of scientific research can be relevant for a responsible development of a culture based on public interest.

References

Aurucci, P. 2023. Il trattamento dei dati personali nella ricerca-biomedica. Problematiche etico-giuridiche, EDS.

Becker, R., D. Chokoshvili and E.S. Dove. 2024. Legal bases for effective secondary use of health and genetic data in the EU: Time for new legislative solutions to better harmonize data for cross-border sharing? International Data Privacy Law, 14(3): 223–246.

Besson, S. 2024. The 'Human Right to Science' *qua* right to participate in science, The International Journal of Human Rights, 28: 497–528.

Brooks, E. and R. Geyer. 2020. The development of EU health policy and the Covid-19pandemic: trends and implications. Journal of European Integration, 42: 8, 1057–1076.

Brooks, E., A. de Ruijter, S.L. Greer and Rozenblum, S. 2023. EU health policy in the aftermath of COVID-19: neofunctionalism and crisis-driven integration. Journal of European Public Policy, 30:4,721–739.

De Montalvo Jaaskelainen, F. 2021. Ethical and legal requirements for biomedical research involving health data in the context of the Covid-19pandemic: is informed consent still playing the leading role? Biolaw Journal, 2: 157–167.

Di Federico, G. 2024. From Dusk Till Dawn. The Case of F. T. v D. W. and the Right to Access Electronic Medical Records in Light of the Future European Health Data Space Regulation, European Papers, 9: 463–477.

Dove, E.S. 2024. The European Health Data Space as a Case Study, Ethics & Human Research, pp. 29–35.

Dove E.S. and J. Chen. 2024. Should consent for data processing be privileged in health research? A comparative legal analysis, International Data Privacy Law, 2020, 10(2): 117–131.

Ferrera, M. 2023. The European Union and cross-national solidarity: safeguarding 'togetherness' in hardtimes. Review of Social Economy, 81(1): 105–129.

Kuiper, E. and D. Brady. 2023. Is the European Health Union ready for the challenges of the21st century? European Policy Centre Discussion Paper, 12 September 2023, https://www.epc.eu/content/PDF/2023/EU_Health_DP_-_DB_and_EK.pdf).

Lilyanova V. 2023. Health-related measures in the national recovery and resilience plans. European Parliamentary Research Service Briefing, September2023; https://www.europarl.europa.eu/RegData/etudes/BRIE/2023/753932/EPRS_BRI(2023)753932_EN.pdf.

Robert, K. 1973. Merton, The normative structure of science. In The Sociology of Science. Theoretical and Empirical Investigations, The University of Chicago Press, pp. 267–278.

Romeo-Casabona, C.M. 2021. The New European Legal Framework on Personal Data Protection and the Legal Status of Biological Samples and Biobanks for Biomedical Research Purposes in Spanish Law. *In*: Slokenberga, S., O. Tzortzatou and J. Reichel (eds.). GDPR and Biobanking. Law, Governance and Technology Series, vol 43. Springer.

Sokol, T. 2024. European Health Data Space, Use of Data and Data Subjects' Control over Their Own Health Data: Can an Opt-Out Restore the Balance? European Journal of Health Law, 31(4): 365–388.

Thomas, J. and S. McNabb. 2019. Principles and ethics of collecting and managing health Data. *In*: Macfarlane, S. and C. AbouZahr (eds.). The Palgrave Handbook of Global Health Data Methods for Policy and Practice. Palgrave Macmillan, London.

Viberg Johansson, J., H.B. Bentzen and D. Mascalzoni. 2022. What ethical approaches are used by scientists when sharing health data? An interview study. BMC Med. Ethics 23, 41.

4

An Overview of GDPR Obligations and Biomedical Research

Guido Scorza

4.1 The Misunderstanding of the Antagonism Between Fundamental Rights

"More technology, less rights, more health", "health first", "without innovation there is no future".[1]

These expressions are so widespread that they now, especially after the dark months of the pandemic, appear true to most.

Yet they are not and, indeed, they conceal a huge error of perspective because they suggest that there may - or, perhaps, even must - be an antagonism between fundamental rights, that there may be or, indeed, must be "tyrant rights",[2] capable of engulfing other rights and capable of preventing the affirmation and exercise of other rights.

And, in most cases, especially in recent years with the progressive affirmation of the data society and the consequent progressive conquest by data - particularly personal data - of a position of centrality in our lives, in our economies and in our democracies, the right pointed at as a tyrant is, more and more frequently, the right to the protection of personal data.

Italian Data Protection Authority, Rome, Italy. Email: g.scorza@gpdp.it

[1] These expressions reflect the growing discourse that often frames technological progress and innovation as inherently conflicting with the protection of fundamental rights, particularly privacy. See, e.g., Resta, G. 2022. Pubblico, privato e collettivo nel sistema europeo di governo dei dati. Rivista trimestrale diritto pubblico. 4: 971.

[2] The idea that certain rights can dominate others in a hierarchical manner has been criticized as a misunderstanding of the nature of fundamental rights, which are intended to be balanced and not in competition. See De Hert, P. and C. Cocito. 2024. *The Added Value of Data Protection within the Framework of Digital Constitutionalism in Europe. In*: De Gregorio, G., O. Pollicino and P. Valcke (eds.). The Oxford Handbook of Digital Constitutionalism, Oxford.

DOI: 10.1201/9781003475927-5

The fault of privacy if doing medical research in Europe is so difficult, all the fault of privacy if doing innovation in Italy is so difficult, and, of course, all the fault of privacy if digitally transforming the national health system is a titanic undertaking.

These are commonplaces as widespread as they are fallacious.

But how are things really or, at least, how should they be?

The starting point is as simple as it is, at least apparently, unknown to most: under the umbrella of our Constitution as well as that of the Charter of Fundamental Rights of the European Union, there are simply no tyrant rights, they cannot and should not exist.

The watchword, among ordered equal rights, is or, at least, should always be "balancing".

And the balancing algorithm[3] is disarmingly simple at least in its enunciation because then, of course, applying it to individual cases can be more difficult: each right can and must be compressed to the minimum extent necessary to guarantee the assertion and exercise of another right.

In short, to put it more simply: the right to privacy must make itself smaller to the minimum extent necessary to allow the right to health to expand as much as it needs to ensure its assertion and exercise, and so, reciprocally, must the latter vis-à-vis the former, and so must both vis-à-vis the right to innovation.

Because, in case the circumstance escapes anyone: the right to privacy, the right to health and the right to innovation, to be understood as a specification of the right to do business, are all fundamental rights.

No antagonism, no rivalry - if only apparent - but, above all, no need to renounce one in order to dispose of the other, not in a democracy, at least.

It is or should be the most sacred of all rights of a citizen and, indeed, of a person: the right not to have to choose between two or more fundamental rights.

And it is or should be the first of the duties of any good government: never ask people to choose between several fundamental rights.

But is it really possible to guarantee this right?

At least in relation to the relationship between the protection of personal data and the right to medical research,[4] which is clearly an essential element of the right to health, the answer seems to be that it can and should be affirmative.

4.2 The Provisions of the GDPR Confirming the Possibility of Balancing

In an interdisciplinary context such as the one in which this paper is being written, it is probably unnecessary to start at the beginning, from the title of GDPR, what is generally referred to as the General Data Protection Regulation.

---

[3] The balancing of rights, particularly in the context of privacy and health, has been discussed extensively in legal theory and practice. See C. Wiese Svanberg. 2020. *"sub § 89". In*: Kuner, C. et al. [eds.]. The EU General Data Protection Regulation, OUP, Oxford. p. 1240.

[4] This issue is central to the ongoing debate regarding data protection in healthcare and the legal frameworks designed to balance data protection with the need for medical research. See Becker, R. and D. Chokoshvili. 2024. Legal bases for effective secondary use of health and genetic data in the EU. Int'l Data Privacy Law 14: 225.

The full title of the Regulation in question is actually more detailed: Regulation (EU) 2016/679 of the European Parliament and of the Council of 27 April 2016 on the protection of individuals with regard to the processing of personal data and on the free movement of such data and repealing Directive 95/46/EC.

Not only, therefore, a personal data protection regulation but also a regulation on the free movement of these data.

This is an important key to understanding the rules in question, which is too often underestimated: if we look carefully at them, the choice of the European Union legislator to adopt a regulation on the subject was not simply dictated by the desire to increase the level of protection of personal data, but also by the conviction that a uniform regulation on the protection of personal data could guarantee an easier and freer circulation[5] of such data within the European Union and, under certain conditions, also outside the territory of the Union.

Not only, therefore, does the title of the GDPR already suggest that the rules contained therein should not and cannot be considered antagonistic to forms of circulation, use or even lawful exploitation of personal data, but, on the contrary, it suggests that promoting such activities with personal data is precisely one of the purposes pursued by the Regulation.

And, on the other hand, in keeping with the title of the meeting that inspired this publication, it is worth considering that a plastic demonstration of the above was recorded precisely during the pandemic when the European Union, managed to design and implement, a sort of "health passport", i.e., the *green pass*, in a very limited time frame, thus allowing the free circulation of extraordinarily sensitive health data.

This was a result made possible precisely by the existence of the GDPR and, therefore, of uniform rules on the protection of personal data, rules that, therefore, if read and interpreted correctly, not only cannot and must not become antagonistic to the exploitation of personal data but, on the contrary, can be a driving force behind it.

In this perspective, it now seems appropriate, before going on with the reasoning, to recall briefly although not exhaustively because of the needs of synthesis in this intervention – the main provisions contained in the GDPR supporting the thesis of the existence of an important and accessible space for a constructive balance between the right to the protection of personal data and the right to health in the declination of the right to medical research.

Among these provisions, Recital 4 of the Regulation deserves, first of all, to be remembered: "The processing of personal data should be designed to serve mankind. The right to the protection of personal data is not an absolute right; it must be considered in relation to its function in society and be balanced against other fundamental rights, in accordance with the principle of proportionality".

---

[5] A key feature of the GDPR is the idea that data protection and data flow are not inherently at odds, but complementary. This view is consistent with the EU's strategy for a Digital Single Market, which aims to enhance the free flow of data while ensuring robust protection for personal information. See European Commission, Shaping Europe's Digital Future. 2020. Available at: https://ec.europa.eu/digital-strategy.

But to remain with the recitals of the Regulation and look more specifically at the balancing[6] issues at issue in this paper, no less relevant is, among others, Recital 33 according to which: "It is often not possible to fully identify the purpose of personal data processing for scientific research purposes at the time of data collection. Therefore, data subjects should be allowed to give their consent to certain areas of scientific research when in keeping with recognized ethical standards for scientific research. Data subjects should have the opportunity to give their consent only to certain areas of research or parts of research projects to the extent allowed by the intended purpose".

Revolving *from the recitals* to the articles, however, it is easy to identify many different indications symptomatic of the effort made by the EU legislator to create the conditions for an effective balance[7] between the right to personal data protection and the right to health also in the form of medical research.

I am considering, for instance, points (b) and (e) of Article 5, which provide for significant limitations of the general rules, inter alia, precisely in relation to treatments instrumental to research:

"(b) collected for specified, explicit and legitimate purposes and not further processed in a manner that is incompatible with those purposes; further processing for archiving purposes in the public interest, scientific or historical research purposes or statistical purposes shall, in accordance with Article 89(1), not be considered to be incompatible with the initial purposes ("purpose limitation");

(e) kept in a form which permits identification of data subjects for no longer than is necessary for the purposes for which the personal data are processed; personal data may be stored for longer periods insofar as the personal data will be processed solely for archiving purposes in public interest, scientific or historical research purposes or statistical purposes in accordance with Article 89 (1) subject to implementation of the appropriate technical and organizational measures required by this Regulation in order to safeguard the rights and freedoms of the data subject ("storage limitation")".

In my view Article 9, which expressly derogates from the prohibition to process special data if: "processing is necessary for archiving purposes in the public interest, scientific or historical research purposes or statistical purposes in accordance with Article 89(1) based on Union or Member State law which shall be proportionate to the aim pursued, respect the essence of the right to data protection and provide for suitable and specific measures to safeguard the fundamental rights and the interests of the data subject".

---

[6] The balancing of fundamental rights under the GDPR is a key principle, articulated both in the recitals and the articles of the Regulation. This principle has been widely discussed in the academic literature on data protection law. See De Hert, P. and C. Cocito. 2024. *The Added Value of Data Protection within the Framework of Digital Constitutionalism in Europe*. *In*: De Gregorio, G., O. Pollicino and P. Valcke [eds.]. The Oxford Handbook of Digital Constitutionalism, Oxford.

[7] The balancing of data protection with the right to health is explicitly addressed in several provisions of the GDPR, particularly Articles 6 and 9, which allow for exceptions to data processing for health-related purposes under specific conditions. See Art. 6 and 9, Regulation (EU) 2016/679, Official Journal of the European Union, L 119/1, 4 May 2016.

Since it is not possible here to analytically remember all the provisions of the GDPR that are relevant for the purposes of the reasoning that is being carried out here, it is worth limiting ourselves to referring to the reading of Articles 14, 17 and 21 and quoting below, because it is relevant with specific reference to the subject under discussion, Article 89, pursuant to which:

"1. Processing for archiving purposes in the public interest, scientific or historical research purposes or statistical purposes, shall be subject to appropriate safeguards, in accordance with this Regulation, for the rights and freedoms of the data subject. Those safeguards shall ensure that technical and organizational measures are in place in particular in order to ensure respect for the principle of data minimization. Those measures may include pseudonymization provided that those purposes can be fulfilled in that manner. Where those purposes can be fulfilled by further processing which does not permit or no longer permits the identification of data subjects, those purposes shall be fulfilled in that manner.

2. Where personal data are processed for scientific or historical research purposes or statistical purposes, the Union or Member State law may provide for derogations from the rights referred to in Articles 15, 16, 18 and 21 subject to the conditions and safeguards referred to in paragraph 1 of this Article in so far as such rights are likely to render impossible or seriously impair the achievement of the specific purposes, and such derogations are necessary for the fulfillment of those purposes.

3. Where personal data are processed for archiving purposes in public interest, the Union or Member State law may provide for derogations from the rights referred to in Articles 15, 16, 18, 19, 20 and 21 subject to the conditions and safeguards referred to in paragraph 1 of this Article in so far as such rights are likely to render impossible or seriously impair the achievement of the specific purposes, and such derogations are necessary for the fulfillment of those purposes.

4. Where processing referred to in paragraphs 2 and 3 serves at the same time another purpose, the derogations shall apply only to processing for the purposes referred to in those paragraphs".

It is difficult, on the basis of the recitals and provisions referred to so far, to deny that the European legislator has indeed taken it upon itself to dig a tunnel through the network of limits and prohibitions placed on the protection of personal data in relation to research matters, so as to lay the foundations for a more effective and accessible exploitation of the same data.

And yet, at the same time, it is undeniable that there are significant problems and issues concerning the processing of personal data for medical research purposes, problems and issues in the context of which, often, the wishes expressed so far seem doomed to remain so.

In the pages that follow, therefore, we will try to investigate some of these issues and provide brief insights into possible solutions, as well as to line up, in a concise manner, the main obligations relating to the processing of personal data in the field of medical research.

4.3 Why Moving from Words to Actions in Balancing Data Protection and Medical Research is so Difficult

The preceding pages, as mentioned above, suggest that the right to personal data protection should not be an obstacle when carrying out processing operations instrumental to research activities.

As anticipated, however, this is what happens in practice.

Identifying some of the reasons behind this phenomenon may, perhaps, help to address the problems and issues related to compliance with the obligations discussed below in such a way as to avert the risk of such compliance being a blocking element with respect to the realization of a medical research project.

In this perspective, based on my first four years of experience at the Italian Data Protection Authority, I think I can say that one of the main reasons for this is the lack of adequate effective interdisciplinarity in medical research project teams.

In particular, in most cases, these teams are composed of excellence in the medical-scientific field in a much higher percentage than skills of a legal nature and related, more generally, to techniques and methods of collecting, storing and managing data, including, of course, personal data.

In this regard, the prerequisite for any balancing exercise between rights such as the one referred to above is that it is based on the conviction that the two or more rights that come into play are equally ordered and equally protagonists of the project in question and that, therefore, one must be concerned to be able to effectively rely on competences of similar importance both on the medical-scientific side and on the computer-legal side with particular reference to the protection of personal data.

This is a conclusion linked, on the one hand, to the need that the rules mentioned in the previous paragraph and the specific obligations mentioned in the next paragraph be interpreted and applied to the concrete case through recourse to appropriate expertise and, on the other hand, to the opportunity that there be members of research teams and, therefore, professionals directly involved in the same, capable of interacting and dialoguing, speaking the same language, with the entire chain of public and private subjects involved in the implementation of the project, including data protection authorities and ethics committees, and, of course, of producing documentation that is accessible to these subjects.

Unfortunately, in fact, conceptual and lexical misunderstandings frequently undermine - and in direct proportion to the complexity and ambition of the research project - the full understanding of the characteristics, dynamics and purposes of personal data processing instrumental to the realization of the project.

With reference to this aspect, it is worth mentioning a frequently underestimated aspect: the composition of the offices of the European Data Protection Authorities is, with few exceptions, significantly unbalanced in the direction of legal expertise vastly superior to any other kind of expertise and a significant deprivation, when not an absence, of medical-scientific expertise and/or real opportunities to recruit such skills, even on a temporary basis, outside organizations.

It is, however, difficult to imagine - given the limitations of recruitment processes in the public sector - that this state of affairs, which inexorably compromises the ability of DPAs to always comprehensively and adequately understand the characteristics of a research project, will change in a short term.

Intellectual honesty, then, requires to recognize that if the foregoing is true, it is equally true that the European regulations as well as the national primary and secondary ones, including, often, the measures of the Data Protection Authorities, betray any ambition of accessibility by insiders to implementing of medical research, being written, both with reference to the structure and the technical-legal language used in such a way as to necessarily require a domain expert for their full comprehension.

In this regard, a significant investment would be required in a dissemination activity of the rules of the individual domains, including, of course, medical and scientific research, aimed at non-jurists but, precisely, at experts and insiders of the specific domains of reference.

Such an activity would, among others, have the merit of allowing a more immediate, virtuous distinction between those who, among data controllers and medical researchers, intend to follow the rules and those who, instead, are looking for easy shortcuts.

At present, unfortunately, the reported complexity of the rules prevents such a valuable distinction by placing on the same level those who do not respect them due to lack of adequate understanding - but would have respected them if they had understood them - and those who do not respect them by choice and, therefore, would not have respected them even if they had understood them.

It is, in this respect, an untenable situation that requires, to put it favourably, more lawyers than lawyers in the management of a medical research project.

Then there is a second issue, no less important than the previous one, which also makes the application of European data protection rules not always easy to apply to the world of scientific research.

As suggested, once again, by the dramatic experience of the pandemic - which, not surprisingly, is the inspiration and occasion for this paper - medical research is now increasingly played out in a global dimension, while regulation is necessarily local, in any case linked to a territory regardless of its size and populousness.

Even the best possible European data protection rules, in such a context, would, however, risk creating asymmetries, frictions and complexities that are not easily manageable - as, in fact, is the case - in the implementation of research projects involving actors also based in non-European countries.

From this point of view, there is probably an urgent need to identify international fora within which standards, guidelines or codes of ethics capable of governing in a transversal and transnational manner the processing of personal data instrumental to the implementation of medical research projects with participants subject to a plurality of legal systems can be discussed and identified, again with a multidisciplinary approach.

This is a regulatory standardization effort that appears to be compatible with the European discipline of the matter and, above all, indispensable for the definitive resolution of problems that, without it, risk being unsolvable simply because of regulatory heterogeneity, even regardless of the level of rigour or complexity of this or that national disciplinary system.

Then there is, and I think it is important to highlight it here, a final issue that, as things stand, often makes the relationship between European data protection rules and medical research more challenging than one would expect.

This is the technological and methodological context.

In this perspective, among others, three profiles are probably worth mentioning, albeit, once again, in summary form.

The first relates to the circumstance that never, in the history of mankind, has such a significant amount of data - including, obviously, personal and health data - been available, even for research purposes, in a natively digital form and, therefore, in a form that lends itself, by definition - more and better than any other - to massive and transnational collection and analysis.

The second is closely related to the first.

As is well known, in fact, big data, the undisputed protagonists of most of the research projects under discussion, represent a so-called "abundant" model in which, essentially, one choses the availability of more data than is actually needed to achieve a given objective and on the situation that, precisely in the executive phase of a research project, one will then be concerned, inter alia, with identifying possible correlations between sets of data and, consequently, the possible actual usefulness of part of the data collected.

In the big data dimension, in other words, it is physiological what for the European data protection framework is pathological, i.e., the processing of more personal data than is strictly necessary to achieve a given objective.

It is evident that this circumstance often complicates certain exercises in the application of the European discipline to matters of medical research.

Lastly, there is a further issue that already concerns the present but is likely to become more and more relevant in the near future: big data and artificial intelligence, in fact, are gradually eroding the boundary between personal data and special data, including health data, as well as that between personal data and non-personal data.

It is becoming increasingly easy and, consequently, increasingly frequent that within the framework of research based on the use of AI and big data, from a given amount of common personal data, one discovers the possibility of tracing even particular data, as well as that certain datasets, although previously anonymized, reveal significant re-identification capacities of the persons concerned in the course of the project.

Such eventualities, of course, significantly complicate the application of data protection rules because they require different protection regimes to be taken into account in the design of the research or, even, to fulfill different obligations from those originally foreseeable at the outset.

But apart from these questions, what are the main obligations that the current European regulations place on those who intend to process personal data, normally of a predominantly special nature, as part of a medical research project?

4.4 The main Obligations under the GDPR for Processing Operations Instrumental to Medical Research

An exhaustive treatment of the obligations that the GDPR provides for in relation to the processing of personal data instrumental to the exercise of the research activity is incompatible with the concise nature of the present work and, moreover, would probably be of little use in any case, on the one hand because a large part of those obligations are not peculiar to the processing operations that are the subject of this exercise and because some of those focal points are dealt with in greater detail in other chapters of the book.[8]

It seems preferable, therefore, to limit ourselves here to proposing some points for reflection on some of the aforementioned obligations either insofar as they are characteristic of medical research-related processing or, insofar as they are likely to give rise to particular questions of application in relation to such processing.

However it is worth pointing out, that the same aforementioned need for conciseness precludes, unfortunately, an in-depth examination of the issues that will be addressed in this paragraph without, however, the conciseness and peremptory nature of certain interpretative proposals being able or necessary to suggest that we are dealing with simple issues or issues of straightforward and unambiguous interpretation.[9]

On the contrary, some of the considerations put forward are far from being shared in the circuit of insiders, with the consequence that a dialogue on them certainly appears useful and opportune.

In this perspective, it is useful to start with the question of legal bases.

As is well known, the European framework provides for a range of different pari-mutuel legal bases and does not link specific treatments to specific legal bases.

The natural consequence of the foregoing is that also in relation to the processing of personal data relating to medical research activities, it will be up to the data controller, on a case-by-case basis, to identify the strongest of legal bases on which to base the processing.

In this regard, allow me to say at the outset that it is my personal conviction that among these legal bases - in general and with specific reference to the case before us of the processing of even particular data intended for use in the context of medical research - the legal basis of consent is often overrated.

This is an understandable overestimation, but one that cannot be shared, above all, because it is not justified by any regulatory provision.

But quite apart from the absence of any reason to give consent primacy among the legal bases of a processing and the consequent need to consider it only one of the

[8] The GDPR sets out various obligations for data controllers in relation to personal data processing, including transparency, accountability, and data minimization, which apply across different sectors, including medical research. For a comprehensive analysis, see Kuner, C. 2020. The General Data Protection Regulation: A Commentary, Oxford University Press, Oxford. pp. 180–195.

[9] The complex and often ambiguous nature of interpreting GDPR provisions in the context of medical research suggests the need for ongoing academic and legal discussion. See Tzouvala, A. 2020. The Evolution of EU Data Protection Law: A Critical Review. European Journal of International Law. p. 215.

conditions legitimizing the processing of personal data absolutely equal to any other, my idea is that in certain contexts - and medical research might be one of them - consent is overrated even with respect to its concrete capacity to guarantee the data subject effective control over his or her personal data.

Trying to summarize much more complex arguments, there are four reasons for this personal conclusion:

a. In most cases consent to the processing of personal data of a medical nature for the purpose of their use for research purposes is sought from patients or relatives of medical patients who are in conditions of natural psycho-physical and emotional fragility. This is a consideration that, personally, I believe weakens the effectiveness of consent as a manifestation of the will of the individual; in situations such as those in which in most cases a request for consent to the processing of certain data for research purposes is proposed, the data subject or the latter's relative has, I believe, more reason to give consent than to deny it, even where he or she is capable of fully exercizing his or her mental faculties;
b. As is well known, in order to be effective, consent must, first and foremost, be informed, and in order to be informed, it must necessarily follow the provision to the data subject of information prepared in accordance with the rules set out in Articles 12 to 14 of the GDPR and such as to enable the data subject to foresee the consequences of his or her choice in the affirmative or negative sense. The foregoing, in my view, always presupposes a clear illustration, inter alia, of the context in which the personal data will be processed in the event of the provision of consent, with particular reference to the quantity and quality of information that, in the course of processing or in the event of the occurrence of any data breach, may be extracted from the data subject's consent. In other words, the data subject, having read the information notice and before deciding whether or not to give consent, should be placed in a position to understand the immediate effects and even the mediated and pathological consequences of his or her possible affirmative choice. Now, it seems to me that this exasperating awareness is all the more difficult to offer to the data subject the more technically or scientifically complex the context of reference is, and my impression - having read a certain amount of information given at the beginning of the processing of personal data instrumental to medical research projects - is that in the cases which concern us, even leaving aside the good faith, scientific rigour and level of ethics of the data controllers, it is particularly difficult to inform the data subjects exhaustively, are not always in an optimal condition from a psycho-physical and emotional point of view. These difficulties, on the other hand, seem, to some extent, to have been taken into account by the EU legislator himself, who, not by chance, in Recital 33 provided that: "It is often not possible to fully identify the purpose of personal data processing for scientific research purposes at the time of data collection. Therefore, data subjects should be allowed to give their consent to certain areas of scientific research when in keeping with recognized ethical

standards for scientific research. Data subjects should have the opportunity to give their consent only to certain areas of research or parts of research projects to the extent allowed by the intended purpose".

c. There is, unfortunately, a widespread practice of promising the person concerned, as a more or less overt incentive to provide the consent in question, diagnostic examinations free of charge, at discounted rates or with priority execution or, more simply, the provision of personalized research feedback. Such a proposition of the request for consent, inexorably, intervenes in the decision-making process, and even where the intervention does not have such features as to suggest negative evaluations as to the existence of the requirement of freedom of consent, inexorably, it is liable to undermine its soundness;
d. The identification in consent of the legal basis of processing of personal data for the purpose of medical research inevitably creates a selection between data subjects whose data may be processed and data subjects whose data may not be processed, a selection based, obviously, on whether or not consent is given. It is, moreover, not infrequent that the inclination to give or withhold consent is conditioned by circumstances that are not without relevance to the performance of the research. It has, for instance, been observed in some contexts that the importance attached to the right to privacy and, consequently, a greater "jealousy" for one's personal data and, therefore, a greater inclination to withhold consent is greater in social classes with greater and easier access to education. And it may, of course, be the case that the same factor - belonging to certain social classes - is also a more or less relevant factor for medical research. From this perspective, it may therefore be the case that basing the processing of personal data on consent ends up affecting the reliability of research results by conditioning them with certain *biases* due to a plurality of different possible elements.

The foregoing considerations should, in my opinion, once again suggest that the assessment of consent as the legal basis to be preferred in contexts such as the ones that are the subject of this writing should be reconsidered.

This, of course, without prejudice to the need to continue to invest every possible resource and energy in making the information provided when requesting consent to process personal data, including particular data, for the purpose of medical research projects, increasingly clear and accessible.

In other words, of course, the alleged particular complexity of the objective cannot be considered in any way exempt from the obligation - which is perhaps more onerous in this case than in other areas - to guarantee the data subject, at a time prior to the commencement of the processing, subject to the known exceptions, effective transparency as to the characteristics of the processing that is the subject of the possible consent.

Partly on account of the foregoing considerations, therefore, my personal opinion - like the previous one formulated in a more peremptory and concise manner than its complexity would suggest - is that for the processing of personal data instrumental to the exercise of research activities, if one ever wanted to identify a legal basis of choice - apart from the aforementioned general principle of the equal coordination of the various legal bases set out in Articles 6 and 9 of the GDPR - one should, perhaps, look to the public interest.[10]

In fact, I find it difficult to imagine a hypothesis in which such an interest could more obviously be shown to exist and to be capable of justifying the processing of personal data even irrespective, at least to a certain extent, of the will of the individual.

And, however, notwithstanding the need for synthesis, it cannot be ignored here to take into account an undeniable limitation linked to the possible recourse to such a legal basis: as is well known, in fact, for the public interest to be usable as a condition legitimizing a specific treatment it is necessary for it to be declined in a specific rule in the law of the Member States, a rule - even leaving aside any question of rank in the various legal systems - that gives concreteness to that balancing exercise between apparently rival interests and rights.

In this context, it is, unfortunately, inevitable - and there is already no shortage of examples in this regard - that this need for normative declination of the general principle contained in the European discipline leads to intra-European fragmentation that may, unfortunately, undermine the real usefulness of recourse to this legal basis as a legitimization of treatments instrumental to the execution of pan-European medical research projects, if not, indeed, of a non-European scope.

A different issue, in addition to those relating to the legal bases set out in Articles 6 and 9 of the GDPR, which seems briefly mentioned so far and which it seems useful to put forward in relation to the application of the obligations laid down by the European framework to the processing of personal data instrumental to medical research projects, concerns transparency, which has been mentioned in passing when addressing the issue of the perhaps excessive centrality often attributed to consent as a legal basis.[11]

In this regard, as is well known, that Articles 12, 13 and 14 lay down the aforementioned general rule of the need to provide the data subjects, whatever the legal basis chosen to underlie the processing, with information containing, at the very least, the minimum information laid down by the legislature itself in the manner laid down in Article 12, a mode that meets the purpose of making the same information effectively comprehensible and accessible.

In this context, paragraph (5) of Article 14 contains one of the most significant exceptions to the above-mentioned general rule.

[10] In medical research, justifying data processing without explicit consent can be difficult but is sometimes necessary when the processing serves significant public interests, such as advancing medical knowledge. This issue is explored further in the case of *Google Spain v. AEPD*, C-131/12, where the ECJ examined balancing personal rights against public interest and the legitimate processing of data.

[11] The centrality of consent as the primary legal basis for data processing in medical research is often questioned, as it may not always be the most practical or feasible option, particularly in large-scale or public health research projects. See Becker. R. et al. 2022. Secondary Use of Personal Health Data: When is it "Further Processing" under the GDPR?. European Journal of Health Law. 30(2): 129–157.

In fact, that provision states that where "the provision of such information proves impossible or would involve a disproportionate effort, in particular for processing for archiving purposes in the public interest, scientific or historical research or statistical purposes, subject to the conditions and safeguards laid down in Article 89(1)...", the information may not be given.

The very nature of this provision as an "exception" immediately suggests that it should always be interpreted restrictively.

Here, on the basis of an increasingly widespread practice and, indeed, it may be said to be a *trend*, it seems useful to remind practitioners of the importance of a correct and strict interpretation of certain expressions contained in the aforementioned paragraph (5) of Article 14.

We refer, in particular, to the terms "impossibility" and "disproportionate effort".

As for the former, the distance separating it from the different expression "difficulty" is worth mentioning.

In order to be clear and to avoid the risk of any misunderstanding, it seems useful to clarify that even the great difficulty of administering an information notice to the persons concerned cannot be considered as an impossibility to do so.

By the expression "impossible", the legislature is evidently referring to a situation in which, irrespective of the use of any means or instruments - subject to what will be said below with regard to the different concept of "disproportionate effort" - it is in any event not possible to provide the information otherwise due.

This is obviously an exceptional hypothesis which, in most cases, will have to be considered to exist when, despite all efforts, although the personal data processed cannot be considered anonymous - since otherwise the processing would be outside the scope of the relevant rules - it is technically impossible to identify the data subjects and, therefore, to reach them with adequate information.

With regard to the expression "disproportionate effort", the interpretative practice here suggests that it should be remembered that such a disproportionate effort cannot be considered an automatic consequence of the onerousness, even significant, of the provision of information, since the cost items required to perform the processing and, therefore, the search in accordance with the current rules cannot be considered secondary or less significant than the other cost items.

In other words, to remain on the economic profile, only a comparative assessment between the overall research budget and the costs necessary to fulfil the obligations under Articles 12-14 of the GDPR could, in theory, lead to the conclusion that it is possible to derogate from fulfilling these obligations or, at least, to fulfil them in a simplified form.

Acknowledgments

The author of this publication is a member of the board of the Italian data protection authority but the considerations above represent only his personal view and not also the view of the Authority.

References

Agraharkar, S. 2019. Data Privacy in Medical Research: A Review of the GDPR's Impact on Clinical Trials. Harvard Journal of Law & Technology. 32(2): 100–120.

Beck, L. 2024. Health Data, Privacy, and Consent: Legal Frameworks in the EU, Journal of European Health Law. 31(3): 142–160.

Becker, R. and D. Chokoshvili. 2024. Legal bases for effective secondary use of health and genetic data in the EU. Int'l Data Privacy Law 14: 223–246.

De Hert, P. and C. Cocito. 2024. The Added Value of Data Protection within the Framework of Digital Constitutionalism in Europe. *In*: De Gregorio, G., O. Pollicino and P. Valcke [eds.]. The Oxford Handbook of Digital Constitutionalism, Oxford.

De Hert, P. and V. Papakonstantinou. 2021. Privacy, Data Protection and Cybersecurity in Europe. Springer, Berlin.

Becker. R. et al. 2022. Secondary Use of Personal Health Data: When is it "Further Processing" under the GDPR?. European Journal of Health Law. 30(2): 129–157.

Edwards, L. and M. Veale. 2017. Slave to the Algorithm: Why a "Right to Explanation" Is Probably Not the Remedy You Are Looking For. Duke Law & Technology Review. 16(1): 18–84.

European Commission. 2020. Ethical and Legal Aspects of Biomedical Research: EU Guidelines on Privacy and Data Protection in Research. European Union.

European Commission. 2021. Data Protection and Biomedical Research: Recommendations for Compliance. European Union.

European Data Protection Board (EDPB). 2020. Guidelines 03/2020 on the Application of the General Data Protection Regulation in the Context of Research. European Data Protection Board.

Fitzgerald, D. 2023. Data Protection and Privacy in the European Health Data Space. Health Law Review. 31(2): 178–202.

Frosini, E. 2024. The Impact of the GDPR on International Medical Research. European Journal of Medical Law. 32(1): 88–105.

Gellert, R. 2021. GDPR and Health Data: Balancing Research and Privacy. Data Protection Journal. 12(4): 198–210.

Geyer, R. and A. Skorupinski. 2023. Personal Data Protection in the Digital Age: EU Law and Global Perspectives. Springer, Berlin.

González, L. 2024. The Future of Data Protection in Health Research. European Review of Private Law. 32(4): 711–730.

Hogarth, S. and O. O'Neill. 2018. Bioethics and Data Privacy in Biomedical Research. The Cambridge Handbook of Bioethics, Cambridge.

Kuner, C., L.A. Bygrave and C. Docksey. 2020. The EU General Data Protection Regulation (GDPR): A Commentary. Oxford University Press, Oxford.

Kuner, C. 2020. The General Data Protection Regulation: A Commentary, Oxford University Press, Oxford.

Lievens, E. 2022. EU Data Protection Law: The General Data Protection Regulation. Cambridge University Press, Cambridge.

López, P. 2024. Biobanking and Data Protection: Navigating the European Landscape, Journal of Health Law and Policy. 12(1): 45–67.

Maughan, D. 2019. Data Protection and Privacy: The Age of Intelligent Machines. Routledge London.

Mitchell, D. 2021. Challenges in Data Privacy Compliance in Biomedical Research: Case Studies and Best Practices. Health Data Management Journal. 22(3): 45–58.

Resta, G. 2022. Pubblico, privato e collettivo nel sistema europeo di governo dei dati. Rivista trimestrale diritto pubblico. 4: 971–996.

Rinehart, K. 2019. Biomedical Research, Privacy, and Ethical Concerns: Navigating the Challenges of Consent and Data Use. Journal of Medical Ethics. 45(4): 234–239.

Rodotà, S. 2021. Tecnologie e diritti, 2nd edn, Il Mulino, Bologna.

Scheltema, A. 2023. Data Governance and the EU: Legal Frameworks and Emerging Issues. Oxford University Press, Oxford.

Tzouvala, A. 2020. The Evolution of EU Data Protection Law: A Critical Review. European Journal of International Law.

Wiese Svanberg, C. 2020. *"sub § 89". In:* Kuner, C. et al. [eds.]. The EU General Data Protection Regulation, OUP, Oxford.

Wachter, S. et al. 2020. Data Protection and Privacy: The Internet of Bodies. Oxford University Press, Oxford.

Wallace, H. and R. Taylor. 2022. Legal Frameworks for Health Data Privacy. Routledge, London.

Zohar, S. 2023. The EU Data Protection Regulation: An International Perspective. Hart Publishing, Oxford.

5

Patients', and Other Experts', Perspectives

Lisa Norén

Understanding expertise requires a shift in perspective, especially for new conditions like long-COVID. As a physician and long-COVID patient with microvascular dysfunction, the author has seen the importance of integrating patient insights into medical practice.

Historically, medical education emphasized learning from the deceased (Mortui vivos docent), but a modern twist is proposed: the ill should teach the healthy. Patients offer invaluable real-time expertise, dedicated focus, and robust support networks that enhance understanding and treatment of long-COVID.

Evidence-Based Medicine (EBM) aims to balance clinical intuition with the best available evidence. The pandemic highlighted the need for adaptable evidence bases, especially for emerging diseases. Combining patient experiences (first-person perspective), clinical interactions (second-person perspective), and scientific research (third-person perspective) provides a comprehensive approach to medical knowledge.

Integrating patient data can accelerate learning and improve healthcare outcomes. Systematically collecting and analyzing the patient's experiences can uncover effective treatments and rehabilitation methods. By evaluating patient contributions and fostering collaboration across different expertise levels, we can enhance our response to long-COVID and similar conditions.

There is an importance of flipping the perspectives and having a broad perspective when it comes to what being an expert means, especially when discovering new grounds. As mentioned, I am both a physician and a patient. I caught COVID in March 2020, and I still struggle with chest pain, shortness of breath, and fatigue. I am one of the 36 million persons experiencing long Covid in Europe, as estimated by WHO in June 2023. Examinations have shown that I suffer from a condition

Swedish COVID-19 patient organization, Stockholm, Sweden. Email: lisa@famnoren.com

DOI: 10.1201/9781003475927-6

called microvascular dysfunction in my heart and in my lungs. I can work part-time from home, but not attend my previous spare time commitments, training, social gatherings, etc.

I have become a patient advocate, committed to both the Swedish COVID Association and a Swedish network for doctors with long-COVID.

When I started studying at medical school back in the 20th century, there was a sign above the entrance to the anatomy hall. It read: *Mortui vivos docent* (the dead teach the living). We were told that we should be humble and hungry for all knowledge that would come our way.

From my experience of being both a physician and a patient, I have come to think about the *Mortui vivos docent* from a patient perspective and would like to suggest a modern way to this old axiom: instead of the dead teaching the living, we should let the ill teach the healthy.

I see three major reasons for that. First, the ill are strongly motivated to find answers and get better. We experience this disease in real-time and are an important body of lived expertise. Also, this is our main focus. We have the time to read articles and pre-prints, watch webinars, etc. Doctors or professors have many interests and learning about long-COVID is just one of their tasks.

Last, we have the infrastructure. It was set up very quickly. Support groups, Twitter, mail groups et cetera, are all parts of the digital society. And patients seem to speak to other patients more often than doctors speak to other doctors!

I am, like all healthcare professionals in Sweden, governed by the law that says that clinicians should provide services according to science and Evidence-Based Medicine (EBM). However, the pandemic and its aftermath have challenged us in our use of EBM.

The point of EBM is to lead us away from pure "intuition". Using EBM does not mean that we have to wait years and years until all riddles and puzzles are solved and cleared. It means that we act on the best possible knowledge available at the time.

At the top of the EBM pyramid are meta-analyses, systematic reviews, and randomized control trials.

When the best possible in the EBM hierarchy is missing - such as in case of a new disease - the evidence base needs to be grounded in other kinds of research to avoid intuition and unsystematic clinical experience and provide safe, ethical, and patient-centred care.

Because randomized controlled trials provide results on a group level, we also need patient-centred approaches to identify individual data and patterns. Sackett, who is one of the modern fathers of EBM, said "Good doctors use both individual clinical expertise and the best available external evidence, and neither alone is enough" (Sackett 1997). This means that to help a patient, as physicians, we need to consider the patient's lived experiences and the scientific evidence at the same time.

Another way to describe this is that there are different experts depending on how far you are from the core of the phenomenon: the first, second and third person experts. All of these perspectives are needed, and they are complementary to each other because they contribute different kinds of expertise.

The third-person expert perspective is held by authorities and scientific bodies. It provides us with objective findings on a group level, it is considered the golden standard and has the advantage of credibility and already established routes of dissemination. The processes are slow but solid, once established. This perspective is very well suited for grounded learning of well-known conditions, but not optimal for new conditions. Partly due to the long delay in feedback when interventions are tested.

The second-person expert perspective is manifested by the clinicians who meet the patients and gain interactional expertise. They can identify patterns in a new group or use previous knowledge in a new patient group.

As Sackett claimed (Sackett 1997), clinical expertise and external evidence go together. We need external evidence to understand and make the best choices for the patient, however, the evidence we use needs to be directly relevant to the patient. As a clinician, you need to be humble and ready to re-assess previous learnings – to not fall into prejudice-based care.

The first-person expert perspective is lived by the patients. Thus, they are the best experts to describe the phenomena that the second- and third-person experts struggle to find and measure with different data points, such as medical tests. Let me give you a few examples of lived expertise in long-COVID:

When the patients describe exercise-related symptoms – implying that examinations need to involve exercise situations. This seems like no rocket science at all but is still not known among healthcare workers.

When the patients describe *adverse reactions to exercise*, recommended by third-person experts based on knowledge from other conditions. Here, Post-Exertional Symptom Exacerbation or Malaise (for short: PESE or PEM) is common when subject to exercise. This example of lived expertise is not available to second or third person experts.

When patients evaluate their self-management, or prescribed symptom-easing treatment of chest pain, shortness of breath, headache, tachycardia, etc. What seems to work, and what doesn't? It's not clear, since data is not collected.

The first-person experts can contribute in different ways with their lived expertise. For example, we have the *Body Politic* group of long-COVID patients in the US, acting as experts to the American authorities and crowdfunding scientific studies. In the UK, a group of *Doctors with long-COVID* has written several articles on the subject and has suggested both a definition and a set of recommendations using a Delphi consensus method together with second- and third-person experts.

Correctly used, the first-, second- and third-person expert model has great opportunities for reaching the best possible learnings and sharing bottom-up and top-down learnings with the clinicians.

Incorrectly used, a third-person expert can take second- and first-person experts as "hostage" in the expert group and still do it in their own third-person way. That usually leads to poor results, in line with Sackett's words.

Literature on research methodology also teaches us that there are different methods we could use in situations when we have little previous knowledge. Traditionally, in scientific models of medical research, the researcher chooses an existing theoretical framework, develops one or more hypotheses derived from that framework, and collects data for the purpose of assessing the validity of the hypotheses. In short: get an idea first, then find data to support it.

For several decades, researchers such as Glaser and Strauss, have argued that we need systematic and empirically-based ways of developing theories and frameworks that we can later test. This argument served as the basis for their method *Grounded theory* which is a systematic methodology that involves the construction of hypotheses and theories through the collection and analysis of data. In short: collect the data first, then generate an idea from it.

These ideas are very relevant for long-COVID. What are the symptoms and findings, and what can their causes be? The more cases you can find, the more power and possibility to capture diversity you get. Data from a large group of patients can be aggregated by big data analysis and can be used to validate each other.

Since many patients have long-COVID, there is a great diversity in backgrounds, professions, and previous health, which leads to good possibilities to observe and describe the reality from different angles.

Many patients started noticing all of this during 2020, but the link to systematically make this knowledge to "Big Data" has been missing, at least in Sweden.

We could also all be more curious about the ones who have got better. What characterizes them? What were their findings before and after, and what actions were taken? This group is lost by both the healthcare and the patient organizations because they disappear when they get better. Very valuable knowledge is lost and, as I write this in 2024, it's already been more than four years. It's time to get structured!

In the emergency phase, the learning process was fast and structured. National quality registers enabled the identification of patterns – what seemed to work and what did not. With or without the blessings from the authorities. Now everybody in intensive care units gets blood thinners, corticosteroids, and antivirals in the emergency phase, and both mortality and morbidity are lower.

The outpatient phase of the pandemic has been more passive. It seems alright to just sit and wait for more knowledge to appear and, meanwhile, healthcare for long-COVID is largely unequal. Had our data been collected, and had we been followed up and reviewed the same way as the COVID-patients in the acute phase, I am certain that we would have known more by now about the dos and don'ts of long-COVID. Both regarding medications and rehabilitation principles. The degree of self-medication among patients is high, but the follow-up is surprisingly unsystematic even with prescribed medication. I have tried several heart medications over the last few years, but the follow-up of them is surprisingly scarce. I'd love to share my data to improve health care.

Gaining new knowledge is a patient safety issue. Wrong guesses might harm the patients instead of helping them. That's why patients' findings are of value when we need to build knowledge fast. Let "the sick teach the healthy"! Ask your first-person experts. Listen, and pay attention to what they say. One of the most important tasks for the third-person experts should be to support the collection of findings of the patients and clinicians.

References

Sackett, D.L. 1997. Evidence-based medicine, Seminars in Perinatology, Volume 21, Issue 1, 3-5, ISSN 0146-0005, https://doi.org/10.1016/S0146-0005(97)80013-4.

6

The Human-Centric Approach in Scientific Research

The AI Act and the New Frontiers of Research Ethics

Alessandro Mantelero

6.1 The Data-Driven Paradigm in Scientific Research

In 2008 the popular technology magazine Wired published an article entitled "The End of Theory: The Data Deluge Makes the Scientific Method Obsolete" (Anderson 2008) that sparked a scientific debate on the impact of Big Data on science and scientific methods. Although it referred to Big Data analytics, based on traditional statistical analysis techniques (e.g., regression and clustering), and did not cover machine learning algorithms, the provocative article highlighted a key element of the trend that is transforming scientific research.

In a nutshell, the author suggested that the new data-driven approach to research would replace the traditional hypothesis-based method, where the latter was a consequence of the scarcity of information. As a result of the datafication of our society and the availability of computing power, it would no longer be necessary to conduct research on a limited sample of the population, formulate hypotheses and extend them to the whole of society, but it would be possible to look directly at the dynamic trends emerging in society, monitored in a comprehensive way by Big Data analytics.

Although criticized for its generalization and thought-provoking approach, the article highlighted an important paradigm shift in research in several fields resulting from the abundance of data and the ability to make intensive use of it. This was just

Associate Professor of Private Law and European Commission Jean Monnet Chair in Mediterranean Digital Societies and Law at Polytechnic University of Turin. He is a member of the Support Pool of Experts of the EDPB and UNDP expert on human rights and AI. Email: alessandro.mantelero@polito.it

DOI: 10.1201/9781003475927-7

the first stage of the recent revolution in research. The second came few years later with AI algorithms that helped researchers not only to navigate large datasets and combine them, but also to replace the role of the researcher in part of the research activities.

In order to address the consequences of these paradigm shifts in data-driven scientific research and their ethical and legal implications, it is useful to outline the three main technological components underpinning them: data generation, data access, and AI.

The longest transition concerns the first component. Since the second half of the last century we witnessed a transition from a paper-based society to a digital society, involving both the digitization of the past, transforming paper documents into digital records, and the digitization of the present, generating new information in digital form, producing new online content, and gradually transforming the physical and social environment into data through the use of sensors, and IoT devices.

The paradigm shift in data access is more recent, dating back to the last decades of the 1990s. It is a consequence of the development of the Internet and the emergence of open data, data sharing, and data re-use trends and policies.

Finally, over the last decade, digital information has been transformed into machine-readable data making it searchable on a large scale and enabling the creation of large datasets, which, together with almost unlimited computational power provided by cloud computing, has made possible the latest wave of AI development centred on machine learning.

All these technological changes have had an impact on the way society and physical world are investigated and represented by researchers. As suggested by technology mediation theory (Verbeek 2011), we necessarily see the world through the lens of the technology we use to look at it. The adoption of a data-driven approach means that data and data processing technologies are increasingly becoming the means to see and understand the world.

In addition, as geographers have been suggesting for many years, we should keep in mind that the map is not territory (Korzybski 1933). The same consideration can be extended to data-driven representations of physical and societal elements, where their digital twins, as well as the dataset generated by processing operations and their elaboration by AI are necessary a selective representation of the world, based on the criteria established by humans and machines.

Combining these two observations, several data ethics issues arise from the use of this technology-mediated representation of the world for research purposes, starting from the new paradigm of the relationship between the researcher and research objects.

Whereas the traditional paradigm was based to a large extent on a two-way relationship centred on direct interaction between researchers and the object of their investigation, through empirical studies or direct access to documents, today's data-driven research is largely characterized by a three-way relationship. The traditional researcher/object relationship has shifted to a researcher/data-AI/object relationship, in which researchers rely on datasets and their AI-driven processing, often paying little attention to the source, generation and collection of data.

This new paradigm has raised specific ethical questions and placed an emphasis on the interplay between data processing and research ethics in the context of AI. As a result, data ethics has emerged as a new field of study and practice in the broader context of ethical evaluation of research activities.

Against this background, it is therefore necessary to (i) frame the ethical issues related to data-driven research, (ii) identify their relationship to AI regulation, and (iii) identify methodological approaches to appropriately address the challenges of AI-powered scientific research. The following sections will address these three issues respectively.

6.2 Framing Ethical Issues in Data-Driven Research

In 2015, in facing the challenges of Big Data, IoT and cloud computing (three key components of the latest AI revolution), the EDPS considered that "an ethical framework needs to underpin the building blocks of this digital ecosystem".[1] To this end, the EDPS set up the Ethics Advisory Group (EAG). In its final document, the EAG emphasized that the challenges posed by these technologies had only been partially addressed by the law, and ethics "offers other insights for conducting an analysis of digital society, such as its collective ethos, its claims to social justice, democracy and personal freedom".[2]

From mid-2010s to 2021 when the European Commission presented the AI Act proposal, "a bewildering array of ethics boards, panels, committees, groups, centres, frameworks, principles, templates, guidelines, protocols, projects and the like have all popped up like woodland mushrooms in a wet Autumn" (Raab 2020). However, this focus on ethics in framing the discourse on the use of AI, including for research purposes, has led to several shortcomings that raise criticisms from a methodological point of view. These criticisms can be grouped into three main categories that will be examined further: (I) improper overlap between ethics and law; (ii) 'transplant' of ethical values; (iii) false ethical problems.

As regards the overlap between ethical and legal values, an example is provided by the main outcome of the Independent High-Level Expert Group on Artificial Intelligence set up by the European Commission,[3] namely the Assessment List For Trustworthy Artificial Intelligence (ALTAI) for Self-Assessment,[4] where such an overlap is evident in the questions concerning fundamental rights, privacy and

[1] European Data Protection Supervisor. 2015. Opinion 4/2015 Towards a new digital ethics, p.12, available at https://edps.europa.eu/sites/default/files/publication/15-09-11_data_ethics_en.pdf.

[2] European Data Protection Supervisor, Ethics Advisory Group. 2018. Towards a Digital Ethics, p.7, available at https://edps.europa.eu/sites/edp/files/publication/18-01-25_eag_report_en.pdf.

[3] Independent High-Level Expert Group on Artificial Intelligence set up by the European Commission 2019. Ethics Guidelines for Trustworthy AI, available at https://ec.europa.eu/futurium/en/aialliance-consultation.1.html.

[4] Independent High-Level Expert Group on Artificial Intelligence set up by the European Commission. 2020. The Assessment List For Trustworthy Artificial Intelligence (ALTAI) for Self-Assessment, available at https://digital-strategy.ec.europa.eu/en/library/assessment-list-trustworthy-artificial-intelligence-altai-self-assessment.

data governance, technical robustness and safety, diversity, non-discrimination and fairness, and environmental impact.[5]

Recalling the results of an analysis conducted on the basis of studies examining the main values underlying the ethical guidelines drafted in this early period of AI regulation (Mantelero 2022), it is possible to identify three main groups of values. While the first group consists of broad principles derived from ethical and sociological theory (e.g., common good, well-being, solidarity) and the second group includes principles from the field of applied ethics and research and medical ethics (e.g., the principle of non-maleficence, the principle of beneficence, and the related broader notion of harm prevention), the last group covers a wide range of legal principles that are already part of data protection and other regulations, such as privacy protection, transparency, accountability, safety and security, fairness, as well as references to human rights principles (human dignity, justice and equity) and specific human rights (non-discrimination and justice).

The inclusion of human rights in the list of ethical principles is critical for two different reasons. On the one hand, this generates confusion among those who want to implement these ethical principles, leading them to regard compliance with certain legal requirements as an expression of soft-law commitments. On the other hand, the 're-codification' of specific requirements already established by the law as ethical values creates uncertainty in the interpretation of legal provisions, as ethical guidelines suggest different nuances in terms of safeguards and scope, with the potential risk of adopting a lower level of protection.

Given the values listed in these guidelines on ethics and AI, the risk of a 'transplant' of ethical values also emerges, i.e. the extension of existing values in one domain to a different and new one. As mentioned, several principles derive from medical ethics and are all centred on Western ethics, without considering different approaches. The risks of a 'transplant' of ethical values thus also entails the risk of an ethics imperialism (Schrag 2017) of Western countries.

This scenario is even more critical because these ethics guidelines do not explicitly discuss their ethics foundations. In many case, the label of ethics is merely used to identify principles that are characterized as not legality binding, rather than being grounded on ethical principles and frameworks.

Finally, the ethical narrative on AI has been also fuelled by inappropriate uses of ethics, as in the cases of false ethical problems. An example is the well-known trolley problem used in the context of AI-based autonomous vehicles, where the discussion on ethical choices (involving children, elderly people, etc., who could be killed) (Awad et al. 2020)[6] missed the crux of the problem.

In a transition phase, in which autonomous cars coexist with human-driven ones, and in the urban context of many cities, characterized by intricate and crowded spaces, safety issues related to autonomous vehicles are not a matter of ethics but

5 The self-assessment checklist also includes questions concerning AI and risk management, such as those dealing with technical robustness and safety, transparency (traceability), and accountability. The latter partially refers to legal issues when accountability questions concern auditing and redress in case of harm.

6 See, e.g., the MIT's Moral Machine project, https://www.moralmachine.net/.

composition and internal organization of the expert committees are not neutral elements, but can influence the outcome of the assessment in terms of quality and reliability of the results, and the independent nature of the evaluation (Mantelero 2022).

Based on the experience of existing ethics committees in scientific research, bioethics and clinical trials, it is possible to draw some lessons for future AI expert committees in terms of key elements, such as independence, multidisciplinarity, and inclusiveness of the committee; transparency of internal procedures and decision-making processes, provisional nature of decisions. However, the committees present a variety of structures and types of organization in terms of member qualifications, rights-holder, stakeholder, and layperson participation, as well as internal or external experts (Ruggie 2007). This shows that there is no one-size-fits-all solution: the differing nature and contextual importance of ethical and societal interests may require different approaches to the role of experts.

In addition, there are several contextual elements that distinguish expert committees in AI from ethics committees in biomedicine: (i) AI applications are much more diffuse and easy-to-develop than medicines; (ii) many AI applications involve forms of automation that have no or few societal consequences; (iii) fundamental rights issues are addressed through the FRIA; (iv) several societal issues are primarily related to large-scale projects (e.g., AI-based automated student evaluation systems).

Since AI developers may be less inclined to seek the advice of experts who have no recognized authority within the institution and who are more likely to be out of touch with them, one solution in organizations focused on AI development could be an internal advisor on societal issues as a permanent contact for day-to-day project development and as a *trait d'union* with the expert committee on AI. This would also help to foster internal participatory deliberation through interaction with the AI developers.

Experts tasked with performing an ethical and social impact assessment operate in a less formalized context, than the fundamental rights framework. They cannot benefit from a quantifiable risk analysis (Mantelero 2024) but rely primarily on an exchange of views within a deliberative process similar to that of traditional ethics committees.

As in the case of the FRIA, ethical and societal assessments have an influence on the design of AI, especially with regard to the necessity and acceptability of proposed AI solutions. They not only examine the AI product/service itself but also look at a wider range of alternative options to meet the identified needs, not necessarily based on AI.

Finally, based on the experience of the ethics committees, the assessment of AI systems/models cannot be left entirely to experts and their interaction with stakeholders. It should also include a participatory dimension, which is essential for effective democratic decision-making in the context of AI. An inclusive approach can also contribute to a better understanding of the societal and ethical issues, as well as context-specific fundamental rights concerns.

From a methodological point of view, the experience of ethics committees both in bioethics and in research suggests that expert committees in AI should adopt a holistic approach, looking at both the scientific-technical aspects and the societal issues of AI projects. This would favour a by-design approach to the societal consequences from the very beginning of product development.

Past experience of ethics committees also suggests other features: (i) independence; (ii) multidisciplinary and inclusive nature; (iii) the role of training and education, both for committee members and for those dealing with social issues within the entities developing and using AI solutions; (iv) the procedural organization and transparency of decision-making processes; (v) access to information on the products/services being assessed; and (vi) the provisional nature of the assessment and the lifecycle monitoring of AI applications, including further assessments.[14]

Despite the fact that these common features are shared by corporate AI ethics boards and provide a starting point for a general strategy, the ethics committees and the corporate ethics boards vary widely in structure, including both internal and external bodies. This demonstrates that issues remain to be addressed and that various solutions are possible.

As in other fields, policy guidance on the non-legal aspects of AI can hardly be black and white and is necessarily context-based. This explains why, especially considering the nature of societal and ethical questions, the possible solutions can be combined and shaped in different ways depending on the specific AI application, its societal impact and the nature of the actors, rights-holders and stakeholders involved in its development and deployment.

In general terms, however, it can be observed that experts play an important role in the ethical and social assessment, which is much more marked in this kind of assessment than in the FRIA. The FRIA does not rely primarily on experts' understanding of the legal framework, which is largely given.

In the FRIA, experts contribute to planning and scoping the evaluation and to defining the level of risk depending on the model. In the ethical and social assessment of AI, the experts' role is much more significant in recognizing the community values, which are context specific and often require active interaction with rights-holders and stakeholders to understand them fully. Here, experts operate in a less formalized context, compared with the fundamental rights framework, and their assessment does not benefit from a quantifiable risk analysis, but is closer to the deliberative processes of ethics committees in other fields.

Finally, a crucial aspect concerns the role of the AI expert committee in civic participation and stakeholder engagement (Agich and Youngner 1991). Participatory issues can be addressed either inside or outside the committees, by including laypersons among their members – representing rights-holders, civil society and stakeholders – or by furthering interaction between the experts, rights-holders, civil society and stakeholders through interviews, focus groups or other participation tools (Taylor et al. 1990).

---

[14] See also Council of Europe, Consultative Committee of the Convention for the Protection of Individuals with regard to Automatic Processing of Personal Data (Convention 108) 2019, para 2.10 ("AI developers, manufacturers, and service providers should adopt forms of algorithm vigilance that promote the accountability of all relevant stakeholders throughout the entire life cycle of these applications, to ensure compliance with data protection and human rights law and principles").

The experience of the ethics committees highlights the value of giving laypersons and stakeholders – and in certain cases rights-holders – a voice directly within the committees, but in the view of the wide impact of many AI systems on society and local communities, it is also necessary to implement specific tools for rightsholder and stakeholder participation. In this respect, AI applications may impact a variety of interests and rightsholders/stakeholders, which in many cases are dispersed and not organized in civil society organizations.

In such cases, the ethical assessment serves to identify these interests and potentially affected groups of people who can only be involved following an initial assessment and not as part of the expert committee from the outset. This does not mean that where homogeneous impacted categories are evident from the earliest stages of an AI proposal (e.g., students in the case of AI-based university admission tools), they cannot be given a voice or included in the assessment teams. Even in this case, however, given the complexity and variety of interests impacted by AI, participatory tools remain an important component of the implementation of impact assessment to identify additional stakeholders and ensuring a wider rights-holder engagement.[15]

6.5 Conclusions

The AI-mediated representation of the world for research purposes raises several data ethics issues that have not been properly framed in the policy debate on AI ethics in recent years. This has led to inappropriate overlaps between ethics and law, 'transplantat' of ethical values, and false ethical problems.

Despite convergence on some common principles, the role of ethics has been overestimated without taking into account the full societal consequences of AI, as well as high level principles and values have shown problems in their concrete implementation, not least because of the role of industry in shaping ethical values according to and subject to business interests.

The shift towards hard law to govern the impact of AI on society, namely the EU AI Act, has not significantly changed this scenario, as the AI Act is largely not applicable to research activities due to specific research exceptions in the Act. As a consequence, specific standards and guidelines from the European Commission and EU research bodies should complement the AI Act by providing an appropriate methodology to address the socio-ethical challenges associated with AI research and development.

In order to contribute to the development of a more robust methodology for a truly human-centric approach to AI-driven research, moving from a high-level discussion of broad common principles to their concrete and effective implementation, the last part of this chapter has outlined some possible approaches for establishing expert

[15] It is also worth noting that direct participation differs from the engagement of spokespersons of selected stakeholders, who often fail to represent the majority of the categories or groups of people involved. Participation tools are therefore vital to an effective democratic decision-making process on AI, an inclusive approach that ensures choice is given to minorities, underrepresented and vulnerable categories.

committees, inspired by ethics committees, to advise AI providers and deployers in their research and development activities.

In spite of the diversity resulting from the contextual nature of ethical and social assessment, it is expected that the relevant European bodies will be able to define some methodological guidelines that will ensure a common approach to ethical issues in AI, based on a concrete participatory assessment of the individual projects and the values existing in the communities where they will be implemented. The purpose of this chapter was to contribute to this crucial debate on research policy in the EU.

References

Agich, G.J. and S.J. Youngner. 1991. For Experts Only? Access to Hospital Ethics Committees. The Hastings Center Report 21(5): 17–25, available at https://doi.org/10.2307/3562886.

Anderson, C. 2008. The End of Theory: The Data Deluge Makes the Scientific Method Obsolete. Wired, available at https://www.wired.com/2008/06/pb-theory/. All cited websites were last consulted in April 2024.

Arda, S. 2024. Taxonomy to Regulation: A (Geo)Political Taxonomy for AI Risks and Regulatory Measures in the EU AI Act. arXiv. Par. 3, available at https://doi.org/10.48550/arXiv.2404.11476.

Awad, E., S. Dsouza, A. Shariff, I. Rahwan and J-F. Bonnefon. 2020. Universals and variations in moral decisions made in 42 countries by 70,000 participants. Proc. Natl. Acad. Sci. USA. 117(5): 2332–2337, available at https://doi.org/10.1073/pnas.1911517117.

Kaspersen, A. and W. Wallach. 2022. Long-Termism: An Ethical Trojan Horse. Carnegie Council for Ethics in International Affairs, available at https://www.carnegiecouncil.org/media/article/long-termism-ethical-trojan-horse.

Korzybski, A. 1933. A Non-aristotelian System and Its Necessity for Rigour in Mathematics and Physics. Science and Sanity. 750.

Mantelero, A. 2022. Beyond Data. Human Rights, Ethical and Social Impact Assessment. pp. 99-101, 76-82: 104–109, Ch. 3. *In*: AI. Springer-T.M.C.. Asser Press, The Hague.

Mantelero, A. 2024. The Fundamental Rights Impact Assessment (FRIA) in the AI Act: roots, legal obligations and key elements for a model template. Computer Law & Sec. Rev. (in press).

Mantelero, A. and M.S. Esposito. 2021. An Evidence-Based Methodology for Human Rights Impact Assessment (HRIA) in the Development of AI Data-Intensive Systems. Computer Law & Sec. Rev. 41: 105561, available at https://doi.org/10.1016/j.clsr.2021.105561.

Newton, C. 2023. Microsoft Lays off Team That Taught Employees How to Make AI Tools Responsibly. The Verge, available at https://www.theverge.com/2023/3/13/23638823/microsoft-ethics-society-team-responsible-ai-layoffs.

Raab, C.D. 2020. Information Privacy, Impact Assessment, and the Place of Ethics. Computer Law & Security Rev. 37: 105404, available at https://doi.org/10.1016/j.clsr.2020.105404.

Ruggie, J.G. 2007. Report of the Special Representative of the Secretary-General on the Issue of Human Rights and Transnational Corporations and Other Business Enterprises: Guiding Principles on Business and Human Rights: Implementing the United Nations "Protect, Respect and Remedy" Framework. United Nations, General Assembly, A/HRC/4/74, available at https://documents-dds-ny.un.org/doc/UNDOC/GEN/G07/106/14/PDF/G0710614.pdf?OpenElement.

Schrag, Z.M. 2017. Ethical Imperialism. Institutional Review Boards and the Social Sciences 1965-2009. Johns Hopkins University Press, Baltimore.

Taylor, L. and L. Dencik. 2020. Constructing Commercial Data Ethics. Technology and Regulation. 197–218, available at https://doi.org/10.26116/techreg.2020.001.

Verbeek, P.-P. 2011. Understanding and Designing the Morality of Things. The Chicago University Press, Chicago/London.

Wagner, B. 2018. Ethics as an escape from regulation: From "ethics-washing" to ethics-shopping? *In*: Bayamlioğlu, E., I. Baraliuc, L.A.W. Janssens and M. Hildebrandt [eds.]. Being Profiled. Amsterdam University Press, Amsterdam.

SECTION 2

Who “Owns” Data?

DOI: 10.1201/9781003475927-8

7

The Data Economy, Health Data and the Lexicon of Property

Comparative Remarks#

Ermanno Calzolaio

7.1 Introduction

This section is titled "Who "owns" data?". Significantly, the term "own" is within quotation marks because data is a *quid* that is relevant to contemporary society and the economy, as it is extremely difficult to frame from a legal perspective (Zeno Zencovich 2021). What is data? Is it ownable? Is the lexicon of property useful for data in general and for health data in particular? The legal debate about these crucial questions is far from having reached final solutions.

Moving from brief remarks on what is data (§1), this paper outlines the different lexicon of property in the civil law and common law legal traditions (§2) and then it discusses the European Union approach, showing that the initial option for data property has gradually shifted toward the access perspective, even if some provisions of the recent Data Act question the common understanding according to which the property discourse has been abandoned (§3). Against this background, the paper focuses on the legal regime of health data according to the GDPR (§4) and the Regulation for the European Health Data Space (§5). After having considered the approach adopted in the UK after Brexit (§6) and the debate about data property in the US and the UK (§7), it finally comes to some conclusive remarks (§8).

University of Macerata. Email: ermanno.calzolaio@unimc.it

This work has been funded by the European Union - NextGenerationEU under the Italian Ministry of University and Research (MUR) National Innovation Ecosystem grant ECS00000041 - VITALITY - CUP E13C22001060006.

DOI: 10.1201/9781003475927-9

7.2 What is Data?

Data has gained crucial importance from an economic perspective. The term "data economy" is now commonly used to denote an economy built on the digital revolution to bridge the physical and cyber worlds.

Data can be distinguished in at least four categories: (i) some are the result of the datafication of real-life objects; (ii) others are related to the datafication of information (news, research results, etc.), thus constituting the outcome of intellectual activity or the representation of nonmaterial entities; (iii) others are related to people taken up through platforms and services available on the Web, allowing news about orientations, opinions, and sensitive aspects (age, name, gender, attitudes, behaviour) to be obtained; and (iv) finally, there are data automatically generated by digital technologies.

Digitalization, therefore, involves both personal data and data that has no direct link to the person. The rapid increase in the amount of data generated worldwide is impressive. It is expected to reach 175 zettabytes in 2025.[1] The European Commission considers that the way data is stored and processed will take place in smart connected objects (cars, home appliances, manufacturing robots) and in computing facilities closer to the user than in classical clouds (this phenomenon is known as "edge computing"). With the advent of the so-called Internet of Things, objects interact and gain "intelligence" because they can communicate data about themselves and access aggregated information from others.

Understanding data is crucial for assessing one's legal entitlement in relation to this new source of value and establishing circulation and access regimes. However, despite their ever-growing economic importance, it is difficult to qualify digital data recurring to traditional legal categories because data has characters of its own, which distinguish it from other material and immaterial objects. In fact, data can be used simultaneously by an indefinite number of subjects, so its use is not "rival". Moreover, data is not perishable, and it does not have to exist at one specific place in the world: the same data could be instantiated in numerous diverse and scattered copies and its use does not alter its consistency in any way. Finally, data do not assume relevance in itself, but rather in aggregation with other data, appropriately collected and processed by computer tools (data analytics), thus soon escaping any possibility of identification.

However, as data has an enormous value, the language of ownership is often used to claim that it is an object of belonging. After all, suffice to remember that the former Business Director of Cambridge Analytica, after giving evidence exposing its data harvesting practices, became a public speaker with the handle #OwnYourOwnData.[2] The term "own" clearly evokes the idea that data belongs or should belong to someone.

---

[1] European Commission, A European strategy for data (2020) COM(2020) 66 final.

[2] See the petition with the strap line "Mark Zuckerberg, change Facebook's rules and give us back control over our data, our digital assets, our property", started in 2018. See Kaiser, B. 2018. Tell Facebook: Our Data Is Our Property #OwnYourData. https://www.change.org/p/tell-facebook-our-data-is-our-property-ownyourdata (accessed Jan. 24, 2024).

7.3 The Lexicon of Property (Civil Law-Common Law)

The word ownership generically indicates that something pertains to a subject. From a legal point of view, it is a very elusive concept. In English, ownership and property are often used alternatively, whereas in other legal traditions, they indicate different concepts. For this reason, to catch the legal issues concerning data ownership, it is worth tracing a brief outline of the differences between the two main models of property in the Western legal tradition.

In the civil law experience (France, Germany, Spain, Italy, etc.), as well as in the definitions contained in the continental codes as in the systematic-conceptual framework, property continues to be shaped as a right tailored to the materiality of its object and connoted by the characters of absoluteness, plenitude and exclusiveness.

The origins of this conception are commonly linked to the influence of Roman law, in particular to the distinction between corporeal things (*res corporales, quae tangi possunt*) and incorporeal things (*res incorporales, quae tangi non possunt*). The former include land, slaves, garments, gold, etc., while the latter include usufruct, the right of inheritance, and claims resulting from a right of obligation but not the right of property, which assumes prominence through a relationship of confusion-identification with the material thing that constitutes its object exclusively. The roman jurists observed that not all material things have relevance from a legal point of view. The moon, for example, is a material thing, but it has no relevance at law, as it cannot be appropriated (at times, at least). In essence, property acts as a filter to distinguish things that exist only from a legal point of view, as its creations, from things in the material sense, which derive their qualification and legal relevance as property insofar as they can be appropriated.

Despite the basic ambivalence of the term *res* (as a physical entity susceptible to appropriation and as an intangible object of ownership in the hands of someone, having the nature of a good insofar as it is part of his patrimony), the corporality of the thing stands in a biunivocal relationship with the commonly accepted notion of property. On the one hand, the right is, so to say, embedded in the thing, in the sense that it is the thing that satisfies the subject's interest (and by this is distinguished from other legal situations). On the other hand, the right of ownership can only have things as its object, i.e., corporal entities.

In contrast, in the common law model (Anglo-American countries), the term property does not take on a unique and precise meaning; instead, it varies depending on the context in which it is used. It can designate a thing physically considered, rights concerning the use and enjoyment of that thing, and rights that are independent of a direct relation to a physical thing, such as a claim. Property designates the ownership of rights. As a result of the feudal model, the common law world has a patrimonialistic conception of the relations of belonging, referring not to things understood in a material sense but to the rights that are exercisable over them and which, insofar as they have patrimonial value, constitute the object of belonging. One of the most relevant consequences of this conception is that more than one right of ownership may exist on the same thing.

Since data is certainly valuable from an economic perspective, the Anglo-American model appears more inclined to include it in the lexicon of property, as is the case for all digital assets, from cryptocurrencies to NFTs. This is one of the reasons why the U.S. data market developed rapidly, as the U.S. immediately adopted a very liberal approach. Data is considered to be freely appropriable, even if related to the individual. As a result, companies in this field have been able to flourish with essentially no restrictions or limited costs.

7.4 The EU Approach to Data: From Property to Access?

The lack of a normative framework for claims in data and the absence of exclusive usage rights on data are at the origin of a deep debate since the beginning of the 2010s in Europe and of intense legislative activity. A full analysis of the EU regulatory landscape is beyond the scope of this contribution. Suffice to mention the well-known Regulation on personal data (GDPR) adopted in 2016,[3] followed by the Regulation of non-personal data in 2018,[4] the Open Data Directive in 2019,[5] the Data Governance Act (DGA) in 2022,[6] the Digital Service Act (DGA) in 2022,[7] the Digital Markets Act in 2022[8] and the recent Data Act, adopted at the end of 2023.[9] This body of laws seems to be in a continuous process, with many announced complementary legislative initiatives, such as the Artificial Intelligence Act and the Revised Product Liability Directive.

Referred to as EU data legislation or EU data law, these data and data-related legislation may be seen as the manifestation of a rising interest of the EU legislator in a comprehensive regulation of digital and data-intensive markets, or in other words, in an EU Data Strategy (Streinz 2021).[10]

3 See the petition with the strap line "Mark Zuckerberg, change Facebook's rules and give us back control over our data, our digital assets, our property", started in 2018. See Kaiser, B. 2018. Tell Facebook: Our Data Is Our Property #OwnYourData. https://www.change.org/p/tell-facebook-our-data-is-our-property-ownyourdata (accessed Jan. 24, 2024).

4 Regulation (EU) 2018/1807 of the European Parliament and of the Council of 14 November 2018 on a framework for the free flow of non-personal data in the European Union [2019], OJ L 303/59.

5 Directive (EU) 2019/1024 of the European Parliament and of the Council of 20 June 2019 on open data and the reuse of public sector information (Open Data Directive) [2019], OJ L 172/56.

6 Regulation (EU) 2022/868 of the European Parliament and of the Council of 30 May 2022 on European data governance and amending Regulation (EU) 2018/1724 (Data Governance Act) [2022], OJ L 152/1.

7 Regulation (EU) 2022/2065 of the European Parliament and of the Council of 19 October 2022 on a Single Market For Digital Services and amending Directive 2000/31/EC (Digital Services Act) [2022] OJ L 277/1.

8 Regulation (EU) 2022/1925 of the European Parliament and of the Council of 14 September 2022 on contestable and fair markets in the digital sector and amending Directives (EU) 2019/1937 and (EU) 2020/1828 (Digital Markets Act) [2022] OJ L 265/.

9 Regulation (EU) 2023/2854 og the European Parliament and of the Council of 13 December 2023 on harmonised rules on fair access to and use of data and amending Regulation (EU) 2017/2394 and Directive (EU) 2020/1828 (Data Act) [2023] OJ L 2854.

10 Communication from the Commission to the European Parliament, the Council, the European Economic and Social Committee and the Committee of the Regions, A European strategy for data (the European Data Strategy), COM/2020/66 final [2020].

At the beginning of this process, legal doctrine and policymakers explored the idea of creating "data ownership" or a "data producers' right".[11] However, it was subsequently found that establishing such a right would be unfeasible due to the difficulty in objectively determining to whom a novel right in data should be attributed among the owner of a device that contains the data, the parties having a contractual relationship with such an owner, or the persons who wrote the data (Gutwirth and Gonzalez Fuster 2018; Strowel 2018). It is also difficult to establish whether this right includes the recoverability of data and exclusivity over access, considering the nonrivalrous nature of data, which can be used by multiple parties without being consumed.

Confronted with the difficulties in solving such relevant problems, the initial perspective aimed at creating *ex novo* property-shaped rights to data was - at least apparently - abandoned in EU data legislation in favour of a governance regime, broadly defined as a system of rights and responsibilities that determine who can take what actions with respect to data, with the focus being placed on the management of such rights and responsibilities (Margoni et al. 2023).

This shift in paradigm is visible, for example, in the wording of Recital 5 of the recent Data Act: "This Regulation ensures that users of a connected product or related service in the Union *can access*, in a timely manner, the data generated by the use of that connected product or related service and that those users can use the data, including by sharing them with third parties of their choice" (emphasis added). Without going into the details, the DA aims to empower users and grant them data control, limiting the *de facto* control of data enjoyed by the data holder in the context of the so-called Internet of Things. It also fosters data reuse for innovative and other public policy purposes, such as the protection of the environment, the transition to the circular economy and better health protection.

The Data Act, in particular, contains two new control rights in data: the right to access the data generated by IoT products and ancillary services (Ch. II) and the right to switch to another service provider (Ch. VI). The assumption that the Data Act testifies the shift from data property to data access is confirmed by the fact that the first of the two rights is not qualified as an exclusive right. However, a closer look at the text of the Regulation seems to contradict this reading. In fact, Article 4(13) clearly prescribes that "A data holder shall *only* use any readily available data that is non-personal data *on the basis of a contract with the user*" (emphasis added). This provision was already contained in Article 4(6) of the proposal, and it attracted a vivid criticism because it unequivocally means that the user has an exclusive right to "his/her" non-personal data and that the data holder cannot use it without a contract (Drexl et al. 2022). This provision is now confirmed in the definitive version of the Regulation, and it introduces a property-oriented right, thus questioning the common understanding according to which the property discourse has been abandoned by the European legislator (Geigerat 2022).

[11] See the EU Commission's Communication "Building a European Data Economy", 10 January 2017, COM(2017) 9 final, availabale at https://eur-lex.europa.eu/legal-content/EN/TXT/PDF/?uri=CELEX:52017DC0009, where a producer's right to use and authorise the use of nonpersonal data is discussed (p. 13).

7.5 Health Data as Personal Data in EU law and the Limits of The Public Interest Exception

Moving now to health data, it should be remembered that the European legislator assumes the distinction between personal and non-personal data. Health data is personal data. In principle, the difference between the two categories is clear: personal data include "any information relating to an identified or identifiable natural person" (Article 4(1) GDPR), while non-personal data are those that cannot identify a natural person. However, identifying a piece of data as exclusively non-personal, is often problematic because information about the person is often intertwined with information from objects.

While the recent Data Act is applicable both to personal and non-personal data ("This Regulation covers personal and non-personal data", Article 1.2), it preserves the distinction. According to Article 1.5, "in the event of a conflict between this Regulation and Union law on the protection of personal data or privacy, or national legislation adopted in accordance with such Union law, the relevant Union or national law on the protection of personal data or privacy shall prevail".

Hence, the key rules are still to be found in the GDPR; these rules conceive personal data as a fundamental right, included among personality rights (name, reputation, image) and give their owner the possibility to demand data to be processed by third parties only in accordance with precise rules and principles (Alpa 2021).

The GDPR's objective of protecting data subjects' fundamental rights from the intrusiveness of new data processing technologies coexists with a further regulatory pillar of European data protection law related to the promotion of the free flow of personal information for the integration and consolidation of the internal market. The fundamental rights and the market integration purposes appear to be placed "on equal footing" (Lynskey 2015).

The two above-highlighted objectives of European data protection law are well reflected in the regulation of health data by the GDPR. The prohibition of processing special categories of data (Article 9.1 GDPR) constitutes a direct regulatory response to the objective of protecting data subjects' fundamental rights against non-consented accesses to very intimate subjective spheres such as the one of health. Some exceptions to this prohibition allow the processing of health data on the basis of different legal grounds, listed under Article 9.2 GDPR:

(i) data subject's consent under Article 9.2 lett. a) and, strictly related to it, the need to protect a vital interest of the data subject under Article 9.2 lett. c) as well as the manifest publicity of the personal data under Article 9.2 lett. e);

(ii) the processing is necessary for reasons of substantial public interest in Article 9.2 lett. g), for the purposes of preventive or occupational medicine, medical diagnosis, the provision of health or social care or treatment or the management of health or social care and systems and services under Article 9.2 lett. h) and for reasons of public interest in the area of public health under Article 9.2 lett. i);

(iii) the processing is necessary for scientific or historical research purposes or for statistical purposes under Article 9.2 lett. j).

The public interest exception could seem to be an easy way to process health data. However, the notion of public interest is not very clear. Moreover, in order to avail of the research exemption, the research must also comply with the framework developed by the Member State in which it is conducted. In fact, according to Article 89.1, the framework must be proportionate to the aim pursued and must respect "the essence of the right to data protection" and provide for "suitable and specific measures" to protect the rights of the data subject. It also prescribes that processing for scientific research purposes must be subject to appropriate safeguards for the rights and freedoms of the data subject, including that technical and organizational measures are in place, in particular to ensure the principle of data minimization. The details of the framework to be developed are left to Member States by Article 89, the result is the fragmentation of data protection standards for health research across the EU (Donnelly and McDonagh 2019).

As an alternative, one could think that the public interest test might easily be avoided by relying upon explicit consent (data processing is possible when "the data subject has given explicit consent to the processing", Article 9.2.a). At a closer look, however, neither does this alternative seem to be without risk. In fact, the data subject can withdraw his/her consent at any time, and in this case, the data will need to be erased, even if this is likely to render impossible or seriously impair the achievement of the objectives of that processing. Moreover, consent is defined by Article 4 to mean "any freely given, specific, informed and unambiguous indication of the data subject's wishes by which he or she, by a statement or by a clear affirmative action, signifies agreement to the processing of personal data relating to him or her". Satisfaction with the adjectives attached to valid consent can be challenging for a researcher seeking to use personal health data.

Another alternative could be anonymization of data, but even this possibility is problematic. While the GDPR has no application in the processing of anonymous data, including for research purposes, Recital 26 describes data as anonymous when where it does "not relate to an identified or identifiable natural person" or is "rendered anonymous in such a manner that the data subject is not or no longer identifiable". According to the definition given by the "Article 29 Data Protection Working Party" (WP29), anonymization is a "technique applied to personal data in order to achieve irreversible de-identification".[12]

In light of Recital 26, personal data that has undergone pseudonymization, which could be attributed to a natural person by the use of additional information, cannot

[12] Opinion, 05/2014 on Anonymisation Techniques, adopted 10 April 2014, 0829/14/EN, 7. The Article 29 Working Party (Art. 29 WP) was established by Directive 95/46/EC and dealt with issues relating to the protection of privacy and personal data until 25 May 2018 (entry into application of the GDPR). It is now replaced by the European Data Protection Board (EDPB), which is an independent European organisation bringing together the national data protection authorities (National Supervisory Authorities) of the countries in the European Economic Area, as well as the European Data Protection Supervisor (EDPS). It ensures that the General Data Protection Regulation and the Law Enforcement Directive are applied consistently and ensures cooperation, including on enforcement. The EDPB constantly refers to the Opinion 05/2014 on anonymisation. See for instance the Guidelines 4/2019 on Article 25 Data Protection by Design and by Default, at https://edpb.europa.eu/our-work-tools/our-documents/guidelines/guidelines-42019-article-25-data-protection-design-and_en.

be described as anonymized. This has relevant consequences because many of the pseudonymization techniques used in health research (such as key coding of data) may not be sufficient to fulfil the GDPR requirements. Research on rare or unusual conditions, research involving long-term follow-up of participants, and research that takes into account social/environmental/economic factors can be anonymized, but this process may deprive these data of their research value (Donnelly and McDonagh 2019).

7.6 The European Health Data Space Regulation

The scenario outlined in the preceding section is evolving due to the Regulation on the European Health Data Space (EHDS) enacted in March 2025.[13] In connection with the Data Act, the EU aims at creating a number of "data spaces", representing areas of increased cooperation in terms of data sharing. Their aims are to deploy data-sharing tools and services for the pooling, processing and sharing of data; include data governance structures compatible with relevant EU legislation; determine, in a transparent and fair way, the rights concerning access to and processing of the data; and improve the availability, quality and interoperability of data both in domain-specific settings and across sectors.

The EHDS is the first one to be addressed by the EU. According to the Regulation, the cross-border exchange of electronic patient records is mandatory, and a critical mass of data is expected to be generated at the EU level; these records would be of adequate quality and could be integrated and reused for policy making, research and regulatory purposes at the EU level. The legislation distinguishes between primary and secondary use of health data. The former is defined as the use of support or direct individual healthcare delivery to the data subject. The latter is defined as the use of individual-level (personal or non-personal) health data or aggregated datasets to support research, innovation, policy making, and regulatory and personalized medicine.

With regard to primary use, the EHDS extends the GDPR data portability right by allowing individuals to transfer their health data in an interoperable format, including inferred data (i.e., valuable health data inferred by a physician or a machine, such as a diagnosis) (Marcus et al. 2022). It is worth remembering that under the GDPR the portability right is ineffective for sharing inferred health data, essentially because it restricts sharing and it does not mandate that providers can easily read data thanks to a common format ("interoperability"). As a result, users can merely access their health data without the ability to use them effectively.

Against this background, the EHDS makes a substantial improvement. It provides that Member States determine which data holders from public and private entities in the health or care sectors (with the exception of microenterprises) would be obliged to make available fifteen categories of data, even when protected by intellectual property rights and trade secrets including EHR and genetic data.

[13] Proposal for a Regulation of the European Parliament and of the Council on the European Health Data Space, COM(2022) 197/2, available at https://eur-lex.europa.eu/legal-content/EN/TXT/?uri=CELEX%3A52022PC0197.

The EU Commission will be allowed to enlarge this list. At the same time, the Regulation prohibits secondary users from making detrimental decisions, including with regard to insurance premiums, advertising or marketing activities, unauthorized access to third parties, and the development of harmful products or services. Finally, the EHDS would also enable sharing of non-personal health data with non-EU countries under certain conditions (Article 64).

Some aspects are unclear, the most relevant one being the problem of consent. With respect to secondary use, the Regulation appears to suggest that national rules requiring consent for the processing of personal data will not apply where they conflict with the operation of the EHDS. In contrast, with respect to primary use, "natural persons shall have the right to give access to or request a data holder from the health or social security sector to transmit their electronic health data to a data recipient of their choice". This ambivalent wording leaves open the question whether consent will still be the legal base of data processing (Li and Quinn 2024). This is a key issue that is open to divergent interpretations.

7.7 The European rules in the UK after Brexit

The EU rules are also applicable in the UK due to the Data Protection Act 2018, commonly referred to as the UK GDPR. In fact, the European legislation adopted in the UK before Brexit is still in force as "retained law" according to the EU (Withdrawal) Act 2018 and now as "assimilated law" under the recent Retained EU Law (Revocation and Reform) Act 2023, in force since 1 January 2024, which consistently reduces the impact of EU rules and principles in the UK.

However, the UK government introduced the Data Protection and Digital Information (No. 2) Bill to Parliament (8 March 2023), approved with amendments by the House of Commons and now before the House of Lords.[14] Final approval is expected in Spring 2024. Its objective is to "update and simplify" the UK data protection laws and to modify the UK GDPR in significant parts. For our purposes, the bill states that scientific research covers "processing for the purposes of any research that can reasonably be described as scientific, whether publicly or privately funded and whether carried out as a commercial or noncommercial activity".[15] Hence, there is an explicit acknowledgment that scientific research can be for commercial purposes. The Bill additionally clarifies that research into public health constitutes "scientific research" only if it is in the public interest, and this largely reflects existing guidance on this topic (Taylor and Whitton 2020).

The Bill also intervenes in the notion of consent. While the general requisites of consent are reaffirmed, according to Article 1.3 "a data subject's consent is to be treated as falling within the definition of "consent" [...] if (a) it does not fall within that definition because (and only because) the consent is given to the processing of personal data for the purposes of an area of scientific research, (b) at the time the

[14] Data Protection and Digital Information Bill, as brought from the Commons (Dec. 6, 2023), https://bills.parliament.uk/bills/3430 (last accessed Feb. 6, 2024).

[15] Sec. 1.2, amending art. 4 of the UK GDPR.

consent is sought, it is not possible to identify fully the purposes for which personal data is to be processed, (c) seeking consent in relation to the area of scientific research is consistent with generally recognized ethical standards relevant to the area of research". These provisions demonstrate that the requisites of consent for scientific research are less rigid than those applicable for consent for other purposes.

In brief, while for the time being, the main principles laid down in the EU GDPR are applicable in the UK, in the near future there will be significant changes, especially concerning health data processing for research purposes, thus enlarging the divide with the European approach (Comandè and Schneider 2022).

7.8 The Debate About Property Rights in Health Data in US and UK

Currently, there is wide debate about whether health data (as well as genetic materials) can be considered to be objects of property rights both in the U.S. and in the UK.

Some eminent scholars defend this view, arguing that a person has possession of data when they control at least one copy. According to their perspective, "the key relationship between a person and information is a relationship of control over the information itself, that someone can access, use, manipulate, and grant to others access to *some instance* of the information somewhere". The value of digital information "exists in having access to and control over the data and in being able to give access and control to others" (Grimmelmann and Mulligan 2023). According to this view, health data is property.

Other legal scholars argue that, as a conceptual matter, the boundaries of property law lie at the limits of control. That which cannot in principle be controlled cannot be owned. As a consequence, "ownable things include anything over which it is conceptually possible to exercise absolute dominion. Unownable things cover anything over which it is impossible, in principle, for a human being to exercise absolute decision-making authority". The outcome of this reasoning is that there are paradigmatic examples of things that can be owned, such as organs, corpses, labour, and unownable things, such as genetic information and personal data (Toomey 2023). In the same vein, some authors reject the possibility of considering that health data can be objects of property rights, arguing that privacy or tort remedies are better suited to protecting individual autonomy and privacy (Contreras 2019).

Sometimes, it is also argued that, at least as far as health data is concerned, the two competing theories reach converging outcomes. Even when considering health data as property, this does not mean that the patient truly owns his/her data once the data is put under the control of a doctor or of a company. If a patient is being treated by a doctor who stores his/her files and (due to misconfigurations and failure to effectively use encryption), the system suffers a data breach, and the patient's health information is obtained by hackers, there is a misappropriation of both the doctor's and the system's data property. Nevertheless, the patient is not in the property picture because his/her data are not under his control and he/she has no data property rights in it. Eventually, the patient can have the right to sue if he/she demonstrates a breach of the doctor's duties to the patient by improperly storing his files or of other privacy rules (Grimmelmann and Mulligan 2023).

7.9 The Potentialities of Ownership

The preceding remarks confirm that the issue of allocating data is challenging in both civil law and common law legal traditions. If the "proprietary" logic continues to appear as a key that is not easy to replace, the lexicon of property seems to be inadequate for intercepting the peculiar characteristics of data, as it is not rivalrous, totally intangible and escaping from exclusive possession.

These difficulties are at the origin of the contradictory European approach, hampered by the slippery distinction between personal and non-personal data. The declared option to abandon the property discourse in favour of the creation of rights of access to data is not concretely achieved. The recent Data Act, for example, does not qualify the new right to access data generated by IoT products and ancillary services as an exclusive right. In addition, the data holder can use non-personal data only on the basis of a contract with the user, thus introducing a property-oriented right. Moreover, with regard to health data, the role of consent continues to be the cornerstone of data processing not only in the GDPR but also in the proposal of the Regulation for a European Health Data Space.

The confusion around the issue of data rights clearly shows that it is difficult to shape the creation of rights of access without solving the problem of allocating the legally relevant object to a person. In other words, before establishing a right to access, the *quid* to which access is provided should be clarified (Szilágyi 2021). The assumption that the property discourse is not appropriate for data reveals the weight of the traditional conception of property in civil law countries, together with misunderstandings about the common law approach. To address this point, it is worth remembering a crucial remark, magisterially illustrated in the following excerpt:

Ownership would seem to the layman to be a simple notion. It is merely a question of *meum* and *tuum*. If the thing is mine, I own it; if it is not, I do not. Often the law conforms to this simple way of looking at ownership. Thus, for example, a person may live alone in a house which he has bought and paid for, no one else has any rights over it. […] he can correctly say, without qualification, that the thing is his. Or he may reasonably say that he is an absolute owner. But even in such simple cases it is evident that the words absolute owner are being used to express two quite different ideas: that his ownership is both indisputable and that it is unshared. The owner is asserting at the same moment that he and no one else is entitled to the house and car, and that his interest in them is exclusive and complete; or, in negative terms, that no other person can regard the things as his and that no other person can do anything which cuts down their use or reduces their value. The one assertion relates to what lawyers call "title", the other to the content of ownership. In English law […] it is perfectly possible to have one without the other (Lawson and Rudden 1982).

In the civil law tradition, ownership is related to physical things, so only one owner is conceivable, whose right is both indisputable and unshared. In the common law tradition, on the contrary, ownership refers to the right to use a chose in possession or a chose in action, with the result that more than one exclusive right can exist in relation to the same asset. Considering these concepts, it emerges that the legal discourse in the civil law tradition seems to be a prisoner of outdated dogmas and concepts. At the same time, in the common law tradition the potentialities offered by a conception of property not anchored on the materiality of its object are sometimes neglected. From this perspective, the "user", the "data holder", and the "data recipient" (echoing the terminology used by the Data Act) can all have an exclusive right in respect of a specific way of using and exploiting data. If ownership does not directly refer to data but rather to the right to use it, it is perfectly possible to find a key to face the challenges of the data economy, while also avoiding the risk of a normative hypertrophy that characterises the European landscape. Working on these basic assumptions could be more fruitful than turning around the delicate issue of allocating data without reaching coherent solutions.

References

Alpa, G. 2021. La "proprietà" dei dati personali. pp. 465–486. *In*: Calzolaio, E. et al. [eds.]. Liber Amicorum Luigi Moccia, Roma Tre Press, Roma.

Bergé, J.S. et al. 2018. The "Datasphere", Data Flows beyond Control, and the Challenges for Law and Governance. European Journal of Comparative Law and Governance. 5: 144–178.

Comandè, G. and G. Schneider. 2022. Differential Data Protection Regimes in Data-Driven Research: Why the GDPR is More Research-Friendly Than You Think. German Law Journal. 23(4): 559–596.

Contreras, J.L. 2019. The False Promise of Health Data Ownership. New York University Law Review. 94: 624–661.

Donnelly, M. and M. McDonagh. 2019. Health Research, Consent and the GDPR Exemption. European Journal of Health Law. 26(2): 97–119.

Drexl, J. et al. 2022. Position Statement of the Max Planck Institute for Innovation and Competition of 25 May 2022 on the Commission's Proposal of 23 February 2022 for a Regulation on harmonized Rules on Fair Access to and Use of Data (Data Act). https://www.ip.mpg.de/en/publications/details/position-statement-of-the-max-planck-institute-for-innovation-and-competition-of-25-may-2022-on-the-commissions-proposal-of-23-February-2022-for-a-regulation-on-harmonized-rules-on-fair-access-to-and-use-of-data-data-act.html.

Geigerat, S. 2022. The Data Act: Start of a New Era for Data Ownership? https://ssrn.com/abstract=4214704 or http://dx.doi.org/10.2139/ssrn.4214704.

Grimmelmann, J. and C. Mulligan. 2023. Data Property. American University Law Review. 72: 829–883.

Gutwirth, S. and G. González Fuster. 2018. L'éternel retour de la propriété des données: de l'insistance d'un mot d'ordre. pp. 117–138. *In*: de Terwangne, C. et al. [eds.]. Law, Norms and Freedoms in Cyberspace. Liber Amicorum Yves Poullet, Bruxelles.

Kaiser, B. 2018. Tell Facebook: Our Data Is Our Property #OwnYourData. https://www.change.org/p/tell-facebook-our-data-is-our-property-ownyourdata.

Lawson, F.H. and B. Rudden. 1982. The Law of Property (2nd ed). Oxford University Press, Oxford.

Li, W. and P. Quinn. 2024. The European Health Data Space: An expanded right to data portability?. Computer Law & Security Review. 52: 1–13.

Liddell, K. et al. 2021. Patient data ownership: who owns your health? Journal of Law and the Biosciences. 8: 1–50.

Lynskey, O. 2015. The Foundations of EU Data Protection Law. Oxford University Press, Oxford.

Marcus, J.S. et al. 2022. The European Health Data Space, Study for the European Parliament's committee on Industry, Research and Energy (ITRE). available at http://www.europarl.europa.eu/supporting-analyses.

Margoni, T. et al. 2023. Data Property, Data Governance and Common European Data Spaces. Computerrecht: Tijdschrift voor Informatica, Telecommunicatie en Recht. https://ssrn.com/abstract=4428364 (§2).

Streinz, T. 2021. The Evolution of European Data Law. pp. 902–936. *In:* Craig, P. and G. de Búrca [eds.]. The Evolution of EU Law (3rd edn). Oxford University Press, Oxford.

Strowel, A. 2018. Les Données : Des Ressources En Quête de Propriété - Regards Sur Quelques Développements Récents En Droit Européen. pp. 251–268. *In*: Degrave, E. et al. [ed.]. Law, Norms and Freedoms in Cyberspace/Droit, Normes et Libertés Dans Le Cybermonde. Liber Amicorum Yves Poullet, Collection Du CRIDS, Larcier.

Szilágyi, F. 2021. The necessity of data allocation: A plea for a private law (property law) perspective. European Property Law Journal. 10 (2-3): 180–240.

Taylor, M.J. and, T. Whitton. 2020. Public Interest, Health Research and Data Protection Law: Establishing a Legitimate Trade-Off between Individual Control and Research Access to Health Data. Laws. 9: 1–23. https://doi.org/10.3390/laws9010006.

Toomey, J. 2023. Property's Boundaries. Virginia Law Review. 109: 131–191.

Zeno Zencovich, V. 2021. Big Data e epistemologia giuridica. pp. 574–586. *In*: Lloredo Alix, L.L. and A. Somma [eds.]. Scritti in onore di Mario G. Losano. Accademia University Press, Torino.

8

Ownership or Control?

Giovanni Maria Riccio

8.1 Introduction

Contemporary society is characterized by a wide range of dichotomies, and internal tensions that permeate every aspect of collective life, from the economy to politics, from ethics to technology. This phenomenon also extends to the issue of personal data protection, which develops within its own system of opposites: on one hand, the need to preserve individual privacy (or control over personal data); on the other, the growing interest in reusing data for commercial, statistical, or research purposes (Schwartz 2004).

This opposition highlights a critical issue raised by many legal scholars, who argue that the reuse of personal data could represent a potential breach in data protection, particularly in a context where such data is increasingly demanded and valued (Litman 2000, Resta 2011). Furthermore, the clash between this opposite sides is reinforced by another individual right, i.e., the right of the owners of the databases which collect and categorize these data.

The present paper analyzes the balance between, on the one side, the control of the data subjects on their own data, and, on the other side, the freedom of research and, at the same time, the rights granted to the health sector operators which make investments for the realization of structured databases in the light of the approval of the European Health Data Space Regulation, that seeks to create a unified regulatory framework for the use of electronic health data across the EU, which balances several key principles to achieve its objectives of promoting innovation, safeguarding data protection, and ensuring equitable access, taking into consideration copyright regulations on the protection of databases.

The dichotomy between, on the one hand, protecting individual privacy and property rules and, on the other hand, reusing data for collective or corporate purposes raises an important reflection: often, attention is focused exclusively on the individual dimension of data protection, overlooking or neglecting the possible

University of Salerno. Email: gmriccio@unisa.it

DOI: 10.1201/9781003475927-10

collective value of effective data protection. In other words, the protection of personal data is not necessarily a matter that concerns only the individual, it should also be considered a public interest or a common (Dekker and Kuchar 2022).

In this sense, data protection has become, over the years, an expression of legal formalism often translated into a mere accumulation of norms and procedures to follow, without a concrete reflection on whether these rules are truly effective in the protection of the rights (Riccio 2024).

The General Data Protection Regulation (GDPR), for example, introduced important principles such as informed consent and the right to access data, which aim to bridge the knowledge gap between those who collect data and those who provide it. The same approach has been followed in other regulations concerning data and, more recently, by the Artificial Intelligence Act.

These legislative attempts assume that reducing informational asymmetries is sufficient to rebalance the relationship between companies that collect data and the consumers who provide it. In other words, the assumption is that knowing how data will be processed, for what purposes, and by whom should enable more effective control over personal information. But is this really the case? Often, it is not (Waerdt 2020). The reality is that privacy notices and documents outlining data processing practices are often written in highly technical language, incomprehensible to many (Solove 2024). Consequently, the process of information and transparency, which should guarantee an informed choice, is reduced to a formalism that overlooks the actual comprehension of the content by the recipient.

This situation underscores the importance of revising both the legal and technological approaches to data protection, as the objective should not only be to inform but to ensure that this information is truly accessible and understandable to everyone. The challenge, then, is to move from a facade of formalism to a real substance that allows individuals to genuinely understand the value and risks associated with sharing their personal data.

At the same time, as I will discuss below, it is necessary to investigate the proprietary relation with personal data, considering also when data aggregated and organized may constitute a protected work under copyright rules.

8.2 Property and Ownership of Personal Data

The General Data Protection Regulation (GDPR) deliberately avoids defining the relationship between an individual (the "data subject") and their personal data in terms of ownership or property rights. Instead, the GDPR focuses on the concept of "control" over personal data, emphasizing the rights individuals have regarding how their information is used, and shared by others. This approach is clear in several GDPR provisions, such as Recital 68, which underline the importance of the individual's ability to manage their personal data, rather than positioning these data as property to be owned or traded (Graux 2024).

The distinction is crucial because personal data (and data in general) are inherently non-exclusive; unlike physical property, they can be shared, replicated, and utilized by multiple entities for diverse purposes without depriving the original "owner" of its utility. For example, an individual's contact information can be used simultaneously by healthcare providers, researchers, or pharmaceutical companies, each for different legitimate purposes. Similarly, the same data may be simultaneously accessed by different persons from different places. Finally, data can be copied without any cost and this exploitation does not affect the data itself (Ostrom 2005).

Analyzing data regulations in proprietary terms may be misleading, as property rules fall under the jurisdiction of individual Member States. In fact, an examination of the rules within different legal systems shows that there are no clear and universally accepted definitions of what constitutes a *thing* (in some cases referred to as a *good*).

Traditionally, ownership has been associated with something tangible, something that could be touched. In Roman law (particularly starting from Gaius, see Gai 2.12-14), intangible assets were categorized in an exclusive manner: they are the *bona qua tangi non possunt*, assets that cannot be perceived with the senses, do not occupy physical space, and merely grant their holder a right (*quae in iure consistunt*). This difference is fundamental, considering that this bipartition between tangible and intangible assets also affected possessory institutions, such as *traditio* (intention of transferring ownership) and *usucapio* (acquisition of ownership through possession), whose scope of application was limited to the former category (Polito 2019).

The distinction would indeed be overcome by Justinian law, but it underwent a profound conceptual transformation only with the advent of the Industrial Revolution, which spurred the emergence of an additional category of assets (in a sense, a subcategory) (Rose 1993, Seignette 1994, Hesse 2002). These were not characterized by the mere granting of a right to their holder but rather by their transcendence beyond the realm of sensory perception, understood in terms of economic entities alone.

As for the definition of the concept of good/thing in legal terms, the civil code that adopts a broader approach is the Austrian one, where things are legally defined by exclusion, meaning that everything differing from persons and that is usable is considered a thing: § 285 ABGB states that "*Everything that is distinct from a person and serves human use is called a thing in a legal sense*" ("*Alles, was von der Person unterschieden ist, und zum Gebrauche der Menschen dient, wird im rechtlichen Sinne eine Sache genannt*"), with a broad definition encompassing both tangible and intangible items, provided they serve human purposes (Boerding et al. 2018, the same definition also influenced the Czech Civil Code: Skřejpek 2016).

Similarly, the Civil Code does not contain an explicit definition, adopting a practical and functional approach and focusing on the nature of things and their relationship to human use and legal systems, where the main distinction is between movable and immovable properties (*biens meubles* and *biens immeubles*) (Terré 2014). The German approach is much narrower, and the term "thing" (*Sache*) refers exclusively to tangible, physical objects that can be perceived by the senses, while intangible items, such as IP rights, or digital data, are not considered "things" under § 90 BGB (Rahmatian 2008).

Common law in the United Kingdom places relatively little emphasis on distinguishing between tangible and intangible things compared to civil law systems. Instead, the legal framework in the UK focuses on the nature and enforceability of

rights themselves, regardless of whether they pertain to physical or non-physical entities and this approach stems from the flexible and pragmatic tradition of common law, which prioritizes the protection and transferability of interests over their physical characteristics (Bridge et al. 2021).

As a result, ownership or property rights can be established over a wide range of assets, including physical items like land, goods, or personal possessions, as well as intangible assets like intellectual property, digital assets, and data. The key consideration is whether the asset in question has value and is capable of being owned, controlled, or traded, rather than its physical presence (Sheehan 2019).

Provided that a common view of the notion of good/thing does not belong to a common core of the European law, as mentioned, the different EU regulations does not qualify personal data in terms of property, even if the control on the data and the role of the consent – as I will try to show in the next paragraph – demonstrates a general legal construction which clearly reminds to the boundaries of the property law.

Furthermore, it has to be remembered that Article 8 of the Charter of Fundamental Rights of the European Union does not recognize personal data themselves as a fundamental right, as rather, it is the protection of personal data that is safeguarded as a fundamental right. This distinction underscores that the right is not to "own" personal data but to ensure they are managed in ways that respect privacy and autonomy and that the regulation grants individuals a framework within which they can exercise control, informed choice, and oversight regarding their data, which aligns more closely with the protection of human rights than with property law. In this sense, the GDPR aims to empower individuals without commodifying their identity, supporting a structure that respects personal autonomy and privacy.

Control, rather than ownership, allows for data to be responsibly shared and utilized while safeguarding individual rights and preventing a reduction of personal data to mere assets in the digital economy.

8.3 The Role of the Consent

As already underlined, data protection regulations are not based on an ownership relationship with personal data but primarily on the ability of the data subject to control the use and circulation of their data. This is achieved by providing the subject with a range of information necessary to balance their position with that of the data controller. Article 6 of the GDPR sets out six legal bases, i.e., six conditions that justify the processing of personal data.

Under Directive 95/46/EC, consent was considered the primary legal basis. However, at least in theory, under the GDPR, it has become one of several legal bases, theoretically even playing a subordinate role compared to the others. In practice, however, consent continues to be regarded as predominant, while legitimate interest and public interest are applied only in residual cases.

Informed consent is a fetish rooted in Enlightenment thinking, stemming from the idea that an informed individual can express a conscious will, thereby legitimizing actions that might otherwise be deemed unacceptable (Hinchman 1996).[1] Although it is not possible to thoroughly address this aspect or the theories that have questioned individuals' true ability to be fully informed, it is evident that in the field of data protection, consent initially experienced significant prominence.

This approach positioned consent as a mechanism that would, on its own, ensure individuals had control over their personal information. As said, under the GDPR consent is still essential, but the Regulation acknowledges other legal bases, such as the necessity of data processing for the performance of a contract, compliance with legal obligations, protection of vital interests, or the pursuit of legitimate interests by the data controller, provided these do not override the fundamental rights of individuals. This shift reflects a broader understanding that consent, while valuable, may not always be sufficient or appropriate, particularly in cases where power imbalances prevent genuine freedom of choice, such as in employer-employee relationships or where services are essential or necessary, as in emergency sanitary situations (Mantelero 2013).

In Europe, the GDPR emphasizes the requirement of "explicit consent", which must be voluntary, affirmative, and informed. This means individuals must clearly indicate their agreement to data processing, typically through an opt-in mechanism. This approach aims to ensure that consent is an intentional, active choice by individuals rather than something that occurs by default. By contrast, in the United States, a different model has prevailed, known as the "notice-and-choice" approach, according to which organizations are required to provide a notice of their privacy practices, usually in the form of a privacy policy, and individuals are deemed to consent if they continue to use the organization's services or fail to opt-out of data collection (Solove 2024).

The European model seeks to ensure that consent is meaningful, protecting individuals from potentially exploitative use of their data. However, in the field of health data, excessive legislative paternalism has led to restrictions on the secondary use of such data. As I will explain further, in this scenario the European Health Data Space could rebalance this trend by facilitating the reuse of data for research purposes, albeit under specific conditions and for clearly defined objectives.

In fact, a minority position (perhaps a confused one, as it was still unaware) was reserved for additional purposes that do not require authorization from the subject concerned: indeed, Article 8, paragraph 3 of Directive 95/46/EC establishes that there is no prohibition on processing health data when the processing is "necessary for medical prevention or diagnosis, the provision of care or management of healthcare services, and when such data is processed by a healthcare professional bound by national legislation's professional confidentiality requirements, including rules set by competent national bodies [...]." However, research activities involving data reuse are not yet mentioned in the first European regulation on data processing.

As above-mentioned, the GDPR was introduced with the ambitious goal of harmonizing data protection laws across the European Union. Its purpose, as outlined

[1] Consent is also the first principle outlined in the Nuremberg Code, drafted following the Nazi war crimes tribunals in 1946. Mindful of the tragic historical experience, the aim was to prevent humans from being used against their will for scientific experimentation.

in Recital 1 and Article 1(1), is to protect natural persons concerning the processing of personal data while ensuring the free movement of such data within the EU, but it is also aimed at facilitating the secure circulation of information within the European Data Space while ensuring consistent protection for all EU citizens. However, despite these intentions, significant discrepancies persist in national legislations concerning the reuse of health data, creating barriers to innovation, policy-making, and crisis management (Kessissoglou et al. 2024).

For instance, Article 9(1) of the GDPR prohibits the processing of special categories of personal data, such as health data, unless specific exceptions outlined in Article 9(2) apply. However, these exceptions, including processing for public interest in healthcare or scientific research (Article 9(2)(h)-(j)), require national implementation and interpretation, leading to divergent approaches among Member States and these inconsistencies often result in situations where individuals cannot access innovative treatments, and policymakers struggle to respond effectively to health emergencies due to restricted access to electronic health data.

Although the GDPR provides a common legal framework, its implementation and application of the single national Data Protection Authorities varies widely across Member States. National legislations differ on critical aspects such as the conditions under which health data can be reused without consent, the scope of permissible data sharing, and the governance structures for managing access. For example, some Member States impose strict consent requirements for secondary use of health data under Article 6(1)(a) and Article 9(2)(a), prioritizing individual privacy over broader societal benefits. While this ensures robust privacy protections, it can delay or obstruct research and innovation. Other Member States, leveraging provisions like Article 9(2)(g)-(j) for public interest or scientific purposes, adopt more flexible approaches that enable faster access to health data but raise concerns about potential misuse or insufficient oversight. This patchwork of rules hinders cross-border collaboration, which is critical for addressing challenges like pandemics or rare diseases that require coordinated EU efforts (Vukovic et al. 2022).

The fragmented regulatory environment creates substantial obstacles for researchers, innovators, regulators, and policymakers (Boyd et al. 2021). For instance, researchers and innovators often face lengthy approval processes and inconsistent access protocols. These challenges slow down the development of new therapies and technologies, as seen during the COVID-19 pandemic when delayed access to aggregated health data impacted the EU's ability to respond quickly and effectively. Policymakers, on the other hand, struggle to obtain comprehensive and timely data needed to address health crises efficiently.

The European Court of Justice (ECJ) has further emphasized the necessity for a balanced approach in cases like *Nowak v. Data Protection Commissioner*,[2] recognizing that data protection frameworks must not stifle innovation or public interest efforts (Jove 2019). However, the lack of a unified approach to health data governance exacerbates the legal uncertainties faced by stakeholders.

By addressing the current fragmentation in health data governance, the EHDS and its associated regulatory framework have the potential to transform how health

2 C-434/16.

data is used in the EU. Building on legal foundations such as the GDPR, the Data Act and the Data Governance Act, the EHDS seeks to harmonize national approaches, streamline data access, and foster cross-border collaboration while maintaining high standards of privacy and data protection. Achieving this vision will require the active engagement of Member States, stakeholders, and the European Commission to ensure that the EHDS becomes a cornerstone of Europe's digital health future.

The EHDS builds on GDPR principles but seeks to address its limitations by enabling individuals to exercise greater control over their health data while facilitating secure and reliable reuse for research, innovation, and policymaking (Eckardt 2024). Data subjects may provide or withdraw their consent for sharing their electronic health data with specific entities or for specific purposes, aligning with GDPR's principle of consent (Article 6(1)(a) GDPR). Furthermore, under GDPR and EHDS, data subjects generally have the right to object to processing based on legitimate interests (Article 21 GDPR).

However, the EHDS includes exceptions to this right where processing serves critical public interests, such as in cases of scientific research and innovation (Article 33(1) o), public health policy-making (Article 33(1)) and compliance with legal obligations as established by EU or national law (Recital 37). In such cases, the regulation emphasizes proportionality, meaning data reuse must be justified, with appropriate safeguards to minimize potential harm to the data subject. Also, the reuse of health data for significant public interest purposes, such as research or public health, may override individual objections under specific conditions, which reflect the EHDS's attempt to balance individual rights with societal benefits [see notably Articles 5 (2), 33 (4), and 35 (1)].

Thus, while the EHDS supports individuals' ability to control their health data, including the right to oppose reuse in certain contexts (e.g., withdrawing consent), this right is not absolute, as the reuse of health data for significant public interest purposes, such as research or public health, may override individual objections under specific conditions.

The regulation requires Member States to establish health data access bodies tasked with managing third-party access to health data. These bodies, which may operate as new entities or within existing institutions, will adhere to principles set out in the GDPR and the Data Governance Act, including transparency, accountability, and proportionality. Notably, they will provide access without requiring individual consent in certain cases, leveraging Article 6 (1)(e) (tasks carried out in the public interest) and Article 9 (2)(j) (scientific research) of the GDPR.

8.4 Protection of Database and Health Data

Although data is difficult to be framed within proprietary rules, the arrangement of the data itself – health data included – may be protected by intellectual property regulations.

The legal framework for the protection of databases under copyright law has been shaped by international conventions, regional directives, and case law. Article 2(5) of the Berne Convention, the European Directive on the legal protection of databases no. 96/9/EC, and the jurisprudence of the European Court of Justice

(ECJ) play pivotal roles in delineating the boundaries of such protection (Pistorius 2024).

Firstly, the Berne Convention for the Protection of Literary and Artistic Works, a cornerstone of international copyright law, ensures the protection of "*collections of literary or artistic works such as encyclopedias and anthologies which, by reason of the selection and arrangement of their contents, constitute intellectual creations*" (Article 2(5)). This provision establishes the principle that databases, insofar as they are intellectual creations, are eligible for copyright protection (Aplin 2006).

Key to this protection is the "intellectual creation" standard, which focuses on the selection and arrangement of content. Mere aggregation of facts without an original method of selection or organization does not qualify. In this context, the European legislation followed the U.S. Supreme Court decision in *Feist Publications, Inc. v. Rural Telephone Service Co.*,[3] which, although outside the Berne Convention's scope, echoes this principle by holding that a telephone directory's mere listing of facts lacks the requisite creativity for copyright protection.

Then, an exclusive right on the data and notably on the health data cannot be granted to his/her owner, while the manner in which facts are compiled can meet the originality threshold if it reflects the author's intellectual effort and may constitute the subject matter of the protection. This approach aligns with Recital 17 of the European Directive 96/9/EC on the Legal Protection of Databases (the "Database Directive"), which reiterates that databases expressing originality in their selection or arrangement are protectable.

For instance, the seminal ECJ decision in *British Horseracing Board Ltd. v. William Hill Organization Ltd.*[4] clarified that databases consisting of a compilation of facts fall within the copyright boundaries if their selection or arrangement constitutes the author's intellectual creation. The Court noted that creativity and originality, rather than effort or skill alone, form the foundation of such protection.[5]

As mentioned, the Database Directive harmonizes copyright protection for databases across EU Member States and it incorporates the Berne Convention's principles and adds a layer of regional specificity by requiring databases to demonstrate a level of creativity and originality. Article 3(1) of the Directive stipulates that databases qualify for copyright if they constitute the author's own intellectual creation by reason of their selection or arrangement.

National courts have also adopted this approach. In *Football Dataco Ltd. v. Sportradar GmbH*,[6] the UK Court of Appeal (at that time still part of the European Union) held that databases containing sports data could be protected under the Directive, provided they satisfied the originality threshold. The court emphasized that creativity in how data is structured or presented is key, rather than the mere labor invested in collecting it.

[3] 499 U.S. 340 (1991).

[4] Case C-203/02.

[5] Para. 30.

[6] Football Dataco Ltd & Ors v. Stan James plc & Ors and Sportradar GmbH and Anor [2013] EWCA Civ 27.

An important consequence of the Database Directive is the potential protection of compilations comprising non-copyrightable materials. In fact, as long as the compilation reflects originality in selection or arrangement, it may attract copyright. The ECJ's decision in *Fixtures Marketing Ltd. v. Organismos prognostikon agonon podosfairou AE (OPAP)*[7] highlighted this principle, ruling that a database containing football fixture lists could be eligible for copyright if its creation involved intellectual effort in organizing the information. However, the database was not protectable solely on the basis of investment in data collection (para. 37).

Despite this statement, the Database Directive introduced a novel dual-layer protection system: (a) *copyright protection*, reserved for original databases reflecting the author's intellectual creation; (b) *sui generis right*, designed to protect databases where significant investment has been made in obtaining, verifying, or presenting their contents, regardless of their originality.

The *sui generis* right, enshrined in Article 7 of the Directive, safeguards the substantial investment behind a database and ensures that third parties cannot extract or reutilize substantial parts of the database without authorization. Unlike copyright on databases, the *sui generis* right is unique to EU law and is not mirrored in international treaties or the domestic laws of non-EU countries (Sganga 2023).

The ECJ's interpretation of this right has been pivotal. In *Database Right Case—Fixtures Marketing Ltd. v. Svenska Spel AB* (Case C-338/02), the Court clarified that the *sui generis* right protects investments related to obtaining, verifying, or presenting database content, but not investments in creating the data itself.

Furthermore, the Database Directive's definition of a database is deliberately broad, encompassing "*a collection of independent works, data or other materials arranged in a systematic or methodical way and individually accessible by electronic or other means*".[8] The ECJ has consistently upheld a wide interpretation of this term: for instance, in *Ryanair Ltd. v. PR Aviation BV* (Case C-30/14), the Court ruled that a collection of flight schedules constituted a database, even if the individual elements lacked originality. Similarly, in *CV-Online Latvia SIA v. Melons SIA* (Case C-762/19), the ECJ reiterated that systematic arrangement and accessibility of data are sufficient to qualify a collection as a database under the Directive.

As said, *sui generis* right is intended to incentivize the creation and maintenance of databases by granting the producer exclusive rights over the database's structure and its non-trivial compilation, though not over the underlying raw data itself. Raw data, by its very nature, is not protected by copyright law because it lacks the original creative expression typically required for copyright eligibility, even if investment, organization, and effort in compiling this raw data into a structured format are protected, as the arrangement and structure reflect intellectual and economic input.

This distinction underscores the Directive's aim to foster innovation while preventing monopolization of raw data: then it is important to highlight that the fact that the data are in the availability of a subject (research institutes, health care institutions, etc.) does not mean that the subject is the owner of the data.

---

[7] F Case C-444/02.

[8] Article 1(2).

Under the Directive, the holder of the *sui generis* right is identified as the "producer" of the database. In the context of research activities, determining who qualifies as the producer can be complex, especially in collaborative or multi-institutional projects where numerous individuals and entities contribute to the database.[9] Typically, the producer is defined as the individual or organization that has made a substantial investment in obtaining, verifying, or presenting the data in a structured manner. In academic and scientific research settings, this role may be attributed to the lead researcher, principal investigator, research institution, or funding body, depending on the specific circumstances of the research project and the contractual arrangements in place.

The Directive also considers exemptions to database protection in specific cases, particularly for educational and scientific purposes. For instance, researchers and educators may have some degree of freedom to use databases for non-commercial research, subject to certain limitations. However, these exemptions vary between jurisdictions within the EU and are designed to balance protecting database producers' rights with the public interest in promoting scientific advancement.

The German Copyright Act (*Urheberrechtsgesetz*) provides exceptions for scientific research and private use (§ 87c), emphasizing non-commercial purposes and requiring acknowledgment of the source, provided that the use does not conflict with the database owner's legitimate interests (Guibault and Wiebe 2013). Similarly, the French Intellectual Property Code (*Code de la propriété intellectuelle*, Article L342-3) specifies that substantial extraction for teaching or research is allowed only for non-commercial purposes, requiring clear source attribution (Guibault and Margoni 2015). In Italy, the Copyright Law (*Legge sul diritto d'autore*, Article 102-ter) adds a further specification, emphasizing proportionality and necessity and requiring users to justify the extent of their extraction or reuse. In these jurisdictions the restriction on non-commercial activities aligns with the directive's provision held in Article 6(2)(b). Divergencies arise regarding teaching and research reuses, where most of the countries limit use to what is necessary for the educational or research goal (e.g., France), while, on the contrary, others require a "strict" necessity (e.g., Poland) or are generally more permissive, allowing broader exceptions under public interest considerations (e.g., Scandinavian countries).

8.5 Access to Health-Data Repositories

The EHDS does not mentions the Database Directive, even if the rights granted by this directive cannot be applied, even if lacking a specific derogation or exception,

[9] With the advent of artificial intelligence (AI) and machine learning technologies, questions arise regarding who holds the rights to databases produced or enhanced by AI systems. When an AI machine autonomously generates a database, the designation of the "producer" becomes less straightforward. While current EU law still requires a human or legal entity to be the database producer, evolving AI capabilities could potentially challenge this paradigm. In cases where AI-generated databases are used in a scientific or commercial context, ownership could lie with the organization that owns or controls the AI, as they would likely be seen as the producer through their substantial investment in the development and operation of the AI system.

in case of the application of the Regulation. The Commission's implicit approach prioritizes collective interests over individual proprietary rights, also striving to achieve a delicate balance between these competing perspectives. Specifically, Section 2 of the Regulation, titled *Governance and mechanisms for the secondary use of electronic health data*, is built on the principle of data access and the requirement that such access serves a higher interest.

Article 41 of the European Health Data Space (EHDS) Regulation establishes obligations for data holders regarding the provision of electronic health data for secondary use. In particular, data holders, which include entities that collect and manage electronic health data (e.g., healthcare providers, public authorities, private organizations), must make data available to health data access bodies upon request, provided certain conditions are met.

The data may be personal or not: in the first case, which may include health-related data, the subjects involved in the transfer of the data are expected to comply with strict obligations held in the GDPR and in the other data protection regulations; in the latter, non-personal data are involved and thus the personal data originally included in the databases should be anonymized.

According to Article 33 EHDS, data holders, except micro-enterprises, must make some categories of data (which are listed in this provision) available for secondary use. These categories include Electronic Health Records (EHRs), human genetic data, data from health registries, clinical trial data, health related administrative data, such as reimbursement data.

Legitimate purposes for which electronic health data can be processed for secondary use beyond patient are regulated by Article 34 EHDS and include scientific research related to health or care sectors, development and innovation activities for products and services, policy-making in health or care sectors, producing official statistics. Even if the data protection standards are maintained, this provision aim to facilitate advancements in healthcare, balancing the conflicting interests of the data-subjects and of those who want to make a secondary use of the health data.

The mechanism designed by the EHDS does not provide a relation between the owner of the database and the subjects which want to use these data. In fact, as above mentioned, EHDS provides that each member State shall "*designate one or more health data access bodies responsible for granting access to electronic health data for secondary use*" or even several health data access bodies, designating "*one health data access body to act as coordinator, with responsibility for coordinating requests with the other health data access bodies*". The application is submitted by any natural or legal person, with the respect of the purposes referred to in Article 34 EHDS, and must specify the purpose and intended use of the data.

However, the analysis of the state of the art in terms of digitalization and administrative organization of healthcare facilities in single Member States highlights the risk of a discrepancy, including in competitive terms, between the European legal systems. This is due to the existing regulatory fragmentation and the fact that the requirements set out in Article 45 EHDS could be interpreted differently by the national health data access bodies. Such a scenario could facilitate a phenomenon of forum shopping, with stakeholders seeking out countries where there are fewer restrictions on data access, thereby partially undermining the goal of harmonization and data circulation within the European Economic Area.

On a positive note, the EHDS (European Health Data Space) does not reference the provisions on database copyright, particularly as it foresees access to individual datasets rather than their organization, even though the boundary between these aspects is far from clear. In facts, taking into account the provision of Article 55 EHDS, datasets may be considered as database exclusively where meets the criteria highlighted on Database Directive – comprising independent data organized systematically and individually accessible – and thus may be eligible for protection under EU law.

Furthermore, the procedure established by the EHDS might appear cumbersome, and it likely is. Indeed, the time required to identify the competent national entities, the technological delays in certain Member States, and the lack of protocols for data transfers could lead to delays that are incompatible with the need to reuse data promptly. However, the decision not to include an explicit derogation from the *sui generis* right under the Database Directive seems reasonable. This contrasts with the recent approach taken in Article 35 of the Data Act, which states that "*the sui generis right provided for in Article 7 of Directive 96/9/EC does not apply to databases containing data obtained from or generated by the use of a product or a related service*".

While this provision has limited applicability to medical databases—except for health data collected through IoT devices—the main issue raised by the Data Act lies in its derogation from the Database Directive. This, in turn, implicitly allows Member States to introduce national regulations that could risk fragmenting the harmonization of the Union's legal framework.

On the other hand, the lack of certainty connected to the concrete application of the EHDS and the missing coordination with the Database Directive may lead to opposition from data holders who might invoke their intellectual property rights to restrict access to data. In fact, Article 41, concerning the duties of the data holders, only provides a general (and generic) obligation to cooperate in good faith with the health data access bodies. The situation is slightly clearer in case of non-personal data, where data holders must ensure "*access to data through trusted open databases to ensure unrestricted access for all users and data storage and preservation*" (Article 41, para. 6).

Article 42 holds that data holders "*may charge fees for making electronic health data available for secondary use*" and these fees may "*include compensation for part of the costs for collecting the electronic health data*". In our opinion, fees are not a compensation for the use of a copyrighted database. First of all, the EHDS does not make any difference between databases and datasets, nor it makes any reference to license agreements, but only to the availableness of the datasets.

The fees must be transparent and proportionate to the cost of collecting and making electronic health data available for secondary use, but, even if the terminology adopted recalls that of copyright regulations (see especially Recital 41 and Article 16 of the Barnier Directive on the negotiations of the collective management organizations), the main purpose of the EHDS is that of reducing the risk of competitive failures, without considering the copyright issues.

The level of fees must be agreed between data holders and data users within one month of the data permit being granted, the decision is entrusted to the national health data access bodies. In the determination of the fees, size, budget and typology of users should be considered, and SMEs, public bodies, Union institutions, bodies, offices and agencies involved in research, health policy or analysis, educational institutions and healthcare providers should be favored.

Furthermore, the data should be provided in a readable format suitable for secondary use. In this regard, it is likely that the relevant European institutions will establish common standards to be adopted.

References

Aplin, T. 2006. The EU Database Directive: Taking Stock. p. 99. *In*: Macmillan, F. [ed.]. New Directions in Copyright Law, Volume 2, Edward Elgar.

Beghini, M. and I. Zambotto. 2022. Res corporales e res incorporales: radici romane e sviluppi contemporanei di una bipartizione, Roma Tre Law Review.

Boerding, A., N. Culik, C. Doepke, T. Hoeren, T. Juelicher, C.V. Roettgen and M. Schoenfeld. 2018. Data Ownership—A Property Rights Approach from a European Perspective. J. Civ. L. Stud. 11: 325.

Boyd, M., M. Zimeta, J. Tennison and M. Alassow. 2021. Secondary use of health data in Europe. Report, Open Data Institute.

Bridge, M., L. Gullifer, G. McMeel and K.F. Low. 2021. The Law of Personal Property. Sweet & Maxwell, London.

Commission of the European Communities, First Evaluation Of Directive 96/9/Ec On The Legal Protection Of Databases, DG Internal Market And Services, Working Paper, 12 Dec. 2005 available at http://ec.europa.eu/internal_market/copyright/docs/databases/evaluation_report_en.pdf.

Dekker, E. and P. Kuchar [eds.]. 2022. Governing Markets as Knowledge Commons. Cambridge University Press, Cambridge.

Eckardt, M. 2024. Data Commons and the EU Data Act. p. 241. *In*: Heine, K. and O. Budzinski [eds.]. Wettbewerb, Recht und Wirtschaftspolitik, Festschrift für Wolfgang Kerber. Nomos, Baden-Baden.

Graux, H. 2024. What is data ownership, and does it still matter under EU data law? An exploration of traditional concepts of data ownership, and of the expected impact of the Data Act. Luxembourg: Publications Office of the European Union.

Guibault, L. and T. Margoni. 2015. Legal aspects of open access to publicly funded research, Enquiries into intellectual property's economic impact. OECD.

Guibault, L. and A. Wiebe [Eds.]. 2013. Safe to be open: study on the protection of research data and recommendations for access and usage. Universitätsverlag Göttingen.

Hesse, C. 2002. The Rise of Intellectual Property, 700 bc to ad 2000. Daedalus. 131(2): 26–45.

Hinchman, L. 1996. Autonomy, individuality, and self-determination. p. 488. *In*: Schmidt, J. [ed.]. What is enlightenment? Eighteenth-Century answers and twentieth-century questions. University of California Press, Berkley.

Jove, D. 2019. Peter Nowak v Data Protection Commissioner: Potential Aftermaths Regarding Subjective Annotations in Clinical Records. European Data Protection Law Review. 175.

Kessissoglou, I.A., S.M. Cosgrove, L.A. Abboud, P. Bogaert, M. Peolsson and N. Calleja. 2024. Are EU member states ready for the European Health Data Space? Lessons learnt on the secondary use of health data from the TEHDAS Joint Action. European Journal of Public Health. 34: 1102.

Litman, J. 2000. Information Privacy/Information Property, Stanford L. Review. 52: 1282.

Mantelero, A. 2013. Competitive Value of Data Protection: The Impact of Data Protection Regulation on On-Line Behaviour. International Data Privacy Law. 1.

Ostrom, E. 2005. Understanding Institutional Diversity. Princeton University Press, Princeton.

Pistorius, T. 2024. The IP Protection of Electronic Databases: Copyright or Copywrong?. The University of Auckland Business School Research Paper (forthcoming), available at SSRN: https://ssrn.com/abstract=5024329 or http://dx.doi.org/10.2139/ssrn.5024329.

Polito, P. 2019. Some Considerations on the Relationship between 'res incorporales' and 'Immaterialgüterrechte'. *In*: Babusiaux and Igini [eds.]. Messages from Antiquity. Roman Law and Current Legal Debates. Koln.

Rahmatian, A. 2008. A Comparison of German Moveable Property Law and English Personal Property Law. J. Comp. L. 3: 197.

Resta, G. 2011. The New Frontiers of Personality Rights and the Problem of Commodification: European and Comparative Perspectives. Tulane European and Civil Law Forum. 26: 33.

Riccio, G.M. 2024. Data protection and appropriate measures: too many uncertainties in the judicial applications? Unio EU Law Journal. 10: 16.

Rose, M. 1993. Authors and Owners. The Invention of Copyright. Harvard Univ. Press. Cambridge, Mass.

Schwartz, P. 2004. Property, Privacy and Personal Data. Harvard L. Review. 117: 2055.

Seignette, J. 1994. Challenges to the Creator Doctrine. Kluwer Law International.

Sganga, C. 2023. Sui Generis Protection of Non-Creative Databases. p. 27. *In*: Bonadio, E. and P. Goold [eds.]. The Cambridge Handbook of Investment-Driven Intellectual Property. Cambridge University Press, Cambridge.

Sheehan, D. 2019. The Principles of Personal Property Law. Hart Publishing, UK.

Skřejpek, M. 2016. The Concept of 'Things' and Their Definition in the New Czech Civil Code from the Perspective of Roman Law. Zeszyty Prawnicze. 14(1): 197.

Solove, D. 2024. Murky Consent: An Approach to the Fictions of Consent in Privacy Law. Boston University Law Review. 104: 593.

Terré, F. 2014. Les biens. 9th ed., Dalloz, Paris.

Vukovic, J., D. Ivankovic, C. Habl et al. 2022. Enablers and barriers to the secondary use of health data. Arch. Public Health. 81: 115.

Waerdt Van de, P. 2020. Information asymmetries: recognizing the limits of the GDPR on the data-driven market. Computer Law & Security Review. 38: 67.

9

Data Ownership and Data Markets

The Chinese Experience

Bingwan Xiong

9.1 Introduction

Suppose a hospital accumulated tons of data consisting of patients' medical records, doctors' diagnoses and prescriptions, etc. The hospital is incentivized to utilize this database in numerous ways. It could use these data to train medical AI to assist future operation, or commission a professional company to do so. It could provide such data to a third party, such as insurance companies. Multiple hospitals could also coordinate to merge their respective data into an enormous data pool. However, the data collected by hospitals contain sensitive information regarding patients' conditions, and it is also quite probable that the data is the fruit of cooperation of the hospital and a third party. When hospitals or other organizations attempt to utilize the data they hold, legal controversies inevitably arise against the backdrop of an unclear property system, inhibiting further exploitation of the commercial value within the data.

On 30 March 2020, the CPC Central Committee and the State Council jointly issued Opinions of the CPC Central Committee and the State Council on Improving the Systems and Mechanisms for Market-based Allocation of Factors of Production, in which data was list as the fifth essential factor of production after land, labor, capital and technology.[1] To implement the guidelines set forth in this opinion, the CPC Central Committee and the State Council later adopted and promulgated Opinions on Building the Data Basic Regime to Better Exploit the Value of Data Factors (also commonly referred to as "Twenty Provisions on Data", hereinafter

Prof. Dr. LL.M. (Harv.), School of Law & Research Center for Civil and Commercial Law, Renmin University of China, Beijing 100872, China. Email:bxiong@ruc.edu.cn

This research is funded by the National Natural Science Foundation of China, a study of Theory and Institutional Design of Public Data Open Utilization and Authorized Operation (72342010).

[1] Available at: https://www.gov.cn/zhengce/2020-04/09/content_5500622.htm, accessed 22 April 2024.

DOI: 10.1201/9781003475927-11

referred to as the "Basic Regime on Data") on 2 December 2022.[2] The Basic Regime on Data aims to provide a fundamental yet comprehensive infrastructure to better exploit the commercial value of data resources, covering topics including a system of data property, an ecosystem of data circulation and services, an allocation system of data profit, and governance of data security.

Designing a property system over data has been a global challenge. The Basic Regime made two important decisions in this regard. First, the Basic Regime on Data has recognized the necessity to endow property rights for data. This decision helps to substantially reduce costs for parties involved in data production and circulation, such as property protection, ownership information identification, contract negotiations, and risk avoidance. Second, the Basic Regime on Data "broke out of the mindset of ownership" (National Development and Reform Commission 2023) and adopted a "bundle of rights" approach to understand and present the rights over data (Xiong and He 2024). To do so, the Basic Regime on Data emphasizes a policy of a "structured allocation of data property rights," defining the property right modules as the right to hold, right to use, and right to manage for profit in Article 3.

The objective of this chapter is to explain the notion of data ownership as the foundation of the "structural allocation" policy, and how such a policy is implemented through a data property regime. The structure and contents of this proposed property regime will be introduced. Although the Chinese government ambitions to cover the ownership issues in all types of data, we shall illustrate the contents and functions of these three rights with examples of health data in order to fit the theme of this book. It should be noted that the Basic Regime on Data does not support a system of data ownership in the technical sense, where complete control over a set of data is assigned to a single owner. In contrast, the Basic Regime on Data allocate different modules and sticks of the rights over data. Thus, the term "data ownership" is only used generally in this Chapter, in the sense that certain rights and interests over a set of data belong to a particular right holder, in contrast with the technical sense of the expression in civil law jurisdictions.

9.2 Rights Modules Within The Bundle of Rights on Data

The Basic Regime on Data proposed a data property system based on the "bundle of rights" theory, although this terminology was not explicitly employed in the document. Choosing this approach is the result of difficulties in identifying a single entity as the sole owner over a certain set of data. The majority of data are produced by multiple processors who collaborate in digital economic activities, where all parties involved have contributed to data production, and no exclusive access to the data was granted (Wang 2022). This presents the risk of concurrency of rights and multiplicity of claims. The future data property regime is bound to satisfy multiple claims from different players in the data production process. Thus, it is improper to depict such data as wholly owned by a single processor, nor is it proper to apply the traditional concept of joint ownership that mainly depicts the right over a set

[2] Available at: https://www.gov.cn/zhengce/2022-12/19/content_5732695.htm, accessed 22 April 2024.

of jointly owned corporeal property. Consequently, it is no surprise that "bundle of rights" approach, which is more flexible than the traditional "ownership-usufruct" approach, has received widespread welcome from the academia (Bao and Xiong 2020, Dai 2021).

When it comes to the expression of such perception in legal doctrines and statutory laws, a clear delineation and a certain extent of standardization of rights over data is required to avoid a "messy bundle" of rights over data. To do so, it is vital to standardize widely seen and justified claims of right sticks over data into right modules at different levels, by focusing on the social relations of data production and circulation based on the aforesaid perception, and by delineating the claims arising from each social relation from key dimensions like the manner in which the right-holder is entitled to act upon the data (Xiong 2023).

According to the Basic Regime on Data and its "structured allocation" policy, possible property right modules enjoyed by data processors include the right to hold data, the right to use data, and the right to manage data for profit. The right to hold data refers to the right to control and manage the controlled data autonomously, and exclude unjustified access, duplication, tampering, destroy, or deletion. The right to use data allows data processors to profit from their own use of data resources. The right to manage data for profit permits data processors to profit from other commercial uses, such as using data as financial assets for investment or as a security interest (Xiong et al. 2023).

Under the such a framework of data property, several problems arise regarding the collection, processing and utilization of medical data.

The first problem concerns the relationship between first-order statutory rights and data property rights. In the scenario of medical data, first order rights mainly comprise privacy rights and personal information rights enjoyed by patients, as provided by the Civil Code and the Personal Information Protection Law (hereinafter as PIPL). Rights to privacy prevents data processors to leak or publicize private information collected in medical operations, under Article 1032 of the Civil Code. Moreover, Article 28 of PIPL expressly states that medical and health information may constitute sensitive personal information, which means that "separate consent" is required for processing activities. Such provisions, while protected the privacy of patients, could pose unnecessarily high costs for processors of medical data, to the extent that a large number of beneficial processing activities are hindered. Thus, it is essential to strike a balance between the protection of privacy and personal information and the utilization and exploitation of health data. In addition, diagnoses and prescriptions may be subject to trade secret protection pursuant to Law against Unfair Competition.

The second problem concerns collaboration in data-processing. The production of data may involve joint contribution from different parties; thus, ownership and legitimacy issues arise as to who is entitled to utilize and manage data for profit. Collaboration could take vertical and horizontal forms in the supply chain. Vertical collaboration usually includes data generated or collected as a byproduct as daily cooperation in production and operation. For example, identical transaction data may be held by e-commerce platform and delivery companies. Horizontal collaboration usually refers to merging data from multiple entities to form a more inclusive data set. For example, several medical institutions pool the medical data into a larger data bank. In the context of medical data, horizontal collaborations are more common in practice. It is necessary to allocate rights and interests over such data properly by differentiating the contractual purposes and social relations of data collaboration.

The third problem concerns data transactions, especially data licensing. Generally speaking, issues arising out of data transactions shall be resolved under the framework of contract law. However, there have been exclusive licensing in practice of data trading. A certain extent of standardization is present in such exclusive licensing. Thus, a regime of recordation is needed to settle ownership controversies in such practices.

The fourth problem concerns commissioned data processing services. Data holders may hire entities with professional processing capabilities to store, filter and analyze the data for them. The rights over fruits of such commissioned processing could be in dispute if the parties failed to specify in the contract. Consequently, a default rule is necessary to allocate rights over data between the parties of such contract. For example, the commissioned party is obligated to hand over the raw data, intermediate data generated in the processing, and data as the result of the processing, unless otherwise provided in the contract. The commissioned party does not enjoy property rights over data under the default rule. In addition, in the cases where the commissioned party is bankrupt, the data shall be returned and shall not be liquidated as debtors' property.

The fifth problem concerns the statutory exceptions of data processing, including fair use and compulsory licensing. While the proposal of establishing a data property regime has received academic and official welcome, it has also raised antitrust concerns, since big-techs holding the data may refuse to share such data with the public. As a result, a system of extensive fair uses should be provided in the data property regime. For example, reasonable scraping of public data, using public data for scientific research, or generative AI training using data could be allowed. In addition, a system of compulsory data licensing could be established to ameliorate monopoly problems.

The sixth problem is the question of public ownership for data. This is largely a problem of Chinese characteristic. Governmental departments collect tons of data in performing their functions, and how the rights and interests over such data should be allocated is under heated discussion. Currently, the provisions regarding scope and utilization of public data are different in each region. A unified regime of public data sharing and licensing is vital for the efficient exploitation of such data.

Amongst these aforementioned issues, licensing and statutory exceptions of medical data show no significant differences with those of other types of data. Thus, in the following chapter, we will not discuss the problem of licensing and statutory exceptions in detail. Issues relating to first-order statutory rights, data collaboration, data processing services, and public ownership of data will be elaborated.

9.3 Modules of First-Order Statutory Rights

The first category of rights modules that comprise the bundle of rights is the first-order statutory rights modules of information subjects, who are the legal entities that supply the information carried by the data. These rights modules are often associated with the protection of personality, copyrights, patents, and trade secrets. According to the Basic Regime on Data, the protection of the first-order statutory rights of information subjects shall be "sufficient".[3] This means that without the consent of information subjects or legal justification, data processors have no right to collect, use, or license the data carrying their information. Moreover, information subjects retain "the right to access or duplicate, transfer the data created with their contribution".[4] Generally, the first-order statutory rights enjoyed by information subjects are considered to precede property rights enjoyed by data processors (Wang 2022, Shen 2023). Nevertheless, given the brevity of the Basic Regime on Data on this issue, reference to other current legislation is needed to further flesh out the content of these rights.

9.3.1 First-order Statutory Rights Module for Natural Persons

In Chinese law, the personal rights module for natural persons mainly derives from the Personality Rights Book of the Civil Code and the PIPL. Such rights mainly contain the right to privacy, and more importantly, the rights over personal information. They are categorized as personality rights in Chinese law, for they are considered to be essential to maintaining a natural person's personal integrity, freedom, and dignity (Cheng 2020).

Consequently, the protection for privacy and personal information was designed to be relatively stringent. The right to privacy prevents unwarranted acts of "prying into, intruding upon, disclosing, or publicizing other's private matters" according to Article 1032 of the Civil Code, which only has limited impact on data processing for medical purposes. However, personal information protection, which restricts "collection, storage, use, processing, transmission, provision, disclosure and deletion"[5] activities, obviously poses a more significant limitation on data processors. For example, Article 23 of PIPL stipulates that provision of personal information to third-parties requires a separate consent from the information subject. Moreover, data collected by medical institutions, which contains medical records, results of medical examinations, diagnoses, and prescriptions, are likely to fall within the scope of sensitive information (Cheng 2021, 262–63), and is subject to stricter scrutiny according to the PIPL[6] (Cai and Chen 2022). The PIPL was largely modeled upon EU's GDPR, yet without a statutory exception for research purposes like Article 89 of GDPR. Thus, requirements for lawful processing of health data collected by medical institutions are difficult to satisfy.

The academia has offered four possible solutions to this problem thus far. The first approach, developed under current statutory framework, is to reinterpret the basic concepts of "personal information," "consent" or "fair use," excluding

3 See Article 5 of the Basic Regime on Data.

4 Id.

5 See Article 4 of PIPL.

6 See Section 2, Chapter II of PIPL.

processing activities which are not targeted against specific individuals (Liao 2022, Liu and Liu 2024, Han 2024). The second approach is to comply with such rules by avoiding provision of personal information. For example, medical institutions and software companies may agree to train medical AI at the medical institution, without transmitting the data to servers owned by the software company. This approach is more suitable for training AI products tailored for a specific institution. However, if the third-party service provider intends to store the data on its own location for its own use, protection of personal information may prohibit providing data without proper consent. The third approach is to explore a more efficient consent mechanism, for example, the Personal Information Management System proposed in EU (European Data Protection Supervisor et al. 2020). The fourth approach suggests that special legislation on medical data utilization should be put in place to allow wider utilization with proper safeguards, following the precedents of Japan (Li and Chen 2020).

9.3.2 *First-Order Statutory Rights Module for Nonnatural Persons*

While the PIPL has specifically addressed the first-order rights enjoyed by natural persons in detail, research regarding the first-order statutory rights of non-natural persons as information sources is still at a beginning stage (Wang 2023, Ding 2023). Apart from common claims such as trade secrets and intellectual property rights, the extent to which non-natural person information sources such as online stores enjoy rights similar to those of natural persons, including query, copy, correction, deletion, and portability remains to be systematically discussed (Xiong and He 2024).

In medical scenarios, diagnosis methods and prescriptions by physicians and medical institutions may fall into the scope of trade secrets. Article 7 of the Basic Regime on Data acknowledged in principle a right to "copy and transfer the data generated by their facilitation," emphasizing the need to "sufficiently protect the legitimate rights and interests of data sources". The policy goal behind this statement is to support information subjects in enhancing their marketing capabilities by aggregating their sales data. Two questions stem from this statement of policy: (a) whether the medical institutions enjoy a legal right of copying and transferring such data to other service providers, and (b) whether a medical institution is entitled to transfer such data to a single professional data service provider, aggregating the data used to be stored by multiple processors. Further discussion on detailed rules is needed to prevent unfair competition, especially in cases involving cross-platform transferring.

9.4 Allocation of Property Rights Over Jointly Contributed Data

Due to its replicability, data generated in certain circumstances can be captured and concurrently held by multiple collaborative processing entities. In social relationships relative to strangers, each concurrent holder, similar to independent data processing entities, enjoys a property right to hold data, which excludes unwarranted external interferences or trespasses. However, determining the data property rights among concurrent holders themselves is a much more complex matter. While such issues could be resolved by prior agreements, there are still many joint processing entities lacking sufficient motivation or capability for such arrangements in practice. Thus, a default rule of allocating rights over such data is necessary.

While Article 5 of the Basic Regime on Data has set out that first-order statutory rights take precedence over the property rights of the data processor, in so far as property rights modules are concerned, there is no clear legislative guidance in respect of claims between data processors that jointly contributed to the generation of the data, especially in the absence of prior data-related agreements. In accordance with the principle of rewarding input as stipulated by Article 7 of the Basic Regime on Data, the primary criterion for determining the priority of data rights claims among different data processors should be the ascertainment of the input and division of labor of each entity in data production (Xiong et al. 2023).

In the case of medical data, multiple medical institutions may have been pooling their data into a much larger data bank for more efficient utilization or AI training. For example, participating institutions can pull out the medical records of a patient generated in another participating institution for references. Such a data bank may be managed by one of the participating institutions, or by a third-party platform. The problem that whether each party enjoys a right to use and a right to manage for profit emerges. In each scenario, it is necessary to differentiate the contractual purposes of data collaboration, in order to allocate rights and interests over such data properly (Xiong 2023).

9.4.1 Scenario A: Data Bank Managed by a Participating Institution

Medical data accumulated in medical institutions' daily operations could be transmitted to and stored at one of the medical institutions with proper capabilities to manage such data technically. Naturally, medical institutions shall exercise their rights according to the provisions in the contract. Nevertheless, in absence of such provisions, conflicting claims over such data may be present. The proper resolution of such disputes requires a more detailed analysis of the contractual relations between the parties.

To begin with, each processor is entitled to hold and use the data set on its own, in order to encourage data processors (in this case, medical institutions) to pool, exchange and exploit their data by forming a comprehensive data set. However, if one party intends to manage the data for profit through means such as transferring it as a whole to external parties, licensing it for third-party use, or making it public, then the consent of the other holders should be obtained. This is because the commercial value of the data held by other collaborating parties may be significantly diluted once some holders manage the data in the aforementioned ways. Consequently, other data holders' reasonable expectations regarding developing, processing, and deriving benefits from the data in the future will be compromized (Xiong 2023).

Questions remain: (a) is there any limitation as to the purpose and scope of the right to use for each party? (b) is it possible for any party to license such data to a third-party without obtaining the authorization of other collaborators? To resolve such difficulties, two general ideas are undoubtedly vital. First, the guiding principles of contract law is indispensable for resolving such disputes, for the relationship amongst processors is in its essence contractual. Second, it is necessary to strike a balance between the interests of each collaborator. Although the law aims to protect each party's interest by limiting the right to manage for profit, such protection is not unconditional. If a holder disagrees with another holder's disposal, they should provide a valid justification, such as specific data management affecting the achievement of cooperative purposes or risking the disclosure of their trade secrets. In the absence of a valid justification or agreement, proper compensation is required.

9.4.2 Scenario B: Data Bank Managed by a Third-Party Platform

Another possible way to form a data bank is for medical institutions to upload their data to a third-party platform, who handles medical data professionally. If medical institutions and the platform fail to specify a part of their respective rights and obligations in advance, whether the platform is entitled to use and manage data for profit is in question.

In practice, a third-party platform holding and managing a comprehensive set of health data professionally is obviously incentivized to develop a product for medical purposes, for example an AI with diagnosing functions. Such utilization of the data falls into the scope of right to use. In addition, a platform holding a comprehensive set of medical data may also be motivated to license such data other interested parties. The legality of licensing depends on whether the platform enjoys a right to manage data for profit.

Two approaches could be proposed to address this dilemma. To begin with, interpreting the contract between medical institutions and the platform more carefully is undeniably vital. Whether such a right to use and manage for profit is conferred to the platform, and the scope of such right, primarily depends on the purpose of the contract. It is natural that the right to use for the platform shall be constrained within the original intent behind the contract. Another possible approach, if contract interpretation alone is not powerful enough, is to view such practice as a trust formed by data. As the trustee, the platform is entitled to manage such dataset for profit, yet also bound by data safeguard duties set forth by data law and the fiduciary duties set forth by trust law at the same time (Sun 2023, Li 2024).

9.5 Protection of Data Ownership in Data Processing Services

The protection issue stemming from professional data processing services is also present in the case of medical data. Although medical institutions are in advantageous positions to collect medical data, they usually lack the capabilities to exploit its value by filtering, managing, and analyzing. Thus, medical institutions are motivated to commission processing companies for their professional services. However, if the parties fail to specify the rights enjoyed by the processing company over data in advance, disputes regarding data ownership will emerge. Especially in the case of insolvency, when a civil enforcement or bankruptcy liquidation is in order, whether the medical institution is entitled to retrieve the data as a property owner is still unclear.

Existing statutes, including Civil Procedures Law and Enterprise Bankruptcy Law, have not provided a clear rule regarding the ownership of data in such cases, since the concept of data property has only been recently proposed (Chen 2023). The Basic Regime on Data, while affirming the positive effects of a property regime to data services in Article 5, did not outline a principle to resolve such disputes in detail, owing to its nature as a general guideline.

In China's current data processing practice, commissioned processing falls into three categories. The first type includes commissioned collection or storage services. For example, companies like "Datatang" collect and sell data according to the requests given by its customers;[7] cloud services stores consumers' data on their own servers. The second type involves data as byproducts of other commissioned services. For example, the electric company sometimes delegates the managing entity of an industrial estate to collect electric bill, thus the latter party continually collects and stores the electricity usage data in the area. The third type includes commissioned processing or commissioned AI training on purpose.

Commissioned data processing in the medical context usually involves the first and third category of aforementioned services. It is undoubtedly essential to delineate the boundaries of the rights of the commissioned processor. In principle, commissioned processors, who provide services such as data collection, storage, processing, and transmission of data, do not possess rights to use or the right to manage for profit, and are not permitted to use the data for purposes beyond the scope of the contract (Xiong and He 2023).

In addition, it has been argued that original data owner is entitled to retrieve the data in the case of processor insolvency. Suppose a hospital, as the rightful owner of a set of data, enters into a contract with a processing company for cloud storage services, and the processor went bankrupt. Data, as the tangible form of information recorded on a medium, is readily separable from both the content and the medium, thus is capable of being retrieved. And even though property rights over data has not been confirmed as rights *in rem* statutorily, it can be argued that the property interests over data such as liquidating data have been reserved according to the purposes of a data service contract (Yu 2021). As the entrusted party does not enjoy a right to use and a right to manage for profit on its own, the data should be returned to the client, and the client's data must not be used for liquidation and distribution purposes, in the event of dissolution or bankruptcy of the entrusted processor. When entrusted processors provide services such as training, labeling, and cleaning of data for others, they should fulfill their obligations in good faith by returning the processed data results, process data, and original data.

This analysis also highlights the importance of introducing a data property regime. Absent the introduction of a data property regime, the medical institution would lack a legal basis to retrieve the data. Entrusted data controlled by the company may be rightfully deemed as the debtor's property, and the medical institution would only be able to claim as an ordinary creditor. To mitigate such risks, data owners

[7] See Linyi Intermediate People's Court of Shandong Province, State v. Chang Kai, Yan Xingwang et al., [2018]. 2 April 2019.

can only resort to alternative measures (such as improving their own data storage technology to enhance their data storage capacity and security capabilities) and would incur additional costs for technical improvements (Xiong and He 2023).

9.6 State as Data Owners

In parallel to the discourse over the propertization of data processed by private firms, the promotion of open access to public data has also occupied the recent academic and policy debates in China, part of which are associated with administrative monopoly in data markets. Up to now, both the Chinese academia and regulators have roughly reached a consensus that, while privacy and personal information protection and data security primarily prevail, extra efforts should also be dedicated to promoting the circulation of data resources so as to better exploit the economic value of data (Xiong and He 2023). One of the essential goals, which is also one of strong Chinese characteristic, is to accelerate the circulation of public data that is directly collected by governmental organs or of a strong public nature.

Currently, multiple government organs collect medical data from medical institutions for regulatory purposes. Does such medical data aggregated by governmental departments qualify as "public data" defined by Article 4 of the Basic Regime on Data? It is therefore vital to delineate the scope of public data that could be accessible to the public.

According to Article 4 of the Basic Regime on Data, public data is data "generated during the performance of legal duties or the provision of public services by government agencies at all levels, as well as by enterprises and public institutions." Differed from "business data" which is subject to property protection set out in Article 5, Article 4 of the Basic Regime on Data emphasizes opening and sharing for the utilization of public data. Public data for public governance and public welfare endeavors will be conditionally shared to the public free of charge, while public data for industrial and sectoral development will be utilized through paid licensing. Thus, whether medical data gathered by government departments are within the scope of public data is a decisive question for its future utilization.

Currently, more detailed definitions of public data are mostly found in regional legislations. Up to March 2023, a total of 30 provinces and cities across the country have begun to legislate on data. At the provincial level, there are over 140 normative documents, of which 33% focuses on public data exclusively (Shi 2023). At the same time, the academia is also debating the scope and governance of public data (Shang 2021, Zhao 2021). However, such legislations and scholarly opinions have reached little consensus regarding the definition of public data (Cai and Chen 2022, Zheng and Tang 2021). Technically speaking, public data should be based upon "public interest" and designated as an extension of the goal of public administration, in order to meet the legislative intent (Zheng and Tang 2021).

If we focus on the particular case of medical data, we would see a more complex picture. According to statistics released by National Health Commission, about half (a percentage of 48.05) of the medical institutions are not of a solely public nature, and about 40 percent (a percentage of 39.58) of all medical institutions are entirely private.[8] There is no reasonable legal basis for including data reported by privately operated medical institutions into public data and utilizing it beyond the original regulatory purposes. Thus, rights over medical data reported by private hospitals should not be turned to responsible governmental departments. The right to use such data of the government is better confined within the legal functions of its office. If the governmental department plans to further exploit such data, the proper legal process is governmental purchase or requisition (Tan 2024).

Compared with data reported by privately operated medical institutions, it is more reasonable that the medical data collected by public medical institutions fall within the purview of public data. However, there is still the issue as to which body is entitled to act as a data owner. Possible options include the public institution directly collecting the data (in our case, medical institutions including hospitals), responsible regulatory organs to which the data is reported, and other related government agencies.

As mentioned above, a number of governmental organs collect data from medical institutions concurrently. The departments of human resources and social security, health departments, insurance institutions, as well as new-type disease control apps, all holds certain parts of medical data to carry out their respective functions, which differs from the data held by medical institutions and is often intended for specific purposes. The data held by the department of human resources and social security primarily focuses on medical insurance settlement data, thus does not include out-of-pocket expenses incurred during medical treatment. Health departments obtain data due to their regulatory responsibilities. Insurance institutions generally have relatively limited accumulation of data (Li and Guo 2021). In addition to these departments who collect and hold the data technically, other departments like state-owned asset administration department, and data administration agencies, are also possible right holders.

While the precise status of such data still awaits further clarification, governmental departments are already beginning to license such data as right holders in practice. Data administrations in some regions have already begun to act as data owners, signing trilateral licensing contracts with medical institutions and third-party companies.

9.7 Conclusions

The policy drafters have consciously decided to depart from the "ownership-usufruct" approach in policy documents, as revealed by experts involved in the drafting of the Basic Regime on Data. They preferred the alternative approach based on the "bundle of rights" theory, which provides a more flexible analytical framework for examining multiple rights claims on data. They considered this approach more appropriate for

[8] See National Health Commission of the People's Republic of China, 2022 China Health Statistics Yearbook, available at: http://www.nhc.gov.cn/mohwsbwstjxxzx/tjtjnj/202305/6ef68aac6bd14c1eb9375e01a0faa1fb.shtml, accessed 6 May 2024.

accurately reflecting the patterns of data entitlements and realizing their full value (Xiong et al. 2023). To efficiently and properly delineate the rights and obligations of each participant in the data economy, it is necessary to focus on the social relations of data production and circulation (Xiong 2023).

In this chapter, we introduced issues arising from exploitation of medical data basing on the aforementioned "bundle of rights" approach. Four of the issues in particular, including the protection of first-order statutory rights, allocation of rights over jointly contributed data, protection of data ownership in data processing services, and the exploitation of public data, are explored in detail in this chapter. We believe that the establishment and development of a data property regime is beneficial for the collection, production, utilization and circulation of medical data.

References

Bao, X. and B. Xiong. 2020. Social capitals in the data of contacts: a research paradigm for the distribution of data properties. China Law Review. 32(2): 146–68.

Cai, P. and L. Chen. 2022. Demystifying data law in china: a unified regime of tomorrow. International Data Privacy Law. 12 (2): 75–92. https://doi.org/10.1093/idpl/ipac004.

Chen, X. 2023. Protection and management of data in the procedure for bankruptcy. Journal of Law Application. 12: 83–92.

Cheng, X. 2020. Research on the fair use of personal information in China's Civil Code. Peking University Law Journal. 32(4): 1001–1017.

Cheng, X. 2021. The Interpretation of Personal Information Protection Law of the People's Republic of China. China Legal Publishing House. Beijing.

Dai, X. 2021. A Relational Approach to Delimiting Data Rights. Peking University Law Journal. 33(6): 1561–1580.

Ding, X. 2023. On the Rights of Data Producers. Journal of Comparative Law. 3: 14–25.

European Data Protection Supervisor et al. 2020. TechDispatch: Personal Information Management Systems. 3, 2020. Publications Office of the European Union. https://data.europa.eu/doi/10.2804/11274.

Han, C. 2024. The Dilemma and Solution of Personal Health Data Protection in Medical AI. Science Technology and Law-Chinese-English Version). 1: 54–60. https://doi.org/10.19685/j.cnki.cn11-2922/n.2024.01.006.

Li, H. and G. Chen. 2020. On conflict and balance between data-driven innovation and personal information protection—a study of japan's medical data regulation experience. Bulletin of Chinese Academy of Sciences. 35(9): 1143–1151. https://doi.org/10.16418/j.issn.1000-3045.20200522002.

Li, M. 2024. The Deconstruction and Reconfiguration of Data Trust. Journal of Northwest Normal University (Social Sciences). 61(3): 110–21. https://doi.org/10.16783/j.cnki.nwnus.2024.03.012.

Li, W. and Y. Guo. 2021. Research on the Legal Issues of Commercial Use of Health and Medical Data. Qilu Journal. 4: 96–103.

Liao, L. 2022. On Path Optimization of weakened application of consent rules in personal information processing: scenario segmentation based on the theory of contextual integrity. Law and Social Development. 28 (6): 156–180.

Liu, G. and Z. Liu. 2024. On the Path of Defining the "Reasonable Range" of Personal Information Processing with Deep Synthesis Technology. Science and Society. 14(1): 85–101. https://doi.org/10.19524/j.cnki.10-1009/g3.2024.01.085.

National Development and Reform Commission. 2023. Accelerate the Construction of a Data Foundation System with Chinese Characteristics to Enable All People to Share the Development Dividends of Digital Economy. Macroeconomic Management. 3: 13–34.

Shang, X. 2021. Construction of Data Usufruct System in Government Data Opening. ECUPL Journal. 24(4): 59–72.

Shen, W. 2023. On the Hierarchy of Data Property Right System: Trichotomy on Data Right Identification. China Legal Science. 4: 26–48. https://doi.org/10.14111/j.cnki.zgfx.2023.04.001.

Shi, J. 2023. Local Legislative Practices of Foundational Architecture for Data: A Preliminary Quantitative Research Based on Policy Documents. Digital Jurisprudence Review. 1: 44–84.
Sun, Y. 2023. Research on the Mechanism for the Ownership and Authorization of Corporate Data Rights. Journal of Comparative Law. 3: 56–73.
Tan, Z. 2024. Theoretical Justification and Institutional Development of Data Requisition. Law and Social Development. 30 (2): 144–163.
Wang, L. 2022. On Data Rights and Interests: From the Perspective of Bundle of Rights. Political Science and Law. 7: 99–113.
Wang, L. 2023. On the Rights of Data Producers. Law and Social Development. 29(6): 36–57.
Xiong, B. 2023. Standardization of Rights to Data. Peking University Law Journal. 35(5): 1145–64.
Xiong, B. et al. 2023. Unpacking Data: China's "Bundle of Rights" Approach to the Commercialization of Data. International Data Privacy Law. 13(2): 93–106. https://doi.org/10.1093/idpl/ipad003.
Xiong, B. and J. He. 2023. Theory, Methodology and Economic Implications of the Delineation of Data Property Rights. Chinese Journal of Law. 45(3): 54–72.
Xiong, B. and J. He. 2024. The System of Basic Legal Regimes for Data Market. Academic Monthly 56(1): 102–14.
Yu, J. 2021. "Separation of Data from the Insolvency Assets. Science of Law (Journal of Northwest University of Political Science and Law. 39(5): 79–88. https://doi.org/10.16290/j.cnki.1674-5205.2021.05.010.
Zhao, J. 2021. Rationality and Legal Significance of Public Data Belonging to Government. Journal of Henan University of Economics and Law. 36(1): 13–22.
Zheng, C. and J. Tang. 2021. On the Normative Meaning of Public Data. Research on Rule of Law. 138(6): 67–79. https://doi.org/10.16224/j.cnki.cn33-1343/d.20211022.010.

10

Parts and Wholes

Developing Innovative Governance Mechanisms for Data Flows Beyond the Paradigms of Privacy and Property

Vagias Karavas

"The whole is always smaller than its parts."

(Latour et al. 2012)

10.1 Introduction

The postfoundational pandemic theory, developed since 2020, presents three prevailing perspectives on the nature of the pandemic: as either insignificant, politically exceptional, or a catalyst for creative change (see on these narratives and their critique Repo and Richter, 2022). While some scholars dispute the pandemic's status as a significant event, arguing it reflects existing inequalities rather than marking a novel rupture (Badiou 2020), others, such as Agamben (2020), interpret it as a politically manufactured crisis serving to reinforce sovereign power. Conversely, a subset of thinkers views the pandemic as a moment of potential transformation, spurred by the material disruptions it has caused to economic and social systems (Žižek 2020a, 2020b).

My aim in this article is not to unravel the complicated theoretical dimensions surrounding the pandemic in order to confirm any of the above theoretical perspectives. Instead, starting from an understanding of the pandemic as a "context-specific assemblage of a myriad of other, longer-standing events – formations, practices, discourses, relations – to which it cannot be reduced" (Repo and Richter 2022), I will argue – and this is one of the premises on which my article builds –

Professor for Legal Sociology, Legal Theory and Private Law at the School of Law of the University of Lucerne and Director of the Institute for Interdisciplinary Legal Studies – lucernaiuris.
Email: vagias.karavas@unilu.ch

DOI: 10.1201/9781003475927-12

that the pandemic, in its highly contingent nature, has served as an accelerator of pre-existing social processes. For example, the Covid-19 crisis has exacerbated pre-existing extractivist practices that target the human body, breaking it down into discrete, explorable and exploitable molecular and biochemical components. The traces of these practices in turn date back to the early twentieth century, when a decisive transition took place, namely the transition from in vivo experiments on whole animals in the nineteenth century to in vitro experiments on isolated body fragments under controlled conditions (Landecker 2007) (regarding the normative effects of the biomedical fragmentation of the human body see Karavas 2020 and Rabinow 1996). From this point of view, the first pandemic of the 21st century appears as a contingent event in a continuum of similar events aimed at reinforcing society's tendency to acquire surplus knowledge, a tendency inherent in modernity's inception.

This becomes evident upon scrutiny of the various research initiatives launched during the pandemic. As noted by Molina (Molina 2022), governmental organizations and private enterprises have endeavored not only to develop "public health measures to address the pandemic and its consequences at multiple social scales: from shelter-in-place ordinances at the state level, mask production and distribution, to vaccine requirements for indoor dining at the city level," but also to pivot their research agendas towards understanding the virus and its impacts.

For instance, Molina (Molina 2022) highlights the UK Biobank's comprehensive initiatives undertaken since the pandemic's onset which have included a serology antibody testing study, a COVID-19 repeat imaging study, a self-test antibody study, an infection study, and lastly a health data linkage study.

Moreover, major private genomics companies such as Verily, Ancestry.com, Helix, and 23andMe have embarked on their own research projects during this period. For instance, Verily, a subsidiary of Alphabet, initiated a COVID-19 immune response study in 2021, while 23 and Me successfully recruited over half a million participants, including over 10,000 non-customers, for its COVID-19 study (for more informations on these studies see Molina 2022).

Lastly, noteworthy collaborations such as the agreement between Pfizer and the Israeli Ministry of Health (MoH) in 2021, titled Real-World Epidemiological Evidence Collaboration, exemplify efforts to expedite vaccine rollout by exchanging vaccine safety and efficacy data for the entire eligible population of Israel (Simana 2021). This agreement underscores the collective response of public and private entities to address the pandemic's challenges and highlights the multidimensional nature of pandemic research and response efforts.

In times of crisis, the imperative to swiftly acquire new insights into the virus and its transmission, potentially leading to life-saving interventions, is both understandable and commendable. However, it is imperative to interrogate the underlying motivations propelling modern society's pursuit of surplus knowledge, particularly when such knowledge holds prospects for commercial exploitation. As Teubner (Teubner 2021) recently elucidated, the drive to generate surplus value is not exclusive to capitalist economies but permeates every functional system in modern society. Each societal domain, in its production of meaning, is oriented toward the generation of surplus value, albeit manifesting diversely across contexts. In this context, a paramount inquiry arises:

Who "owns" medical big data, the very bedrock facilitating the production of surplus value in modern society?

As I will explain, the legal doctrine's answer to this question vacillates between two alternatives: data protection law on the one hand and (intellectual) property law on the other. In the discussion that follows, therefore, I will first provide a succinct assessment of the existing legal landscape in relation to data rights, while highlighting the limitations that arise from the application of both data protection law and (intellectual) property law in the context examined here. Drawing on insights from the scholarship of Gabriel Tarde, I will then question a basic assumption underlying the prevailing legal framework: the notion that data can be uniquely attributed to a particular source, either as a part (single datum) or as a whole (big data). Finally, I will argue for the need for alternative models of data governance that go beyond the traditional legal paradigms of ownership and privacy. Based on the principle of data solidarity, I will argue for the creation of data trusts as a way out of the paradoxes and dead ends into which the current legal system leads us.

10.2 Oscillations

10.2.1 Data protection Law

In the realm of data concerning living persons, data protection legislation typically comes to the forefront. According to the European Union's General Data Protection Regulation 2016/679 (GDPR), personal data – defined as "any information relating to an identified or identifiable natural person" (Art. 4(1)) – is accorded legal protection. At the heart of data protection law lies the principle that only the individual concerned can determine the circumstances under which their data may be shared, including with whom, when, how, and under what conditions.

Given that research institutions frequently rely on the exchange of personal data to advance understanding of the virus, a pertinent question arises: does this data-sharing, often involving cross-border transfers, align with GDPR?[1] This issue has been the subject of numerous discussions in the relevant literature during and after the pandemic (Bentzen et al. 2023, Kędzior 2021, Staunton 2021, Becker et al. 2020, Kuner 2020). Accordingly, the prevailing opinion argues that the GDPR provides three possible legal frameworks for data transfers (see on this view Kuner 2021): (1) formal adequacy decisions by the European Commission concerning third countries or international organizations;[2] (2) appropriate safeguards like contractual clauses or legally binding agreements between public authorities;[3] or (3) exceptions for specific circumstances.[4]

1 The topic is admittedly complex and cannot be analyzed in detail here. Instead, only cursory reference will be made to certain legal options.

2 Article 45 GDPR.

3 Article 46 GDPR.

4 Article 49 GDPR.

Of the three options, the latter, i.e., opting for derogations, is perhaps the most promising way to create a legal basis for data transfers, provided that the prescribed conditions are met. The two derogations that are most relevant to the fight against the pandemic are that the data transfer is necessary for important public interests[5] or to protect the vital interests of the data subject or other persons (see again Kuner 2021). The recitals of the General Data Protection Regulation confirm that these exemptions cover situations such as epidemic monitoring and contact tracing for infectious diseases. However, one has to add in this context that these exemptions are intended for use only under very strict conditions. For example, the public interest exception only applies to interests that are recognized by the EU or Member State law.[6] Furthermore, according to the European Court of Justice, the exceptions to data protection rights must be interpreted narrowly. Furthermore, data protection authorities have suggested that exemptions should only be used for "occasional" and "non-repetitive" transfers. Considering that neither the text of the GDPR nor the CJEU rulings use the terms "occasional" or "non-recurring" for these exemptions, and that "combating serious cross-border threats to health" is a constitutional interest of the EU under Article 168(1) TFEU,[7] it can be concluded that the above exemptions provide a legitimate ground for such data transfers.

Nevertheless, whether data transfers can succeed depends not only on legal premises, but also on the removal of corresponding technical barriers, such as the heterogeneity of data sources (clinical studies, observation cohorts, patient records, etc.) and data collection tools, for which a meta-harmonization of common data, metadata standards and standardized processes is currently lacking (Tacconelli et al. 2022).

Despite the technical complexities involved, the legal response to the intricacies of data transfers is notably sophisticated, yet it prompts a deeper inquiry into the status of data subjects who have contributed their personal data to research endeavours. Does data protection legislation effectively foster their inclusion and fair involvement in the management and utilization of their aggregated data? Unfortunately, the response to this question leans towards the negative, stemming from an inherent limitation within the framework of data protection law, rooted in its foundational objectives.

At its core, data protection law is designed to uphold the right to privacy, construed as the freedom to lead an autonomous life shielded from intrusions by external parties (see above all Rössler 2017). Consequently, it governs the dissemination and acquisition of information by others, intricately tied to the principle of safeguarding one's autonomy from external interference. Therefore, it primarily concerns itself with data, i.e. semantic information relating to identifiable natural persons, offering protection against risks to identification. Notably, it does not extend its jurisdiction to situations where such risks are absent, such as instances involving anonymized data, notwithstanding the vulnerability of anonymity in light of advancements in de-anonymization and re-identification techniques.

---

5 Article 49(1)(d) of the GDPR.

6 Article 49(4) GDPR.

7 Consolidated Version of the Treaty on the Functioning of the European Union, O.J. C 326. 26.10.2012. 13390.

Furthermore, data protection law inadequately grapples with the implications of data analytics, which empower third parties to glean insights about individuals beyond the scope of the individual subjects it primarily addresses (Calacci and Stein 2023). This deficiency arises from an individualistic orientation inherent in data protection law, which fails to acknowledge that data utilization in the era of big data and advanced analytics often yields collective impacts beyond individual control (see Purtova 2018, who argues that due to the impact of data analytics, data protection law could become a law for everything). To effectively counteract exclusionary trends within the realm of scientific research, which further undermine the autonomy of protected subjects, there arises a necessity for data protection law to be sensitized to threats against collective autonomy (see on this notion Graef et al. 2023) and to devise appropriate mechanisms for empowerment accordingly (see, e.g., see Calacci's 2022 suggestions regarding platform workers).

10.2.2 Data as Property

Due to the limitations of the data protection law paradigm, some scholars advocate for treating data as property, irrespective of their personal nature. Amstutz (2018) is among these scholars, presenting a compelling argument for recognizing property rights to data (see further Thouvenin 2017, Wiebe 2016, Bräutigam and Klindt 2015) (Zech 2015a, 2015b).[8] He asserts that the growing influence of big data on civil society threatens our capacity to shape societal dynamics according to individual preferences. Consequently, he asks whether the recognition of property rights to data could prevent these dangers.

Ownership, in Amstutzs′ perspective, transcends mere protection from external interference; it empowers individuals to shape their civil sphere in alignment with personal preferences (Amstutz 2018). As big data predominantly consists of user data from online activities, granting ownership of this data could potentially safeguard individual freedom. This proposition stems from the distinctive nature of digital technology, which segregates communication and information, making user data susceptible to unnoticed collection and analysis by big data companies (Amstutz 2018).

This pivotal consequence of the digital era enables mass data collection and predictive analytics, leading to algorithmic governmentality, where individuals are subtly influenced within predetermined social contexts (Amstutz 2018). Directly assigning data ownership to users could disrupt the current state of cyberspace, thwarting unauthorized data mining and endowing users with control over their data through legal means.

By shifting responsibility from big data corporations to data owners, the introduction of data ownership could thus spark a transformative shift in cyberspace, empowering civil society to regain control within digital networks. This legal paradigm shift would signify a significant evolution in internet governance, akin to the initial revolutionary advancements with the emergence of digital network technologies.

---

[8] The discourse on data ownership relates primarily to so-called "raw" data.

Despite Amstutz' well-meaning intentions, his view is problematic for a number of reasons: Supporters of a property right in data imagine the act of generation of data as a solitary act of property acquisition, which - similar to Locke's theory – takes place in a *terra nullius*. However, as the legal anthropologist Maurer (Maurer 2015) correctly notes: "Data [...] needs at least two to tango." The act of generation requires not only the contribution of the data subject, who initiates the process of data generation by using the relevant media or infrastructures, but also the infrastructures provided by third parties such as private companies or governmental agencies. So, instead of imagining the generation of data as a solitary act, carried out by an individual actor, we should rather think of it as an act of co-production that presupposes a network of different actors, infrastructures and media. Therefore, issues of property should, if any, be thought in terms of common or joint-property (Dulong de Rosnay and Stalder 2020, French and Barchers 2017), rather than in terms of personal property.

Finally, as far as Amstutz' argument regarding the normalizing effects of algorithmic governmentality is concerned, the following should be added: It is not algorithmic governmentality *per se* that is problematic. Due to its high degree of complexity, modern, functionally differentiated society is very much dependent on the algorithmic analysis of big data for the purpose of creating forecasts and patterns of interpretation (Nassehi 2019). By the way, the pandemic has highlighted this need. What is more problematic is the fact that these patterns are created without the participation of the subjects who provide the "raw" data. What is needed, therefore, are not new rights of exclusive attribution, but new rights of co-determination in the places where such analyses take place, namely – in the context of this study – in the governance structures of the databases managed by the researchers (Karavas 2018: 257).

10.2.3 Data as Intellectual Property

Now, as Birnhack (Birnhack 2021) mentions, the transition from "single datum" to "big data" or aggregated data raises the question of whether intellectual property law can be invoked to protect the later.

Indeed, the applicability of intellectual property law to big data faces inherent challenges, stemming from three key factors (Birnhack 2021). Firstly, big data does not originate from individual authorship. It thus cannot qualify as an original work in the realm of art and literature. Secondly, data serve as foundational material for subsequent creative endeavors, aligning with the concept of human knowledge as a collaborative effort and advancing societal progress. Thirdly, the unfettered flow of information plays a vital role in democratic discourse, and the imposition of property rights on raw data could impede the open exchange of information and ideas.

However, the Intellectual property law in the EU provides, in the form of a sui-generis-property right, i.e., the so-called database right,[9] that aggregated data may

[9] European Directive. Directive 96/9/EC of the European Parliament and of the Council of 11 March 1996 on the legal protection of databases.

be protected if they are structured according to some pre-defined selection criteria and if each datum is tagged. But even in the case of unstructured collections of data, data collectors can achieve control over their collection through other means, such as technological locks or trade secret law.

However, intellectual property law is also characterized by its own blind spot, which in turn consists in assigning rights only to the creator of the data collection. The process of data collection, however, is an immensely lengthy and complex task that is collective by its very nature; something which intellectual property law, in turn, can hardly accommodate. Intellectual property law is therefore ill-prepared, first, to recognize the labour that data subjects have also put into creating these data collections and, second, to give these subjects rights of control over their collected data (Karavas 2018). Incidently, the latter is also reiterated in the GDPR which states that the rights of the data subject should not adversely affect the rights or freedoms of others, including trade secrets or intellectual property and in particular the copyright protecting the software.[10]

10.2.4 Scenarios

Birnhack (Birnhack 2021) summarizes the four primary scenarios relating to the legal regime applicable to medical big data as follows:

"(a) Structured, anonymous datasets: The controller enjoys copyright, and data subjects enjoy de facto privacy protection.

(b) Unstructured, anonymous datasets: The controller does not enjoy copyright protection and may revert to other means, such as trade secret law. Data subjects enjoy de facto privacy.

Both (a) and (b) carry the risk of de-anonymization, especially if the controller shares the dataset with others.

(c) Structured, identifying datasets: The controller enjoys copyright protection but is subject to various obligations to protect data subjects' privacy. Sharing the dataset would pose a high privacy risk. Data protection law reinforces copyright.

(d) Unstructured, identifying datasets: The controller does not enjoy copyright protection and is under an obligation to protect data subjects' privacy, pushing the controller to seek other restrictive practices of not sharing the data."

None of the above scenarios contain provisions that recognize data subjects as the co-authors of the datasets in question. Instead, data subjects are primarily involved in providing consent and facilitating data collection and processing procedures. In addition, as mentioned above, data transfers can take place without the explicit consent of the data subjects if this is in the public interest. Nevertheless, a mere juxtaposition of individual rights and general interests in this regard can only lead to imprecise distinctions. How can one mainly differentiate between individual and public interests without inevitably encountering paradoxes? The only promising way out of this dilemma therefore seems to be the productive further development of the

---

10 Recital 63 GDPR.

paradoxes in which we are trapped towards concepts of collective data management that fit into the existing legal framework.

10.3 Tarde's Monadology or How to Overcome the Dichotomy Between Parts and Wholes

But what alternatives could there be to the options provided by the current legal regime? To put this question differently: How can the inclusion and egalitarian participation of data subjects in the use and control of their aggregated data be made possible without reproducing the exclusionary logic inherent in the current legal regime?

In order to achieve this goal, we first need to question the main distinction that underlies the current legal regime, namely the distinction between single data supposedly attributable to an individualized (human) source regardless if the latter is identifiable or not and aggregated data or datasets to which, as we have seen, different rights and obligations apply. To put it differently: Data protection law as well as property law refers to data that is attributable to a single source, whereas intellectual property law refers to aggregated data attributable to one "author". It goes without saying that "data are aggregative. They pile up. They are collected in assortments of individual, homologous data entries and are accumulated into larger or smaller data sets" (Gitelman and Jackson 2013). However, this fact, which is due to the numerical nature of data, should not lead us to believe that there is a significant difference between single data and aggregated data. This becomes clear when we take a detour via Tarde's monadology.

Against the backdrop of the digital turn in the social sciences, Latour et al. (Latour et al. 2012) offer us in their text with the title "The whole is always smaller than its parts – a digital test of Tarde's monads" a fascinating reinterpretation of Tarde's social theory (Tarde 2009); a long-forgotten social theory that famously dispensed with concepts such as micro and macro analysis or individual and society. According to Latour et al. (Latour et al. 2012), social theory today is concerned with the explanation of the emergent properties of social structures by means of an operation of addition or subtraction, depending on the relevant starting point. As they note: "[It thus establishes] a first level of individual entities, then [adds] to them a few rules of interaction, in order to observe whether the dynamics of interaction lead to a second level, that of aggregation, which has generated enough new properties to deserve to be called a 'structure', that is, another entity for which it is possible to say that 'it is more than the sum of its parts'" (Latour et al. 2012).

This operation can eventually be used to explain all kinds of social phenomena, from crowds to swarms, corporations, markets, states, and so on. For Latour et al. (Latour et al. 2012), however, social theories that proceed in this way provide us with a very limited vision of the social; the social is namely imagined as a linear and successive sequence from single atoms to interactions between these atoms and then to whole structures. Latour et al. in turn offer us an alternative to this common-sense version of social theory. To this end, they turn to Tarde's social theory, which eschews such leaps of aggregation and instead proposes a flat vision of the social based on a concept borrowed from Leibniz, namely the concept of "monad".

But what is a monad? As Marrero-Guillamón (Marrero-Guillamón 2015) notes: "The term monad, from the Greek monas, a 'unit' or a 'one', has a long tradition in philosophical writings, dating back to the Pythagoreans, but it was Leibniz who popularized it in the early 1700s. In the modern sense it is used to describe: '(1) a simple, irreducible, and sometimes indestructible entity; and (2) the minimal unity into which the cosmos and all composite things in it can be resolved; yet (3) containing with in itself, in contrast to material atoms, powers and relations of which it is itself the source'."

Tarde, however, departs in his social theory clearly from such an understanding of the term monad. He rather argues, using insights from the natural sciences, that only because we observe things from a distance that we see them as distinct, undifferentiated elements, i.e., as monads in the abovementioned sense. In reality all things in the world constitute nothing else than "a multitude of different elements that are connected to each other in the same way that they are connected to the elements of other aggregates" (Marrero-Guillamón 2015, with direct references to Tarde's work). And this applies not only to the entities of the physical world, but also to all social things. The smallest a social unit, the more complex and social it is.

According to Tarde, monads are animated by a striving for dominance and expansion. In order to follow this urge to spread, they also require a connection with other monads. Because in isolation they are powerless. However, this urge does not lead to the formation of an alliance of different units. Their association is instead "symbiotic" (Folkers and Opitz 2020), in the sense that they "interpenetrate each other instead of being strictly separated from each other" (Tarde 2009). Latour et al. (Latour et al. 2012) explain this symbiotic form of association as follows: "A monad is not a part of a whole, but a point of view on all the other entities taken severally and not as a totality." And in another place of their text, they also state: "Such is what is meant by a monad, a point of view, or, more exactly, a type of navigation that composes an entity through other entities and, by doing so, particularizes all of them successively – 'all of them' being an open ended list the size and precision of which always depend on more inquiries and never from the sudden irruption of a superior level accessible through a brutal shift in methods."

Now, what can we gain by applying this *monadological* principle to our case at hand?

Following Tarde, we could argue that the assumption that we can perform legal attribution operations along the dichotomy between a part (single datum) and the whole (dataset) is misleading. After all, there is no such thing as single data understood as individual elements, since such a view would deprive data of one of their constitutive features, namely their associative, relational character (Zeno-Zencovich 2020), as well as there is no aggregated data understood as the sum of their parts.

Consequently, association is not what comes after single data have been collected and analyzed, but what characterizes data in the first place. Accordingly, the difference between single and aggregated data is nothing else than a difference of perspective. Single and aggregated data are thus both different types of navigating this vast sea of data. Accordingly, with Latour et al. (Latour et al. 2012) we could argue, "that to be part of a whole is no longer to 'enter into' a higher entity, but for any given monad it is to lend part of itself to other monads without either of them losing their multiple identities."

10.4 A Quest for Alternatives: The Concept of "Data Trust"

If, as we have seen, data never refers to a single source, neither in its semantic nor in its syntactic dimension, the act of attribution carried out by the current legal system is then without a reference point. In other words, there is no single source to which the data can be exclusively attributed neither in their individualized nor in their aggregated form (see for the same argument with a specific reference to personal data Zeno-Zencovich 2009).

Instead of arguing about the exclusive allocation of rights, I would therefore like to suggest a different approach at this point; a path that argues for the introduction of participatory design models in which data collectors and data subjects jointly redesign the way in which data are collected and managed; models that should furthermore be implemented in the places where such data are collected and stored, i.e., in databanks (see for a similar argument in the biomedical context Karavas 2018).

At this juncture, a pertinent inquiry arises: "Why should the empowerment of data subjects be of concern?" The assertion posited herein presupposes that the prevailing crisis of public trust in research emanates from a discernible sense of disenfranchisement and disempowerment within the contemporary research regime. Individuals contributing biological samples and data to research find themselves relegated to passive roles, thereby lacking recognition as active agents in the process. Consequently, the imperative of inclusion transcends mere justice; it assumes the character of an existential quandary that undermines the very legitimacy of scientific research. The exclusion of one of its pivotal constituents – the donors of bodily samples and data – exerts a dual impact, as substantiated by numerous empirical studies: firstly, it erodes social trust in the research enterprise, and secondly, it compromises the enterprise's capacity for introspection concerning its societal obligations, notably the provision of socially responsive innovation services. Put differently, the inclusion of data subjects becomes a prerequisite for enabling the research regime to confront its reliance on its human environment. Acknowledging this interdependence, in turn, becomes imperative for the sustenance of said regime.

The necessity to involve data subjects as stakeholders in the management of their aggregated data through collaborative efforts with researchers has gained prominence within scholarly discourse, often discussed under various terms such as "data sovereignty" (Augsberg and Gehring 2022), "data justice" (Taylor 2017) or "data solidarity" (Prainsack et al. 2022).

Proponents of the latter concept argue, for instance, that data solidarity encompasses scenarios where individuals recognize and act upon shared commonalities with others, thereby transcending the dichotomy between individual interests and collective well-being (Prainsack et al. 2022). They emphasize that solidarity does not entail individuals sacrificing their own needs for the benefit of others, nor does it prioritize collective needs over individual ones. Rather, it acknowledges the interconnected nature of individuals' needs, interests, and identities within broader social, environmental, and technological contexts (*idem*). Within this framework, solidarity underscores and advocates for practices, mechanisms, and structures that empower individuals to recognize and address the needs and interests of others, thereby fostering the development of institutional and legal frameworks supportive of such solidarity (*idem*).

In the endeavor to actualize the concept of data solidarity, proponents advocate for a framework of solidarity-based data governance grounded on the following three fundamental pillars (Prainsack et al. 2022). The first pillar should aim to facilitate data utilization that yields substantial public value while minimizing the risk of causing undue harm. Under the second pillar, the focus should shift to the prevention and alleviation of harm across all data applications, by developing robust mechanisms for mitigating harm that are readily accessible to all impacted parties, independent of legal complexities or fault attribution. Lastly, the third pillar should endeavor to bolster mechanisms for equitable benefit distribution, ensuring that a portion of the profits stemming from commercial data exploitation is reinvested into the public domain.

The concept of data trusts (Blankhertz 2020, Delacroix and Lawrence 2019, Ruhaak 2019) represents a promising legal mechanism for actualizing the principles of data solidarity, thereby addressing the concerns outlined in the aforementioned three pillars.

While concrete examples of data trusts remain scarce, their potential is great and should be further expanded. From the outset, the overarching goal of data trusts has been to foster the development of sustainable and compliant data practices that align with the interests of diverse stakeholders. Serving as intermediaries, data trusts facilitate the broader access and sharing of data while ensuring the protection of rights for all involved parties. However, the primary motivation behind the inception of data trusts has been to empower data subjects, mitigate power differentials, and facilitate participatory data governance.

As a result, these entities can play a crucial role in harmonizing individual interests with collective oversight and public value, enabling the articulation and enforcement of their collective interests against both public and commercial research entities. In this regard, their function bears resemblance to that of labor unions, as they aggregate individual influence to advocate for collective interests.

In view of this last comparison, we do not have to start from scratch when developing participatory design models such as data trusts, but can draw on our rich historical experience with co-determination and works councils in collective labor law. While some might question the applicability of collective labor law in our contemporary context, especially amidst the dominance of neoliberal ideologies, Dukes' (Dukes 2014) nuanced perspective offers valuable insights.

Dukes (*idem*) reminds us that labor law serves as a crucial battleground where conflicts between social classes over economic benefits and the balance between market competition and social welfare are negotiated. By recognizing the ongoing importance of such conflicts, we can appreciate the potential of labor law principles to inform contemporary debates about data governance.

Indeed, parallels between the realms of data subjects and workers abound (see Fumagalli 2024). Both groups navigate power dynamics in their interactions with institutions and face questions about the extent to which their lives should be commodified or controlled by external forces, whether economic or technological/scientific. By exploring historical experiences of workers' participation in decision-making processes within the workplace, we can glean valuable lessons about fostering meaningful engagement and representation in data governance discussions.

However, as we embark on this journey, it's essential to underscore a fundamental principle: the voluntary nature of participation. Unlike the coercive mandates of communist regimes, where individual participation is often enforced as a social duty, any participatory data governance model must uphold the right of each individual data subject to choose whether or not to engage. This commitment to voluntary participation is not just a matter of individual autonomy; it's a safeguard against reproducing authoritarian structures within data governance frameworks.

So is a kind of "data communism" being advocated here, in which the individual has the opportunity to opt out? The analogy may be provocative, but in principle it is about nothing less than the pursuit of a democratization of the science system, i.e., the creation of data-based research structures in which individual subjects have a say in decisions and in which collective dialogue replaces top-down control. By embracing democratic participation structures in the collection and management of big data, we can create fertile ground for transformative discussions about the ethical, social, and economic implications of data use, ultimately fostering a more equitable and inclusive research landscape.

Acknowledgments

I would like to thank all participants of the EuCARE workshop on "Privacy, data ownership and data sharing: an interdisciplinary dialog for post-pandemic transnational research" (Foundation Brocher, Geneva, 19-20 June 2023) for the insightful discussions and comments.

References

Agamben, G. 2020. The invention of an epidemic. Coronavirus and philosophers. European Journal of Psychoanalysis, available at: https://www.journal-psychoanalysis.eu/articles/coronavirus-and-philosophers/.

Amstutz. M. 2018. Dateneigentum. Funktion und Form. Archiv für die civilistische Praxis (AcP) 218: 521, 495, 509.

Augsberg, S. and P. Gehring [eds.]. 2022. Datensouverenität. Positionen zur Debatte. Campus Verlag, Frankfurt am Main.

Badiou, A. 2020. On the Epidemic Situation. Verso Blog. 23 March 2020, available at: https://www.versobooks.com/blogs/4608-on-the-epidemic-situation.

Becker, R., A. Thorogood, J. Ordish and M.J.S. Beauvais. 2020. COVID-19 Research: Navigating the European General Data Protection Regulation. J. Med. Internet Res. 22: e19799, available at doi: 10.2196/19799.

Bentzen, H.B., H.K. Olav and G. Ursin. 2023. Maximizing the GDPR potential for data transfers: first in Europe. Lancet Reg Health Eur. 27: 100600, available at doi: 10.1016/j.lanepe.2023.100600.

Birnhack, M. 2021. Who Controls Covid-Related Medical Data? Copyright and Personal Data. IIC 52: 823, 824, available at https://doi.org/10.1007/s40319-021-01067-5.

Blankertz, A. 2020. Designing Data Trusts - Why We Need to Test Consumer Data Trusts Now. Stiftung Neue Verantwortung, available at https://www.stiftung-nv.de/sites/default/files/designing_data_trusts_e.pdf.

Bräutigam, P. and T. Klindt. 2015. Industrie 4.0, das Internet der Dinge und das Recht, Neue Juristische Wochenschrift. 1137–1142.

Calacci. D. 2022. Organizing in the End of Employment: Information Sharing, Data Stewardship, and Digital Workerism. CHIWORK '22: The Symposium on Human-Computer Interaction for Work, June 08–09, 2022, Durham, NH. ACM, New York, NY, USA. 1–9, available at https://doi.org/10.1145/3533406.3533424.

Calacci, D. and J. Stein. 2023. From access to understanding: Collective data governance for workers European Labour Law Journal 14. 253–282, available at https://doi.org/10.1177/20319525231167981.

Delacroix, S. and N.D. Lawrence. 2019. Bottom-up Data Trusts: Disturbing the "One Size Fits All" Approach to Data Governance. International Data Privacy Law 9: 236–252, available at https://doi.org/10.1093/idpl/ipz014.

Dukes, R. 2014. A Global Labour Constitution? Northern Ireland Legal Quarterly 65. 283–301, available at: http://eprints.gla.ac.uk/76158/1/76158.pdf.

Dulong de Rosnay, M. and F. Stalder. 2020. Digital Commons. Internet Policy Review 9, available at https://doi.org/10.14763/2020.4.1530.

Folkers, A. and S. Opitz. 2020. Symbiosozialität – Zwischen Leben und Gesellschaft. Zeitschrift für theoretische Soziologie 2: 243.

French, S.P. and C. Barchers. 2017. Designing a Data Commons for Big Data. Computers in Urban Planning and Urban Management Conference Proceedings 83, avalaible at https://scholarworks.umass.edu/larp_faculty_pubs/83.

Fumagalli, A. 2024. The process of valorization in the platform capitalism. *In*: Mezzadra, S., N. Cuppini, M. Frapporti and M. Pirone [eds.] Capitalism in the Platform Age. Springer Studies in Alternative Economics. Springer, Cham, available at https://doi.org/10.1007/978-3-031-49147-4_3.

Gitelman, L. and V. Jackson. 2013. Introduction. p. 8. *In*: Gitelman, L. [ed.]. "Raw Data" is an Oxymoron. The MIT Press, Cambridge, Mass.

Graef, I., T. Petročnik, and T. Tombal. 2023. Conceptualizing Autonomy in an Era of Collective Data Processing: From Theory to Practice. DISO 2, 19, available at https://doi.org/10.1007/s44206-023-00045-3.

Karavas, V. 2020. Biomedical collective labour: politics, sovereign subjects and empowerment in biobank research. *In:* Jacob, M.A. and A. Kirkland [ed.]. Research Handbook on Socio-Legal Studies of Medicine and Health. Edward Elgar Publishing, Cheltenham, UK.

Karavas, V. 2018. Körperverfassungsrecht: Entwurf eines inklusiven Biomedizinrechts. Dike Verlag, Zürich.

Kędzior, M. 2021. The right to data protection and the COVID-19 pandemic: the European approach. ERA Forum 21. 533–543, available at https://doi.org/10.1007/s12027-020-00644-4.

Kuner, C. 2020. Data Crossing Borders. Data Sharing and Protection in Times of Coronavirus, available at: https://verfassungsblog.de/data-crossing-borders/.

Landecker, H. 2007. Culturing Life: How Cells Became Technologies. Harvard University Press, Cambridge, Mass. 14.

Latour, B., P. Jensen, T. Venturini, S. Grauwin and D. Boullier. 2012. 'The whole is always smaller than its parts' – a digital test of Gabriel Tardes' monads. The British Journal of Sociology 63: 595, 598, 607.

Marrero-Guillamón, I. 2015. Monadology and ethnography: Towards a Tardian monadic ethnography. Ethnography 16: 243, 244.

Maurer, B. 2015. Principles of Descent and Alliance for Big Data. p. 77. *In*: Boellstorff, T. and B. Maurer [eds.]. Data Now, Bigger and Better! Prickly Paradigm Press, Chicago.

Molina, S.J. 2022. Biological Citizenship and Surveillance in the COVID-19 Pandemic. Berkeley Journal of Sociology 63, available at: https://berkeleyjournal.org/2022/06/24/biological-citizenship-and-surveillance-in-the-covid-19-pandemic/.

Nassehi, A. 2019. Muster. Theorie der digitalen Gesellschaft. C.H. Beck, München.

Purtova, N. 2018. The law of everything. Broad concept of personal data and future of EU data protection law, Law, Innovation and Technology 10. 40–81, available at DOI:10.1080/17579961.2018.1452176.

Prainsack, B., S. El-Sayed, N. Forgó, L. Szoszkiewicz and P. Baumer. 2022. Data solidarity – A White Paper. Geneva: The Lancet & Financial Times Commission on Governing Health Futures, 6, 9, available at: https://www.governinghealthfutures2030.org/wpcontent/uploads/2022/12/DataSolidarity.pdf.

Rabinow, P. 1996. Severing the Ties: Fragmentation and Dignity in Late Modernity. *In*: *idem*. Essays in the Anthropology of Reason. Princeton University Press, New Jersey.

Repo, J. and H. Richter. 2022. An evental pandemic: thinking the COVID-19 'event' with Deleuze and Foucault. Distinktion: Journal of Social Theory. 220–237, 234. Available at DOI: 10.1080/1600910X.2022.2086595.

Roessler, Beate. 2017. X—Privacy as a Human Right, *Proceedings of the Aristotelian Society*, Volume 117. 187–206, available at https://doi.org/10.1093/arisoc/aox008.

Ruhaak, A. 2019. Data Trusts: Why, What and How, available at https://medium.com/@anoukruhaak/data-trusts-why-what-and-how-a8b53b53d34.

Simana, S. 2021. Is Israel Trading Medical Information for Vaccines? Ethical and Legal Considerations, available at: https://blog.petrieflom.law.harvard.edu/2021/01/25/israel-covid-vaccine-health-data/.

Staunton, C. 2021. Data Sharing and the Public Interest in a Digital Pandemic, available at: https://verfassungsblog.de/data-sharing-and-the-public-interest-in-a-digital-pandemic/.

Tacconelli, E., A. Gorska, E. Carrara, R.J. Davis, M. Bonten, A.W. Friedrich, C. Glasner, H. Goossens, J. Hasenauer, J.M.H. Abad, J.L. Peñalvo, A. Sanchez-Niubo, A. Sialm, G. Scipione, G. Soriano, Y. Yazdanpanah, E. Vorstenbosch and T. Jaenisch. 2022. Challenges of data sharing in European Covid-19 projects: A learning opportunity for advancing pandemic preparedness and response. Lancet Reg Health Eur. 21:100467, available at doi: 10.1016/j.lanepe.2022.100467.

Taylor, L. 2017. What Is Data Justice? The Case for Connecting Digital Rights and Freedoms Globally. Big Data & Society 4: 205395171773633.

Tarde, G. de 2009 [1895]. Monadologie und Soziologie. Suhrkamp, Frankfurt am Main, 48.

Teubner, G. 2021. The Constitution of Non-Monetary Surplus Values. Social & Legal Studies, 30. 501-521, available at https://doi.org/10.1177/0964663920952531.

Thouvenin. Florian. 2017. Wem gehören meine Daten? Zu Sinn und Nutzen einer Erweiterung des Eigentumsbegriffs. Schweizerische Juristen-Zeitung 113. 21–32.

Zech, H. 2015a. Industrie 4.0 – Rechtsrahmen für eine Datenwirtschaft im digitalen Binnenmarkt. GRUR. 1151–1160.

Zech, H. 2015b. Daten als Wirtschaftsgut – Überlegungen zu einem Recht des Datenerzeugers, Computer und Recht. 137–146, available at https://doi.org/10.9785/cr-2015-0303.

Zeno-Zencovich, V. 2020. Free-Flow of Data. Is International Trade Law the Appropriate Answer? p. 182. *In*: Fabbrini, E. Celeste, J. Quinn [eds.]. Data Protection Beyond Borders. Transatlantic Perspectives on Extraterritoriality and Sovereignty. Hart Pub, London.

Zeno-Zencovich, V. 2009. La 'comunione' di dati personali. Un contributo al sistema dei diritti della personalità. IL DIRITTO DELL'INFORMAZIONE E DELL'INFORMATICA. 5-22.

Žižek, S. 2020a. Pandemic! COVID-19 shakes the world. OR Books, New York/London.

Žižek, S. 2020b. Pandemic! 2. Chronicles of a time lost. OR Books, New York/London.

Wiebe, A. 2016. Protection of industrial data – a new property right for the digital economy? GRUR Int. 877–884.

11

Data Ownership

Empowering Students in the Digital Age

Francesca Osima[1,#] and *Panagiotis Chatzimichail*[2,#,*]

The advent of the digital era has revolutionized many sectors, and education is one of the most profoundly affected. Central to this transformation is the concept of data ownership and its use by students. As students generate vast amounts of data through their interactions with digital educational platforms, the ownership, control, and use of this data can significantly empower them, shaping their learning experiences, enhancing their autonomy, and fostering a more personalized and effective educational journey.

11.1 The Importance of Data Ownership

Data ownership in education refers to the rights of students to control and manage the data they produce through their educational activities. This encompasses academic performance data, engagement metrics, and even personal reflections captured through digital tools. Empowering students with ownership of their data can lead to several benefits:

- ***Enhanced Personalization***: When students have control over their data, they can use it to personalize their learning experiences. Personalized learning paths, adaptive learning technologies, and individualized feedback become possible, catering to diverse learning styles and needs. Research has shown that personalized learning can improve student engagement and outcomes significantly (Pane et al. 2015).

[1] Head of Projects and Operations, Organising Bureau of European Student School Unions (OBESSU), Brussels, Belgium.
[2] Head of External Affairs, Organising Bureau of European Student School Unions (OBESSU), Brussels, Belgium.
# Both authors have contributed equally.
* Corresponding author: panagiotis@obessu.org

DOI: 10.1201/9781003475927-13

- ***Increased Autonomy***: Data ownership fosters a sense of autonomy and responsibility. Students who own their data can set personal learning goals, track their progress and make informed decisions about their educational journey. This autonomy is linked to higher motivation and self-regulation skills, which are critical for lifelong learning (Zimmerman 2002).
- ***Improved Engagement and Motivation***: When students understand and use their own data, they become more engaged in their learning process. Access to performance metrics and progress reports can motivate students to improve and achieve their goals. Studies indicate that student engagement is a key predictor of academic success (Fredricks et al. 2004).
- ***Transparency and Trust***: Data ownership promotes transparency between students and educational institutions. It ensures that students are aware of how their data is being used and can hold institutions accountable for their data practices. This transparency builds trust and encourages a more collaborative relationship between students and educators.

11.2 The Role of Educational Institutions

Educational institutions play a pivotal role in facilitating data ownership for students. They must implement policies and technologies that support data access, portability, and control. Key strategies include:

- ***Developing Robust Data Policies***: Institutions should establish clear policies regarding data ownership, privacy, and usage. These policies must comply with legal frameworks such as the General Data Protection Regulation (GDPR) in the European Union, which grants individuals significant rights over their personal data (European Commission 2020).
- ***Implementing Data Portability***: To empower students, institutions should ensure that data is portable. This means that students should be able to easily access, download, and transfer their data across different platforms and institutions.
- ***Educating Students About Data Literacy***: Institutions should provide education on data literacy, helping students understand how to interpret and use their data effectively. This includes training on data privacy, security, and ethical considerations. Data literacy is crucial for enabling students to make informed decisions and protect their privacy (Mandinach and Gummer 2013).
- ***Leveraging Technology***: Educational technology (EdTech) tools that support data ownership and personalization should be integrated into the learning environment. Learning Management Systems (LMS), e-portfolios, and other digital tools can help students track their progress and utilize their data for personal growth.

11.3 Challenges and Considerations

While the benefits of data ownership are clear, several challenges must be addressed:

- ***Privacy and Security***: Protecting student data from unauthorized access and breaches is crucial. Institutions must implement robust security measures and educate students on best practices for data protection.
- ***Equity and Access***: Ensuring equal access to the tools and resources students need to own their data. This means bridging digital divides and supporting students with different levels of digital literacy.
- ***Ethical Use of Data***: Ethical considerations around data use must be at the forefront. This includes ensuring that data is used to support student learning and well-being, rather than for punitive measures or excessive surveillance.

11.4 Conclusion

Data ownership in education has the potential to profoundly empower students in the digital era. By granting students control over their data, educational institutions can enhance personalization, foster autonomy, ensure engagement, and build trust. To realize this potential, it is crucial to implement thoughtful policies, robust technologies, and a commitment to equity and ethical data practices. Moving forward, data ownership in education can help both educators and students co-create more effective and inclusive learning environments.

References

European Commission. 2020. General Data Protection Regulation (GDPR). Retrieved from https://ec.europa.eu/info/law/law-topic/data-protection_en.

Fredricks, J.A., P.C. Blumenfeld and A.H. Paris, 2004. School engagement: Potential of the concept, state of the evidence. Review of Educational Research 74(1): 59–109.

Mandinach, E.B. and E.S. Gummer. 2013. A systemic view of implementing data literacy in educator preparation. Educational Researcher 42(1): 30–37.

Pane, J.F., E.D. Steiner, M.D. Baird and L.S. Hamilton. 2015. Continued progress: Promising evidence on personalized learning. RAND Corporation.

Zimmerman, B.J. 2002. Becoming a self-regulated learner: An overview. Theory Into Practice 41(2): 64–70.

12

Regulating Health Data

A Comparative Perspective

Nicolai Søderlund Kofod[a] and *Wen Xiang*[b,*]

12.1 Introduction

On November 1st 2021 the most significant piece of personal data protection legislation within the People's Republic of China (PRC) came into effect. The Personal Information Protection Law, also known as the PIPL, is a codification of the right to privacy and the protection of personal information as it is laid out under the newly instituted Civil Code of the People's Republic of China.[1] The PIPL determines the rights of individuals and duties of personal information handlers. In doing so it supplements the extensive legal regime of privacy regulation already present in the PRC such as the 2017 Cybersecurity Law (CSL) and the 2021 Data Security Law (DSL). Up until the introduction of the PIPL, the data protection regime in the PRC had been criticized as being fragmented and scattered throughout several regulations, which are still finding their footing (Geller 2020, 1202). This fragmentation has further been exacerbated by the introduction of five central pieces of privacy legislation over the course of three years.[2] The regulations are meant to complement and clarify the abovementioned laws, which make up the core of the PRC's attempt at digital governance. But sectors which rely on steady personal data flows continue to be outspoken about the lack of structure and the lack of transparency which persists even under the PIPL.[3]

[a] Region Hovedstaden, Copenhagen, Denmark. Email: N.S.Kofod@hotmail.com
[b] University of Copenhagen, Copenhagen, Denmark.

[1] Civil Code of the People's Republic of China 2020. Art. 1034.

[2] See the Personal Information Security Specification, the Critical Information Infrastructure Security Protection Regulation, the Draft Online Data Management Regulations, the Cybersecurity Review Measures and the Outbound Data Transfer Security Assessment Measures.

[3] Ministry of Foreign Affairs of Denmark, Sino-Danish Center for Education and Research, and Danish Chamber of Commerce, Survey of Danish Companies in China (2022). 33–34.

* Corresponding author: xiang.wen@hotmail.com

DOI: 10.1201/9781003475927-14

A principal outspoken industry is the life science sector which relies heavily on access to personal data for its research and development of medicinal products. The functionality and profitability of the life science sector is directly affected by any incoherence which may exist within the PRC's attempt at privacy regulation. Looking therefore to the EU where the value of the data economy is set to exceed 820 billion euro in 2025 lawmakers (Løffler Nielsen and Orthmann Grønbech 2021, 1–2), through various legal measures such as the General Data Protection Regulation (GDPR) and the EU Data Act, have tried to accommodate a widespread access to data for key industries such as the life science sector. In light of this on the one hand recently developed legal discipline, and on the other hand a decades long legal tradition of privacy regulation, it becomes interesting to comparatively analyze how these two prominent jurisdictions have approached the challenge of regulating the use of health data and how interplay between the jurisdictions is facilitated by emerging legislation.

This chapter is therefore structured around analysis and discussion of PRC and EU privacy frameworks. Section 2 provides an analysis of the privacy policies which have driven legal regulation in the PRC and the EU. Section 3 compares the frameworks currently in place with an emphasis on the GDPR and PIPL in relation to the use of health data within biomedical research. Section 4 gives a brief analysis of future harmonization prospects, and Section 5 offers concluding remarks.

12.2 Health Data Privacy Policies

In order to give a voice to its policy on digital governance, as well as promoting interconnectivity of internet infrastructure and the prosperity of an internet economy, China held its first World Internet Conference in Wuzhen in November of 2014.[4] At the summit, Chinese President Xi Jinping had prepared a statement, saying that "China is ready to work with other countries to deepen international cooperation, respect sovereignty on the Internet, uphold cyber security, and jointly build a cyberspace of peace, security, openness and cooperation".[5]

This statement warrants analysis. Firstly, it provides an insight into China's overall strategy of promoting cyber sovereignty, and secondly, between those separate points of reflection, it offers an indication as to which thoughts have preoccupied Chinese legislator's minds: peace, security, openness and cooperation. These thoughts have shaped extensive national security legislation as well as cyber security and personal information protection legislation. Consequently, to understand Chinese legislation on this topic, cyber sovereignty and the underlying regards of said legislation serve as two departure points for analysis.

---

4 World Internet Conference, Wuzhen Summit: China Holds First World Internet Conference, Urges Better Governance. <http://www.wuzhenwic.org/2014-11/20/c_46286.htm> accessed 11 February 2024.

5 Chinese President Xi Jinping Calls for International Cooperation on Cyberspace Security. The Economic Times (19 November 2014) <https://economictimes.indiatimes.com/news/international/world-news/chinese-president-xi-jinping-calls-for-international-cooperation-on-cyberspace-security/articleshow/45205445.cms> accessed 11 February 2024.

Following the 2014 World Internet Conference, the Standing Committee of the PRC passed the 2015 National Security Law (NSL), the purpose of which is to maintain national security and to defend the interests of the Chinese people. In this vein the NSL places an emphasis on closely tying national security to a broad interpretation of the principle of sovereignty.[6]

Important to the Chinese concept of sovereignty is the power to rule supreme in a state's own territory. And new to the Westphalian view of *summa potestas* propagated in the NSL, is the demand that this supreme power carry over into more intangible territory, e.g., cyberspace where the PRC seeks an increasing oversight. Sovereignty in cyberspace, however, is not a new conception, nor was it news when China presented its vision at the 2014 Wuzhen World Internet Conference. In the years leading up to the conference a desire to exercise a greater control over the internet had been at the heart of Chinese cyber strategies (Creemers 2020, 8-11). Similarly, the international community had long endeavoured to strike a balance between states' control over their own territories and cyber means and a free internet. The initial bid to accommodate this control had been made in 2013 via the first Tallinn Manual. However, the first manual was criticized, amongst others by the PRC, for militarizing cyberspace and for avoiding the topic of sovereignty in cyberspace (Schmitt 2017). So a second manual was commissioned for release in 2017 which finally enshrined the application of sovereignty in cyberspace as one of its black letter rules and as an expression of customary international law.

Looking back at the 2015 National Security Law, and the 2014 statement, it is evident that China sought to further the position that the principle of sovereignty is applicable in cyberspace, which was still contested at the time. The Manual, however, now accommodates the degree of "control" (Segal 2020, 92-95) that China internationally sought over its internal affairs as it has become the prevailing view that the principle of sovereignty is applicable in cyberspace (Schmitt and Vihul 2017, 31-32). Consequently, the focus of China's legislature has now been enabled to shift from international policies to domestic internet and data policies, which can now be constructed freely, although still subject to international obligations.

Shifting the focus to the second part of the Wuzhen statement, China's domestic policy on digital governance begins to materialize. The implementation of the PIPL, CSL and DSL has been a strong tool in demonstrating to the EU and the United States, that China also has a robust model for data and security (Segal 2020, 105). The model in itself, however, being more comprehensive at times than the EU and U.S. strategies, offers an alternative route in light of the general modus operandi of the Chinese government, and the interests such governance cultivates.

The full legal effect of the Charter of Fundamental Rights of the European Union, was secured by the entry into force of the Treaty of Lisbon. As a result, since 2009, every citizen of the EU has enjoyed the fundamental right "to the protection of personal data concerning him or her".[7] Consequently it makes sense for the EU to shift its focus into niche areas of data regulation and prioritize a health data space where "natural persons can easily control their electronic health data" and which

[6] National Security Law of The People's Republic of China (NSL). Arts. 1 and 2.

[7] The Charter of Fundamental Rights of the European Union. Art. 8(1).

enable "researchers, innovators and policy makers to use… electronic health data in a trusted and secure way that preserves privacy".[8] The EU has thus chosen to make the rights of its citizens its regulatory focus and to increasingly weigh those interests over that of government or industry interests.[9] On the other side of the Atlantic, tech giants such as Google and Intel have triumphed in expressing dissatisfaction with increasing regulations on the commercial use of data – and indeed tried to hinder such efforts – leading to an industry-centric data policy (Zuboff 2019). The regulatory focus of the EU and the U.S. stand in stark contrast to each other. In the coin toss of data protection interests the EU has chosen the side which favours that of the subject vis-á-vis the U.S.-favoured side advantaging commercial use instead.[10] Building upon this analogy China has chosen a third approach: to own the coin (Chander and Schwartz 2023, 25-32). This road less travelled by in data protection circles is fostered by three primary objectives: to control information, to legitimize the Chinese regime and to secure the continued rule of the Chinese Communist Party (CCP) (Segal 2020, 97). The Chinese strategy of digital governance is therefore preoccupied with controlling public discourse and streamlining political opinion for which a vast array of authoritarian tools has already been used as seen under Section 4.

Understanding these policies serves as an interpretative outset for an analysis into how the PRC and EU legal frameworks for the use of health data differ. Indeed, the EU builds upon a three decades long tradition of privacy regulation which provides for a head start that allows for established user-centric policies to be translated into regulation of niche areas such as health data and life science. An effort which has only just begun for the PRC. But it must still be noted that the EU and PRC policy outsets for regulating health data are rooted in vastly different views on privacy and may therefore have vastly different outcomes.

12.3 Comparing Legal Frameworks

Heralded as the "Chinese GDPR" it is apparent that Chinese lawmakers sought inspiration in the European data protection regime when forming the PIPL.[11] However, as discussed, the main outset of Chinese privacy governance differs from that of the EU, and as a result the PIPL and GDPR differ in the comprehensiveness of their scope. Focusing on how both regulations have tackled the use of health data key differences emerge.

Throughout the GDPR concessions are made as to the processing of personal data for the sake of scientific research purposes. To ensure that no developmental

---

[8] Proposal for a Regulation of the European Parliament and of the Council on the European Health Data Space (2022) section 1.

[9] China Cybersecurity and Data Regulation: What Multinationals Should Know (Teneo, 25 October 2021). <https://www.teneo.com/china-cybersecurity-and-data-regulation-what-multinationals-should-know/> accessed 11 February 2024.

[10] China Cybersecurity and Data Regulation: What Multinationals Should Know (n 16).

[11] ELRA, Comparative Analysis of GDPR & PIPL (2021) http://elra.info/en/dissemination/legal-issues-papers/comparative-analysis-gdpr-pipl/.

activities or sectors are excluded scientific research purposes must be interpreted in a broad manner, encompassing privately funded research, technological development and fundamental research, etc.[12] In light of the use of health data the preamble furthermore mentions clinical trials, once more signalling that special thought has been offered to the life science sector and the vast amount of health data consumed within that industry.[13] Moreover the GDPR is lenient with the fundamental personal data processing principle "purpose limitation", when interpreted in light of research activities. Whereas no data subject can be asked to expect that their personal data are processed for incompatible purposes for which the data were originally collected, the principle is construed so as to always allow further processing when done for scientific research purposes.[14] A similar leniency is offered to those who seek to store personal data for longer periods than what is simply necessary when done so for scientific research purposes.[15] Lastly, European lawmakers have dedicated a legal basis for the processing of personal data to research. Although subject to national implementation this concession is the final piece of a legal puzzle which through and through facilitates the continued flourishing of the life science sector and its research activities.[16]

Admittedly both the GDPR and PIPL are strikingly similar. Both use a similar language and structure, both set up fundamental data protection principles such as purpose- and storage limitation, both call for a legal basis for the processing of personal data and both define health data as a special category of personal data. But absent from the PIPL are the special mentions and indeed concessions offered under the GDPR to personal data processing done for scientific research purposes.

Why this discrepancy exists may be due to differences in legal culture and the contrasting ways in which EU and PRC lawmakers view privacy regulation. Looking at the PIPL and GDPR's purpose statements the GDPR seeks to facilitate the "free movement of personal data" and the PIPL seeks to "promote the rational use of personal information".[17] Although the variation in language is slight, there is a degree of control to the PIPL formulation, and once more the abovementioned view in Section 2 of why each jurisdiction deems privacy regulation necessary shines through. As a result, the PIPL has a more general and narrower outlook. When trying to discern what basis' may prove helpful in allowing for researchers to access health data art. 13 leaves two choices: 13(1) consent and 13(7) lex specialis. As will be discussed below, consent – and the overarching emphasis the PIPL places on consent – is sometimes an unviable basis for the processing of someone's personal data for research purposes. Researchers are therefore left with trying to find any allowance under any other laws or regulations for their access to health data. This is seemingly

---

[12] Regulation (EU) 2016/679 General Data Protection Regulation (GDPR). Preamble 159.

[13] ibid. Preamble 161.

[14] ibid. Art. 5 (1)(b).

[15] ibid. Art. 5 (1)(e).

[16] ibid. Art. 9 (2)(j).

[17] Art. 1 of the GDPR and PIPL. In this case the Stanford translation: Personal Information Protection Law of the People's Republic of China. (2021 version, DigiChina translation) https://digichina.stanford.edu/work/translation-personal-information-protection-law-of-the-peoples-republic-of-china-effective-nov-1-2021/.

innocuous, but where the GDPR allows researchers to process health data in all cases where the law is not directly opposed to it, the PIPL has conversely narrowed its scope and limited access for researchers to what is specifically provided for by the law. Consequently, accessing health data for research purposes within the PRC may be subject to more limitations than it is in the EU as it is up to the lawmakers to choose which concrete activities are permitted.

12.3.1 Consent

If no other legal basis for the processing of personal data for the sake of conducting research can be found in any other law, the PIPL allows the researcher to simply obtain the consent of the individual whose personal data they are aiming to process. In fact, due to the consent-centric nature of the PIPL it is a prerequisite for the processing of health data, being a special category of personal data, that the researcher has at all times obtained the individual's separate consent[18] along with having found another suitable basis for the processing under art. 13. That suitable basis in addition is, as shown above, often reduced to consent.

Consent, to be valid under the GDPR, must among other criteria be freely given.[19] The Working Party 29 Guidelines on consent explain that "free" "implies real choice and control for data subjects".[20] In other words data controllers, often times private life science companies or public research institutions, must pay close attention to any imbalance of power which may exist between them and the data subject. The imbalance is most pronounced when the research institution is a public service provider which often holds some leverage over the data subject or when the refusal to give consent by the data subject is detrimental to the subject's situation.[21] However, private companies which provide crucial services may also have enough leverage for the relationship to be imbalanced.

One particular instance within research where this nuance proves relevant is during clinical trials. Often a clinical trial of a pharmaceutical or medical device is performed on patients who are in dire need of novel treatments. The private company offering that treatment or the clinic in which the patient is receiving such treatment may therefore inadvertently hold that treatment as leverage over patients in need. The European Data Protection Board (EDPB) has commented on exactly this scenario expressing concern that an imbalance of power in clinical trials may imply that consent is not freely given by patients.[22] The EDPB considers that those seeking to perform clinical trials and processing personal data must examine this imbalance thoroughly. An imbalance which is once more most prevalent where the patient is "not in good health condition".[23] In the end the EDPB, based on the Working Party

18 PIPL art. 29.
19 GDPR art. 4(11).
20 Article 29 Working Party Guidelines on consent under Regulation 2016/679 (2018). 5-11.
21 ibid. 5.
22 EDPB, Opinion 3/2019 Concerning the Questions and Answers on the Interplay between the Clinical Trials Regulation (CTR) and the General Data Protection Regulation (GDPR) (Art. 70.1.b). 3-6.
23 ibid. Para 20.

29 guidelines, has concluded that "consent will not be the appropriate legal basis in most cases". [24] This conclusion was recently subjected to a more wide-ranging interpretation in a verdict set down by the Danish supervisory authority, Datatilsynet. A hospital had posted pictures and text pieces about admitted patients on its social media accounts. The posts were of an informative nature about the activities going on at the hospital, with personal excerpts about recovering patients and their experiences. The hospital had taken every precaution to obtain the written explicit consent of patients participating on social media, and had made plain to patients that their treatment would not be affected by their participation or lack thereof. Datatilsynet, however, found that the consent had not been valid as it had not been freely given. Datatilsynet focused in its conclusion increasingly on the imbalance of power that exists between a public authority providing health services and a vulnerable patient – even when the personal data is only used for social media – citing the exact concerns highlighted above by the EDPB.[25]

Looking to the PIPL's brand of consent it is evident that a likened criterion to that of "freely given" exists, as consent must be voluntary.[26] Whether this holds the same value as "freely given" under the GDPR is yet uncertain, as case law and supplementing regulations are yet to provide guidance in this area. However, three years prior to the enforcement of the PIPL, the Personal Information Security Specification took effect, which hints at the early thoughts on the components of a valid consent. The main component of consent is here similarly that consent must be "freely given"[27] wherefore "voluntary" and "freely given" are used synonymously throughout the specification.[28] That "freely given" did not make its way into the PIPL is therefore not of much importance when "voluntary" did, at least when looking at the implications this may have for conducting clinical research.

It remains to be seen whether the interpretation of "freely given" under clinical circumstances that the EDPB and Datatilsynet have applied will similarly exist under the PIPL. But that the same imbalance of power may become present under the PIPL bodes ill for the coherence of the Chinese data protection regime when it comes to clinical research. Under normal circumstances a legal basis to deviate from the criterion of obtaining consent could simply be introduced in any other regulation by lawmakers.[29] But in relation to art. 29 of the PIPL, there exists no such remedy within the PIPL or outside it to avoid having to obtain the separate consent of the patient whose health data one may try to process for research.[30] Ultimately this finding further

[24] ibid. Para 20.]

[25] Hospital kan ikke bruge samtykke til at offentliggøre billeder af patienter på Instagram. 2023. Datatilsynet J.nr.: 2023-432-0016.

[26] PIPL art. 14.

[27] Translation: China's Personal Information Security Specification (Mar. 2020) (DigiChina) https://digichina.stanford.edu/work/translation-chinas-personal-information-security-specification accessed 11 February 2024. Para. 5.5a.

[28] ibid. Para. 5.5b.

[29] PIPL art. 13(7).

[30] Looking at the National Institutes of Health ClinRegs database on China: https://clinregs.niaid.nih.gov/country/china the PIPL is indeed the primary regulator of the processing of personal information within clinical trials.

underlines the argument that the PRC data protection regime was initially formulated with the main goal of controlling public discourse and streamlining political opinion, with little thought of how this primary goal may affect industries which are reliant on access to health data.

12.3.2 National Security

So far, the interplay between Chinese and European privacy regulation has been suboptimal when it comes to utilizing health data. Turning now to national security, which has been the primary focus of the establishment of privacy regulation within the PRC, a final critical nuance emerges which may potentially dissuade at least parts of the European life science sector in pursuing some health data endeavors in the PRC.

In 2015 the National Security Law offered up the following definition of what, to the PRC, constitutes national security: "National security refers to the relative absence of international or domestic threats to the state's power to govern, sovereignty, unity and territorial integrity, the welfare of the people, sustainable economic and social development, and other major national interests, and the ability to ensure a continued state of security".[31] The definition is overwhelmingly broad which means it has been difficult to anticipate how the National Security Law is to be enforced and in what scenarios it will be utilized. The law, however, makes frequent reference to the completion of the rule of law and the establishment of national security rule of law.[32] Consequently, it is not coincidental that the Cybersecurity Law, Data Security Law and PIPL, which came into effect after 2015, contain several references to national security.

The CSL is formulated on the basis of ensuring national security via cybersecurity. It does so by prohibiting any use of the internet in activities endangering national security. In the same vein it prohibits any subverting incitement of national sovereignty, or any incitement of separatism.[33] This rule in particular was invoked in 2018 when the US-based multinational hotel operator, Marriott International, was indicted, by the Shanghai government, offering a glimpse into the practical application of the safeguarding of national security. Marriott International had sent out a Mandarin-language questionnaire to its customers in which Tibet, Hong Kong, Macau and Taiwan were listed as separate countries.[34] In the end the questionnaire was viewed as a subversion of Chinese sovereignty and the hotel operator apologized for the incident by proclaiming that "Marriott International respects Chinese sovereignty and its territorial integrity".[35]

31 NSL art. 2.

32 ibid. Arts. 15 and 70.

33 Cybersecurity Law of the People's Republic of China (CSL) (2017) https://digichina.stanford.edu/work/translation-cybersecurity-law-of-the-peoples-republic-of-china-effective-june-1-2017/ Art. 12.

34 Marriott International Apologises for Questionnaire Gaffe in China *Reuters* (11 January 2018) <https://www.reuters.com/article/china-marriott-idUKL1N1P609V> accessed 11 February 2024.

35 ibid.

Under the CSL certain companies may gain the label Critical Information Infrastructure Operator (CIIO) if they form part of "important sectors and areas such as public telecommunications… public services, e-government, etc., as well as other critical information infrastructure that, whenever it is destroyed, loses its ability to function or encounters data leaks, may gravely harm national security, the national economy, the people's livelihood and the public interest".[36] Encompassed by this definition is also thought to be the healthcare sector (Segal 2020) and as such parts of the life science sector can be subject to rules concerning national security, as the compromise of critical information infrastructure per definition has the ability to endanger national security.[37]

In relation to CIIO's the CSL has been predominantly focused on platform operators, who must be aware of the material which is published through their platforms as art. 70 of the CSL prescribes, that the publication or transmission of information, which is prohibited by CSL art. 12, para. 2, can lead to legal liability under the law. Art. 12 being the foremost provision for indictments such as the Marriott incident above. And even if the illegal content is not explicitly produced by the company itself, it is the CIIO's responsibility to police their own platforms according to art. 47 of the CSL. This means that the operator must be aware of the content produced by users, and set out to immediately stop transmission of such material, delete such material and report the alleged perpetrator to the relevant authorities.[38] The Cyberspace Administration of China has made clear its resolve to enforce the law in this manner, by putting the three tech giants Tencent, Sina and Baidu under investigation in 2017.[39] The three internet giants control WeChat, Weibo and Tieba respectively which are used as a messaging service, a microblogging website like Twitter and a search engine forum. The companies had failed to properly manage their social media platforms as some of their users "spread misinformation of violence and terror, false rumors, pornography, and other information that jeopardizes national security, public safety, and social order".[40] As a result all three platforms were fined in accordance with art. 68, wherefore the fines, although not public, could be as high as 500,000 RMB.[41]

The sharing of false rumors or pornography seems innocuous compared to insinuating that Chinese-claimed territories are indeed sovereign, but it is far from the only example of the limits of the terms "national security, national honor and national interests" being tested. In Gao v. Tiedong Public Security Bureau, the plaintiff was arrested for allegedly posting messages such as "Let the Chairman come after me, sooner or later I will oppose him", in a WeChat group, referring to President Xi

[36] CSL art. 31.

[37] CSL arts. 31 and 35.

[38] ibid. Art. 47.

[39] The Diplomat, China Fines Its Top 3 Internet Giants for Violating Cybersecurity Law (2017) <https://thediplomat.com/2017/09/china-fines-its-top-3-internet-giants-for-violating-cybersecurity-law/> accessed 11 February 2024.

[40] ibid.

[41] ibid.

Jinping.[42] According to the public security bureau, the plaintiff signed a confession after interrogation and was sentenced to fifteen days of administrative detention. The verdict was in part based on CSL arts. 12 and 70. In a similar case, Lu v. Sandu Shui Public Security Bureau, the plaintiff received ten days of administrative detention for posting a video to Baidu depicting an individual being injured during a property demolition conducted by the local government.[43] The public security bureau claimed that the video was fake, and that it was damaging to the reputation of the local government, which the plaintiff appealed. During the appeal the court insisted that internet users should exercise caution when posting about "party and government agencies [and] mass incidents… that may disrupt social order" (Pyo 2021, 251-252). The plaintiff was sentenced either way.

The above cases site instances where large social media platforms are used to speak negatively about the Chinese Communist Party or the State as a whole. However, no platform operator is spared from the level of enforcement or the broad definition of national security. Comments on health-related forums, or public online forums which facilitate the contact between the investigator in a clinical trial and the clinical trial participant may be subject to such policing. The result is that the CIIO, possibly an actor within the life science milieu, must ensure that such forums are moderated to avoid liability.

The PIPL has joined the CSL in setting out rules against any personal information handling activities which may harm national security or public interest. In addition, any foreign organization or individual that engages in such activities may be blacklisted from handling personal information under the PIPL.[44] Although the PIPL is yet to reveal its own particular brand of national security as case law is still scant, it is not far-fetched to imagine that the extra-territorial application of the law, which is exactly similar to that of the GDPR, can potentially serve as a tool for curbing any criticism abroad. Or that the blacklisting of foreign organizations and individuals may harm life science companies trying to establish themselves in the PRC.

When viewed through a GDPR lens, the combination of this interpretation of national security and laws protecting personal data appear especially foreign. The GDPR explicitly excludes questions of national security from its scope. Instead, such questions fall to the individual member states to regulate separately, in proportion to the provisions of the free movement of data propelled under the GDPR. The broad interpretation of national security, being unusual in European data protection circles,

[42] Gao Guangjun and Anshan City Public Security Bureau, Tiedong Branch, Public Security Management:, Other (Public Security) First Instance Administrative Judgment, China Judgements, and Online, Liao 0381 Xingchu 30 Hao (2020). As referenced in Pyo, G. 2021. An Alternate Vision: China's Cybersecurity Law and Its Implementation in the Chinese Courts. Columbia Journal of Transnational Law. 60: 243–250.

[43] Lu Wenyou, Sandu Shui Autonomous County Public Security Bureau Public Security Administration: Public Security Management (Public Security) Second Instance Administrative Judgment China Judgements Online, (2018) Qian 27 Xingzhong 167 Hao (2018). As referenced in Pyo, G. 2021. An Alternate Vision: China's Cybersecurity Law and Its Implementation in the Chinese Courts. Columbia Journal of Transnational Law. 60: 251–252.

[44] PIPL arts. 10 and 42

must therefore be borne in mind at all times for any part of the life science sector trying to establish itself in, or do business with actors in the PRC.

12.4 Concluding Remarks

In this chapter two increasingly different approaches to the regulation of health data have been explored. We argue that the EU approach is ultimately more open and facilitating towards the sharing of and access to health data for the expansive life science sector and for researchers. But it must be noted that both regulations build upon differing legal cultures under which are placed a value on different outcomes. Due to the infancy of the PRC's legal culture regarding privacy it is not yet certain whether this inability to adapt to the specifics of health data sharing will persist. The PRC has throughout the past ten years introduced more privacy regulation than the EU, and it remains to be seen what regulation will be the product of a matured PRC effort. Evidently the central PRC view that privacy regulation is foremost about control runs counter to the efforts underway at the moment in the EU to facilitate health data spaces. Therefore, the EU will most likely be more effective in attracting life science companies and start-ups, as well as promoting public health overall through research. But the reformation of the Chinese legal system has only accelerated over the past few decades, and as we have seen its privacy regulation can be formed swiftly. Inspired by EU privacy regulation it becomes interesting to see what the PRC digital governance programme has to offer next.

References

Chander, A. and P.M. Schwartz. 2023. Privacy and/or Trade. University Chicago Law Review. 90(1): 49–135.

Creemers, R. 2020. China's Conception of Cyber Sovereignty: Rhetoric and Realization. pp. 107–142. *In*: Broeders, D. and B. Van den Berg [eds.]. Governing Cyberspace: Behavior, Power, and Diplomacy. Digital Technologies and Global Politics. Rowman & Littlefield, Lanham.

Geller, A. 2020. How Comprehensive Is Chinese Data Protection Law? A Systematization of Chinese Data Protection Law from a European Perspective. GRUR International. 69(19): 1191–1203.

Løffler Nielsen, J. and M. Orthmann Grønbech. 2021. Den Stigende Regulering Af Data - Med Særlig Fokus På dataejerskab. UfR.

Pyo, G. 2021. An Alternate Vision: China's Cybersecurity Law and Its Implementation in the Chinese Courts. Columbia Journal of Transnational Law. 60: 228–275.

Schmitt, M.N. and L. Vihul. 2017. Respect for Sovereignty in Cyberspace. Texas Law Review. 95(7): 1639–1670.

Schmitt, M.N. 2017. Tallinn Manual 2.0 on the International Law Applicable to Cyber Operations (2nd edn). Cambridge University Press. Cambridge.

Segal, A. 2020. China's Vision for Cyber Sovereignty and the Global Governance of Cyberspace. NBR Special Report #87 The National Bureau of Asian Research.

Zuboff, S. 2019. The Age of Surveillance Capitalism (1st edn). PublicAffairs. New York.

SECTION 3

Sustainable Health Data Sharing

DOI: 10.1201/9781003475927-15

13

Data-Related Ethics Issues in EU-Funded Research Projects

Edward S. Dove

13.1 Introduction

For over 40 years, the European Union (EU)—the supranational political and economic union currently comprized of 27 Member States—has a provided significant amount of funding for research and innovation projects (see Table 1) under a scheme known as Framework Programmes for Research and Technological Development (Framework Programmes, or FPs for short). The recipients of these grants—whom I will refer to as "beneficiaries" in this chapter—may be situated within or outwith the EU,[1] although the primary mission of the FPs is to support and foster research in the European Research Area (ERA), which is the ambition to create a single, borderless market for research, innovation, and technology across the EU. The current FP scheme, known as Horizon Europe and which runs through 2027, has a budget of €95.5 billion.

The specific objectives and actions vary between the FPs and funding periods. Horizon Europe currently is divided into three pillars and one part, corresponding to its main priorities:[2] Pillar I (Excellent Science) aims to increase the EU's global scientific competitiveness. It supports frontier research projects defined and driven by top researchers themselves through the European Research Council (ERC), funds fellowships for postdoctoral researchers, doctoral training networks and exchanges

Maynooth University, Maynooth, Ireland; Email: edward.dove@mu.ie

[1] Non-EU countries may choose to "associate" to FPs by "paying to play". Under Horizon Europe, for example, at least 18 non-EU countries are associated countries. This includes the United Kingdom, Türkiye, Israel, Serbia, and North Macedonia. For the full list, see European Commission, "List of Participating Countries in Horizon Europe" (V3.4 – 10.04.2025), available at: https://ec.europa.eu/info/funding-tenders/opportunities/docs/2021-2027/common/guidance/list-3rd-country-participation_horizon-euratom_en.pdf.

[2] See European Commission, "Horizon Europe", available at: https://research-and-innovation.ec.europa.eu/funding/funding-opportunities/funding-programmes-and-open-calls/horizon-europe_en.

DOI: 10.1201/9781003475927-16

Table 1. EU Framework Programmes for Research and Technological Development.

Framework Programme name/ID	Period	Budget (€)
FP 1	1984–1987	3.8
FP 2	1987–1991	5.4
FP 3	1990–1994	6.6
FP 4	1994–1998	13.2
FP 5	1998–2002	15.0
FP 6	2002–2006	16.3
FP7	2007–2013	50.5 over seven years + 2.7 for Euratom over five years
FP 8/Horizon 2020	2014–2020	77
FP 9/Horizon Europe	2021–2027	95.5

for researchers through Marie Skłodowska-Curie Actions, and invests in world-class research infrastructures. Pillar II (Global Challenges and European Industrial Competitiveness) supports research relating to societal challenges and reinforces technological and industrial capacities through clusters. It also includes activities pursued by the Joint Research Centre (Lövestam et al. 2025),[3] which supports EU and national policymakers with independent scientific evidence and technical support. Pillar III (Innovative Europe) aims to make Europe a frontrunner in market creating innovation via the European Innovation Council. It also helps to develop the overall European innovation landscape through the European Institute of Innovation and Technology (EIT), which fosters the integration of the knowledge triangle of education, research, and innovation. The part, entitled Widening Participation and Strengthening the European Research Area (ERA), aims to increase support to EU Member States in their efforts to make the most of their national research and innovation potential. It is also worth noting that Horizon Europe is implemented also through the European Defence Fund[4] and complemented by the Euratom Research and Training Programme.[5]

As with funding schemes around the world, whether public or private, ethics is a core component of a successful grant application. EU-funded research projects are expected to fully comply with relevant national and international laws and research ethics norms. While the responsibility for ethics self-awareness and compliance lies with the individuals carrying out the research, formal accountability lies with the signatory of the Grant Agreement, which in EU parlance is known as the Host

[3] See Joint Research Centre, available at: https://commission.europa.eu/about-european-commission/departments-and-executive-agencies/joint-research-centre_en.

[4] European Defence Fund, available at: https://defence-industry-space.ec.europa.eu/eu-defence-industry/european-defence-fund-edf-official-webpage-european-commission_en.

[5] Euratom Research and Training Programme, available at: https://research-and-innovation.ec.europa.eu/funding/funding-opportunities/funding-programmes-and-open-calls/horizon-europe/euratom-research-and-training-programme_en.

Institution. As the Horizon Europe Framework Programme Grant Agreement template states:

> The action must be carried out in line with the highest ethical standards and the applicable EU, international and national law on ethical principles. [...] The beneficiaries must commit to and ensure the respect of basic EU values (such as respect for human dignity, freedom, democracy, equality, the rule of law and human rights, including the rights of minorities).[6]

As the chapters in this book have well illustrated, increasingly all kinds of research projects are defined by data—the growing volume and variety collected *ab initio* or reused or linked from previous collections (also known as data repurposing), and the growing focus on development of artificial intelligence (AI), which requires significant amounts of (training) data. Data-driven research generates its own ethics issues, however, as academic literature has long noted (O'Keefe and O Brien 2023) (Martin 2022, Shukla et al. 2022, Machon 2022, Hasselbalch 2021). Recent initiatives by the EU to develop a "European data strategy"[7] and data infrastructures, such as the European Health Data Space,[8] raise a number of issues relating to (among other areas) privacy and human rights, data protection, power, trust, and dignity.

It is impossible within a single chapter to cover the broad panoply of funding schemes provided by the EU, including those for both civil and defence-related purposes, much less the full range of ethics issues research projects making use of data can raise. I will therefore focus on data-related ethics issues commonly seen within Pillar 1 of Horizon Europe applications, i.e., those related to applications to the ERC, Marie Skłodowska-Curie Actions, and research infrastructures. I flag those ethics issues that are commonly identified by beneficiaries themselves in their Ethics Self-Assessments,[9] and those that I have come across most commonly in my experience as an "ethics expert" for the European Commission when asked to screen and advise on applications or projects that raise ethics questions of varying type and degree.

---

6 European Commission, *Horizon Europe: Model Grant Agreement*, Arts 14.1-14.2, available at: https://ec.europa.eu/info/funding-tenders/opportunities/docs/2021-2027/horizon/agr-contr/unit-mga_he_en.pdf.

7 European Commission, "European data strategy", available at: https://commission.europa.eu/strategy-and-policy/priorities-2019-2024/europe-fit-digital-age/european-data-strategy_en.

8 European Commission, "European Health Data Space", available at: https://health.ec.europa.eu/ehealth-digital-health-and-care/european-health-data-space_en.

9 Applicants for EU funding schemes are expected to inform the EU if their research proposal raises one or more of the issues listed in the pro forma ethics issues table, in which case they must also complete an "Ethics Self-Assessment". This self-assessment must include a description of the identified ethics issue(s), the ways in which the applicant intends to deal with it or them in order to ensure ethics compliance, and any relevant documents (such as authorisation or permissions that are already available). The Ethics Self-Assessment is integrated in the applicant's Grant Agreement if they are successfully funded (in Annex 1, as description of the action, ethics requirements, etc.) and thus give rise to binding contractual obligations.

As this volume focuses on aspects relating to data collection and data sharing, this chapter likewise focuses on ethics issues that may be considered data-related, including the processing of personal data from research participants (or data subjects more broadly) and the use of AI. However, the issues identified below should not be interpreted as applying exclusively to data-driven projects; they may well arise in projects that involve very little in the way of data collection or use.

13.2 The Ethics Review Procedure in Horizon Europe

Before proceeding to unpack some of the more commonly identified data-related ethics issues, it is worth explaining the process of the ethics review procedure in Horizon Europe, which has instituted an Ethics Appraisal Procedure. This procedure has a component before the start of the project (if successfully funded) and then components conducted during the project (what are termed Ethics Checks, Ethics Reviews, and Audits). The aim of the Ethics Appraisal Procedure is to identify any ethics issues and ensure that the project complies with expected standards of ethical research and that risks to research participants and researchers alike (not to mention society more broadly) are minimized.

Horizon Europe funding agencies carry out an Ethics Appraisal process with the assistance of independent and highly qualified ethics experts, who generally are situated within an EU Member State or European country. Following scientific evaluation of proposals, those that are considered meriting funding are then passed over for the ethics evaluation. Here, the ethics experts examine the proposals' documentation, and in particular the completed Ethics Table and Ethics Self-Assessment found in Part A of the proposals.[10] The Ethics Appraisal process includes: (a) the ethics screening process conducted before the start of the project (both pre-screening and screening); and (b) the ethics review process and audits conducted during the lifetime of a project. The ethics screening process starts right after the scientific evaluation and is concluded by ethics clearance before the signature of the grant agreement. Proposals can receive ethics clearance during the pre-screening, screening, or assessment phase.

It is important to note that only proposals selected for funding are subject to an ethics pre-screening and screening process. A pre-screening of the proposals is done by the internal ethics officers at the relevant funding agency in order to identify proposals that can be cleared. The rest of the proposals go to screening, which are conducted by independent external ethics experts. In this phase, the ethics experts are asked to identify the proposals that have "serious or complex ethics issues" (discussed further below) that will be the subject of a more in-depth analysis in an additional independent ethics panel (called an Ethics Assessment). The proposals

[10] The European Commission provides guidance to applicants on this. See European Commission, *EU Grants: How to complete your ethics self-assessment* (Version 2.0, 13 July 2021) (hereafter, "Ethics Self-Assessment Guidance Note"), available at: https://ec.europa.eu/info/funding-tenders/opportunities/docs/2021-2027/common/guidance/how-to-complete-your-ethics-self-assessment_en.pdf.

that are not considered ethically serious or complex are cleared at screening without ethics requirements or deliverables.

As already indicated above, in Horizon Europe, a clear distinction is made between "normal" and "serious or complex" ethics issues. How proposals are managed is then based on this distinction. The majority of Horizon Europe grants will raise normal (or otherwise non-material) ethics issues; by this mean I mean issues that are known to and commonly experienced by researchers in their daily work, such as processing of any kind of personal data and involving research participants in some capacity (e.g., survey or interview). These relatively minor and standard ethics issues are managed by the project participants themselves, such as the Principal Investigator, the Host Institution, and any others contributing to the research (or in EU parlance, "grant beneficiaries"), and without involving contractual deliverables and without having to send documentation to the relevant funding agency (e.g., ERC, European Research Executive Agency). Instead, the beneficiaries are expected to "keep on file" all documentation pertaining to the ethics compliance of the research, and they may be requested to submit this documentation to demonstrate compliance and good standing, although routine monitoring does not take place. In these cases, beneficiaries are sent an "ethics clearance for grant signature" and an Ethics Summary Report. The latter contains the general analysis of the issues, any observations to which the funding agency wishes to draw particular attention to, any requirement or recommendation to appoint an ethics advisor or a board, and (as discussed below) only for complex or serious cases only, any contractual ethics deliverables.

A small proportion of research projects will feature what are termed "serious or complex ethics issues".[11] Projects with these kinds of ethics issues are subject to more extensive assessment prior to grant agreement signature (specifically, a further Ethics Assessment by a panel of independent ethics experts). The outcome of this assessment may include requiring specific "ethics deliverables", such as a data protection impact assessment, a human rights impact assessment, or copies of template informed consent forms, to be included in the grant agreement as part of the contractual obligations.

As already noted above, it is open to the relevant funding agency under Horizon Europe to require or recommend that some beneficiaries, including those that do not raise "serious or complex ethics issues", appoint an independent ethics advisor or ethics board to help the project participants achieve good ethics management and provide assurance of this to the funding agency. Likewise, it may be appropriate for a funded project to have an "Ethics Check" or "Ethics Review" during the lifetime of the project (there may be several arranged for projects that have a longer duration of four or five years); generally, these Checks or Reviews will only occur for projects deemed to raise "serious or complex ethics issues". An Ethics Check is a semi-elaborate procedure carried out by at least one independent external ethics expert. An Ethics Review is a more elaborate and in-depth procedure carried out

[11] See generally European Commission, *Identifying serious and complex ethics issues in EU-funded research* (5 July 2021), available at: https://ec.europa.eu/info/funding-tenders/opportunities/docs/2021-2027/horizon/guidance/guidelines-on-serious-and-complex-cases_he_en.pdf.

by up to five independent external ethics experts and may include a panel meeting with the research team. Ethics Checks or Reviews often take place when decisions as to the research design with ethical implications will be taken after the project started, and where it therefore is not possible to fully review the project prior to the start, or when closer ethics oversight is deemed necessary. In Ethics Reviews, often independent ethics experts work with the funding agencies' Project Officer (staff members who help coordinate projects and liaise with the projects' investigators and Host Institutions) to assess compliance with the ethics requirements or deliverables; in the case of Ethics Checks, this tends to be undertaken exclusively by an ethics expert, although occasionally also with feedback from the Project Officer.

Ethics experts for Horizon Europe proposals are expected to analyze the ethical aspects of the proposal based on their subject-matter expertise and on an independent basis (i.e., not as a representative of their country, organization, religion, or philosophy). Experts are bound by confidentiality obligations. Their ethics analysis is expected to take into consideration all the documents provided by the applicant, including the full scientific proposal (Part B1 and B2), Part A (containing the Ethics Issues Table and the Ethics Self-Assessment), as well as any attached ethics Annexes. In their analysis, among other matters, ethics experts are asked to consider: (1) potential ethics issues regarding the objectives of the proposal, the methodology to conduct the research (from an ethical rather than scientific perspective), and the possible ethical implications of the results; (2) conformity with legal and ethical principles (at EU and national level); (3) any issues with the completed Ethics Self-Assessment, such as the capacity of the researchers or participating institutions to effectively mitigate the risks arising from the project's execution, or awareness of (or lack of) key ethics issues in the proposal. Ethics experts are asked to not question the scientific evaluation of the proposal, although the epistemic divide between ethics and science has long-been known as artificial and difficult to maintain (Dove 2020).

Having now briefly discussed the ethics review procedure in Horizon Europe, I now turn to explore several of the main data-related ethics issues that can arise in research projects.

13.3 Data-related Ethics Issues

In this section, I discuss several data-related ethics issues that tend to arise in Horizon Europe funding applications. Sometimes these issues will be considered "normal"; other times, they may meet the threshold of "serious or complex". Commonly, data-related ethics issues raise questions pertaining to compliance with data protection law, such as the EU General Data Protection Regulation (GDPR),[12] but may also raise questions relating to respect for persons and their privacy (as a component of research ethics and human rights) as well as, for example, research misuse. In what

[12] Regulation (EU) 2016/679 of the European Parliament and of the Council of 27 April 2016 on the protection of natural persons with regard to the processing of personal data and on the free movement of such data, and repealing Directive 95/46/EC (General Data Protection Regulation).

follows, I identify four core areas of ethical focus: (1) research involving humans; (2) personal data; (3) AI; and (4) research misuse.

13.3.1 Research involving humans

A significant amount of Horizon Europe-funded research involves human research participants in some capacity. As noted by European Commission guidance, "Humans must be considered as 'research participants' whenever they are recruited, observed, actively engaged, or in any other way may be influenced, manipulated, or directed by the research".[13] Often this can take the form of people being surveyed, interviewed, participating in a focus group, or being observed ethnographically, as well as through involvement in a clinical trial or psychological experiment. Much of their involvement will entail collection of their data in some form and can also include tracking them with consent (e.g., via GPS on a smartphone) or the secondary use of their data provided initially for other purposes (e.g., consent to access health records about them for a different or non-health-related purpose).

From a data-related perspective, research involving humans can raise ethical concerns about (among other aspects) respect for human dignity and wellbeing, a fair distribution of benefits and burdens (what might be termed justice-related concerns), and where appropriate, the securing of the participants' informed consent to participate in the research activity. Depending on the nature of the activities planned, it may be appropriate for the beneficiary to apply for and keep on file evidence of ethics approval obtained from a competent research ethics committee (REC), which in turn would be expected to assess whether the beneficiaries are able to obtain the free and fully informed consent of the participants. While ethics approvals may be readily obtainable from an institution such as a university, hospital, or (government-funded) region-based REC, entirely private organizations (e.g., start-up companies) may find it difficult to secure ethics approval. In such cases, it would be appropriate for them to incorporate independent ethics advice via the appointment of an ethics advisor or board.

Where beneficiaries are collecting and using data relating to research participants (and whether it constitutes "personal data" or not),[14] they should also be mindful of issues related to privacy, such as compliance with Article 8 of the European Convention on Human Rights and Article 7 of the EU Charter of Fundamental Rights, data protection (discussed more below), and data management (sometimes also called data governance). Privacy is a fundamental right in EU law (and broader human rights law) that differs from data protection. It encompasses the qualified right of an individual to not be interfered with against their will and the ability for them to exercise some form of autonomy over their life (including information relating to them). At least in its informational dimension, privacy is a state of affairs whereby data relating to a person are in a state of non-access or controlled access. It embodies a broad range of rights and values, such as the right to be let alone,

[13] Ethics Self-Assessment Guidance Note, n 11, p. 5.

[14] For which, see the section below on "Processing of personal data".

intimacy, seclusion, and personhood. In their research activities, beneficiaries ought to be mindful of how to engage in methods of data collection and use in ways that respect privacy (including use of non-identifiable data where possible); this may involve obtaining the informed consent of participants prior to any such data use, or otherwise using data in ways that accord with participants' reasonable expectations of such use. For the data governance aspect, this means ensuring that the data are kept securely in line with the beneficiaries' organizational safeguards. More generally, in relation to privacy, this means ensuring that any statements in the information sheet (as part of the research ethics norm of securing informed consent to participate in a project) that the participant's privacy and confidentiality will be respected, including ensuring that any publication of research results do not lead directly or indirectly to a participant's identification.

As noted in the European Commission guidance,[15] data-related aspects of research involving human participants that can be considered to raise "serious or complex" ethics issues include use of high-risk or inappropriate techniques and methods that may expose participants to unacceptable risk (such as insecure online surveys addressing sensitive issues or processing sensitive/special category data, methods that encourage participation by vulnerable groups without regard to their vulnerability); or studies whose findings carry a high risk of exposure or identification of individuals and/or stigmatization of a minority or marginalized group (e.g., on the basis of poverty, ethnicity, religion, gender/gender identity, disabilities, age, sexuality or perceived involvement in illegal activities). In such cases, beneficiaries are encouraged to seek ethics guidance from an ethics expert within their institution (if one exists), as well as guidance from their institution's Data Protection Officer, or DPO (which the majority of institutions involved in Horizon Europe bids have in place, and which are discussed in more detail below). Beneficiaries are also encouraged to ensure there is detailed information on the informed consent procedures for the research participants, alongside project-specific templates of the information sheets and informed consent or assent forms (in language and terms understandable to the participants) with all relevant information regarding the protection of personal data, including the DPO contact details for host institutions required to appoint a DPO under the GDPR.

13.3.2 Processing of personal data

It is fair to say that the most commonly identified and most intensive of the ethics-related aspects of EU-funded research projects is the processing of personal data. This has become a heightened area of concern since the coming into force of the GDPR in May 2018. The GDPR is a complex piece of legislation, comprising 99 articles (and over 55,000 words of text). It provides a comprehensive legal framework for the protection of Europeans' personal data, as well as for the promotion of responsible data processing for a range of legitimate purposes, including scientific research and healthcare delivery. The GDPR is fundamentally about protecting and promoting

[15] Ethics Self-Assessment Guidance Note, n 11, p. 6.

the fundamental right to data protection. Data protection aims to protect the rights, freedoms, and interests of individuals whose personal data are collected and used in various ways. The objective is to ensure fairness in the processing of data and, to some extent, fairness in the outcomes of such processing. As the European Commission guidance note on ethics and data protection states:

> Data protection is both a central issue for research ethics in Europe and a fundamental human right. It is intimately linked to autonomy and human dignity, and the principle that everyone should be valued and respected. For this principle to guide the development of today's information society, data protection must be rigorously applied by the research community.[16]

In short, any projects with research activities that involve processing of personal data,[17] regardless of the method used, will be caught in this regulatory and ethics web. This means that activities that might involve a person's name, address, identification number, pseudonym, occupation, e-mail, CV, their location data, their Internet Protocol (IP) address, cookie ID, phone number, any data provided by their smart meters, and so on is likely to be considered personal data, bringing data protection law within scope. Similarly, "data processing" under the GDPR is defined broadly to encompass any operation (or set of operations) performed on personal data, either manually or by automatic means. This includes: collection; recording; organization, structuring, and storage; adaptation or alteration; retrieval and consultation; use; disclosure by transmission, dissemination, or otherwise making available; alignment or combination; or restriction, erasure, or destruction.

What the above indicates is that data processing applies in almost all research activities. Moreover, data sharing commonly is envisaged both during the lifetime of the project as well as afterwards, which means that the ethics issues associated with personal data can endure long after the research activities within the funded portion of the projects are completed.

The GDPR requires processing personal data to be undertaken with a "legal basis", of which there are six kinds under Article 6(1) GDPR (I note that consent is one available legal basis, but it is not a strict requirement for all data processing activities). If there is processing of "special category" personal data, which is sensitive data concerning ethnicity, sexual orientation, health, genetics, biometrics, and so on, then there must also be, alongside a legal basis, an appropriate exemption under Article 9(2) GDPR, of which there are ten kinds (I note that explicit consent is one kind of exemption, but again, is not a strict requirement for all data processing activities). For example, under Article 9(2)(j), data controllers may process special category personal data (e.g., health data) where the processing is necessary for

16 European Commission, "Ethics and data protection" (5 July 2021), p. 4, available at: https://ec.europa.eu/info/funding-tenders/opportunities/docs/2021-2027/horizon/guidance/ethics-and-data-protection_he_en.pdf.

17 Personal data is information relating to an identified or identifiable natural person, who in turn is known in data protection law parlance as the "data subject". Personal data may contain confidential information, but the two are not synonymous (e.g., a person's publicly available tax records comprise personal data but may not be considered confidential).

scientific research in accordance with Article 89(1) GDPR and that appropriate safeguards are provided for. The processing must also be consistent with EU or Member State law (such as laws relating to clinical trials). Appropriate safeguards would include ensuring that the research will not cause substantial damage or distress to the data subject; ensuring that the scientific research project has approval from a competent and independent REC; and ensuring that the data controller has technical and organizational safeguards in place that ensure respect for the principle of data minimization and ensure that exemptions to data subjects' rights are not exercized unless the rights are likely to render impossible or seriously impair the achievement of the purposes of the processing. Examples would include pseudonymizing personal data where compatible with the research purpose and avoiding processing identifiable data where the research purpose can be fulfilled by further processing with anonymized data. Other examples include having robust IT security and data protection policies in place, as well as risk-assessment exercises of data systems, regular review and audit of data protection impact assessments, and developing efficient IT systems, policies, and procedures to handle data subjects' requests (Dove 2018).

For many EU-funded research projects, it may be sensible to conduct what are called data protection impact assessments (DPIAs), which are outlined under Article 35 GDPR. Where a type of data processing—in particular one using new technologies—is likely to result in a high risk to the rights and freedoms of data subjects (taking into account the nature, scope, context, and purposes of the processing), the data controller[18] must, prior to the processing, carry out an assessment of the impact of the envisaged processing operations on the protection of personal data. DPIAs are a process to help controllers identify and minimise the data protection risks of a project. DPIAs generally (1) describe the nature, scope, context, and purposes of the processing; (2) assess necessity, proportionality, and compliance measures; (3) identify and assess risks to individuals; and (4) identify any additional measures to mitigate those risks. Many websites of the national data protection authorities across EU Member States provide guidance on how to complete a DPIA.[19]

Beneficiaries of EU-funded research projects must be mindful that the GDPR (and national data protection laws) provide a number of "data subject rights" to individuals, although these may be qualified for scientific research purposes, as I discuss below. Data subject rights include:

- Rights of data access, e.g., the right for data subjects to obtain from the data controller confirmation as to whether or not personal data concerning them are being processed, where and for what purpose(s); and the obligation for controllers to provide a copy of the personal data, free of charge, in an electronic format;

[18] The data controller is the natural or legal person, public authority, agency, or other body which, alone or jointly with others, determines the purposes and means of the processing of personal data. For EU-funded research projects, often this will be investigator and/or the Host Institution.

[19] See, e.g., Data Protection Commission (Ireland), "Data Protection Impact Assessments", available at: https://www.dataprotection.ie/en/organisations/know-your-obligations/data-protection-impact-assessments.

- The right of data subjects to have the data controller erase their personal data, cease further dissemination of the data, and potentially have third parties halt processing of the data (the so-called "right to be forgotten");
- The right for a data subject to receive the personal data concerning them, which they have previously provided in a "commonly used and machine-readable format," and to transmit those data to another controller (the so-called "right to data portability");
- The right to object to processing "on grounds relating to his or her particular situation" where processing is based on Article 6(f) "legitimate interests" or Article 6(e) "task carried out in the public interest", including profiling based on those provisions; and
- The right for a data subject not to be subject to a decision based solely on automated processing, including profiling, which produces legal effects concerning them or similarly significantly affects them (i.e., a right not to be subject to pure AI-driven decisions, or machine learning or other technologies that result in a decision about the individual such as a medical diagnosis), provided the decision is not (1) necessary for entering into, or performance of, a contract between the data subject and a data controller; (2) authorized by EU or Member State law to which the controller is subject and which also lays down suitable measures to safeguard the data subject's rights and freedoms and legitimate interests; or (3) is based on the data subject's explicit consent.

Article 89(2) GDPR allows EU Member States to create derogations from (i.e., exceptions to) the otherwise core data subject rights of data access, to rectification, to erasure, to restriction of processing, and to object, whenever these data subject rights "are likely to render impossible or seriously impair the achievement" of scientific research purposes. Such a derogation must be essential for the fulfilment of these purposes. These derogations are subject to the conditions and safeguards laid out in Article 89(1), such as data minimization and pseudonymization (Recital 156 GDPR). The "right to object" is curtailed in a slightly different manner: Article 21(6) GDPR gives data subjects an unconditional right to object in the context of scientific research but can be overridden where "the processing is necessary for the performance of a task carried out for reasons of public interest".

Beyond legal bases, special category data processing exemptions, and data subject rights, it is also worth noting the strict international data transfer restrictions under Chapter V of the GDPR. Broadly speaking, there are four avenues for lawful transfer of personal data outside the EU: (1) an adequacy decision (Article 45); (2) appropriate safeguards (Articles 46 and 47); (3) specific derogations (Article 49); and (4) exceptions for one-off (or infrequent) transfers (second subparagraph of Article 49(1)). It is outwith the scope of this chapter to provide details on each of these avenues; readers are encouraged to consult specific guidance elsewhere,[20] although

[20] See, e.g., European Data Protection Board, "International Data Transfers", available at: https://www.edpb.europa.eu/sme-data-protection-guide/international-data-transfers_en; Health Research Board (Ireland), " Transferring, processing or storing personal data outside of Ireland", available at: https://www.hrb.ie/funding/gdpr-guidance-for-researchers/gdpr-overview/international-data-transfers/.

it bears noting that none of these legal avenues are particularly straightforward to satisfy and raise particular challenges for data sharing for a variety of research-related purposes (Dove 2025).

For many EU-funded research projects, and given the complexity of data protection law, it may be sensible for beneficiaries to discuss data protection-related ethics issues with a DPO. Under the GDPR, DPO appointment will be mandatory for controllers and processors which are public authorities or whose core activities, among other things, consist of processing "large scale" special categories of data (e.g., health-related data or genetic data). Thus, DPOs are required for universities and scientific research and medical organizations. Importantly, the DPO must be appointed on the basis of professional qualities and, in particular, expert knowledge on data protection law and practices. DPOs must be provided with appropriate resources to carry out their tasks and maintain their expert knowledge; must report directly to the highest level of management; and must not carry out any other tasks that could result in a conflict of interest.

In sum, the various ways beneficiaries could look to address common data protection-related ethics issues in EU-funded projects include:

- Verifying compliance with the applicable national legal framework(s) in case special derogations pertaining to the rights of data subjects or the processing of genetic, biometric, and/or health data have been established in the country where the research takes place.
- Thinking through the justification for the processing of sensitive ("special categories") of personal data, as per Article 9 of the GDPR.
- Confirming that all the personal data that will be processed in the project are relevant and limited to the purposes of the research project (in accordance with the "data minimization" principle).
- Thinking through the technical and organizational measures to safeguard the rights and freedoms of the data subjects.
- Thinking through the security measures to prevent unauthorized access to personal data or the equipment used for processing.
- Applying generally accepted or state-of-the-art anonymization or pseudonymization techniques, where possible.
- Confirming that for any personal data to be transferred from the EU to a non-EU country or international organization, such transfers are in accordance with Chapter V of the GDPR.
- Confirming that for any personal data to be transferred from a non-EU country to the EU (or another third country), such transfers comply with the laws of the country in which the data was collected.
- Providing suitable justification for any research involving profiling, alongside careful explanation on how the data subjects will be informed of the existence of the profiling, its possible consequences, and how their fundamental rights will be safeguarded.

- For the further processing of previously collected personal data, identifying a suitable legal basis under Article 6(1) GDPR, alongside appropriate technical and organizational measures to safeguard the rights of the data subjects.
- Evaluating the ethics risks related to the data processing activities of the project, together with an opinion if a Data Protection Impact Assessment (DPIA) should be conducted as per Article 35 GDPR.

13.3.3 Artificial intelligence

A growing number of EU-funded research projects integrate some form of AI-based systems or techniques. While there is no general definition of AI that enjoys widespread consensus, it may be taken to mean (as per the AI Act's definition) a machine-based system designed to operate with varying levels of autonomy, that may exhibit adaptiveness after deployment and that, for explicit or implicit objectives, infers, from the input it receives, how to generate outputs such as predictions, content, recommendations, or decisions that can influence physical or virtual environments.

The EU's AI Act[21] has had further effect on how applicants and beneficiaries address aspects of AI in their project activities. This Act, which was proposed by the European Commission in April 2021 and passed in March 2024, it aims to establish a common regulatory and legal framework for AI across the EU. Specifically, it classifies and regulates AI applications based on their risk to cause harm to humans. This classification includes four categories of risk: unacceptable, high, limited, and minimal, plus one additional category for general purpose AI (GPAI).

Applications deemed to represent unacceptable risks (e.g., evaluating or classifying individuals or groups based on social behaviour or personal traits, causing detrimental or unfavourable treatment of those people; deploying subliminal, manipulative, or deceptive techniques to distort behaviour and impair informed decision-making, causing significant harm; compiling facial recognition databases by untargeted scraping of facial images from the internet or CCTV footage) are banned. High-risk AI applications (e.g., a CV-scanning tool that ranks job applicants) must comply with security, transparency, and quality obligations and undergo conformity assessments. Limited-risk AI applications (e.g., chatbots) only have transparency obligations, and those representing minimal risks (e.g., many AI-enabled video games and spam filters) are not regulated. For GPAI, transparency requirements are imposed (e.g., technical documentation, instructions for use, comply with the Copyright Directive, a summary about the content used for training), with additional and thorough evaluations when representing particularly high risks.

The AI Act largely imposes obligations on providers (developers) of high-risk AI systems. This applies both to developers that intend to place on the market or put into service high-risk AI systems in the EU, regardless of whether they are based in the EU or a third country, and also developments in third countries (e.g., USA, UK) where the high-risk AI system's output is used in the EU.

---

[21] AI Act, available at: https://eur-lex.europa.eu/legal-content/EN/TXT/?uri=OJ:L_202401689.

The AI Act will be implemented through the establishment of an "AI Office", sitting within the European Commission, which will monitor the effective implementation and compliance of GPAI model providers. Among other activities, the AI Office will: develop tools, methodologies, and benchmarks for evaluating capabilities and reach of GPAI models, and classifying models with systemic risks; draw up state-of-the-art codes of practice to detail out rules, in cooperation with leading AI developers, the scientific community and other experts; investigate possible infringements of rules, including evaluations to assess model capabilities, and requesting providers to take corrective action; prepare guidance and guidelines, implementing and delegated acts, and other tools to support effective implementation of the AI Act and monitor compliance with the regulation; provide advice on best practices and enabling ready-access to AI sandboxes, real-world testing and other European support structures for AI uptake; and foster international cooperation and governance on AI, with the aim of contributing to a global approach to AI.[22]

After entry into force in May 2024, the AI Act will apply after:

- 6 months for prohibited AI systems;
- 12 months for GPAI;
- 24 months for high-risk AI systems under Annex III;
- 36 months for high-risk AI systems under Annex II; and
- Codes of practice must be ready 9 months after entry into force.

Separate from (but connected to) their legal compliance obligations under the AI Act, beneficiaries are strongly encouraged to abide by the *Independent High Level Expert Group on AI's Ethics Guidelines for Trustworthy AI*.[23] According to the Guidelines, "trustworthy AI" has three components. It should be: (1) lawful—respecting all applicable laws and regulations; (2) ethical—respecting ethical principles and values; and (3) robust—both from a technical perspective and also taking into account its social environment. As the Ethics Guidelines note, "Laws are not always up to speed with technological developments, can at times be out of step with ethical norms or may simply not be well suited to addressing certain issues. For AI systems to be trustworthy, they should hence also be ethical, ensuring alignment with ethical norms".[24]

Further, the Ethics Guidelines promulgate a set of 7 key requirements that AI systems should meet in order to be deemed "trustworthy". These ethical obligations largely complement the legal obligations imposed by the EU's AI Act. A specific assessment list aims to help verify the application of each of the key requirements:

- *Human agency and oversight*: AI systems should empower human beings, allowing them to make informed decisions and fostering their fundamental

---

[22] See European Commission, "European AI Office", available at: https://digital-strategy.ec.europa.eu/en/policies/ai-office.

[23] Independent High Level Expert Group, *Ethics Guidelines for Trustworthy AI* (8 April 2019), available at: https://digital-strategy.ec.europa.eu/en/library/ethics-guidelines-trustworthy-ai. The Independent High Level Expert Group was set up by the European Commission in June 2018.

[24] Ibid., pp. 6-7.

rights. At the same time, proper oversight mechanisms need to be ensured, which can be achieved through human-in-the-loop, human-on-the-loop, and human-in-command approaches.

- *Technical robustness and safety*: AI systems need to be resilient and secure. They need to be safe, ensuring a fall-back plan in case something goes wrong, as well as being accurate, reliable and reproducible. That is the only way to ensure that unintentional harm can be minimized and prevented.
- *Privacy and data governance*: besides ensuring full respect for privacy and data protection, adequate data governance mechanisms must also be ensured, taking into account the quality and integrity of the data, and ensuring legitimized access to data.
- *Transparency*: the data, system, and AI business models should be transparent. Traceability mechanisms can help achieving this. Moreover, AI systems and their decisions should be explained in a manner adapted to the stakeholder concerned. Humans need to be aware that they are interacting with an AI system, and must be informed of the system's capabilities and limitations.
- *Diversity, non-discrimination, and fairness*: Unfair bias must be avoided, as it could have multiple negative implications, from the marginalization of vulnerable groups, to the exacerbation of prejudice and discrimination. Fostering diversity, AI systems should be accessible to all, regardless of any disability, and involve relevant stakeholders throughout their entire lifecycle.
- *Societal and environmental well-being*: AI systems should benefit all human beings, including future generations. It must hence be ensured that they are sustainable and environmentally friendly. Moreover, they should take into account the environment, including other living beings, and their social and societal impact should be carefully considered.
- *Accountability*: Mechanisms should be put in place to ensure responsibility and accountability for AI systems and their outcomes. Auditability, which enables the assessment of algorithms, data and design processes plays a key role therein, especially in critical applications. Moreover, adequate an accessible redress should be ensured.[25]

In general, ways to address AI-related ethics issues and to establish trustworthy include:

- Being transparent with participants and users that they are interacting with an AI system and are informed (in a language and terms understandable by all) about the AI's abilities, limitations, risks, and expected benefits.
- Avoiding any breach of human rights and freedoms (e.g. freedom of expression, access to information, freedom of movement). It might be appropriate to undertake a human rights impact assessment covering the development, deployment, and post-deployment phases, including detailed information on how respect for fundamental human rights and freedoms will be ensured.

[25] Ibid.

- Designing the AI in ways that avoid objectification, dehumanization, subordination, discrimination, stereotyping, coercion, manipulation of people, or creation of attachment or addiction.
- Designing the AI in ways that avoid bias in both input data and algorithm design, as well as potential discrimination, stigmatization, or any other adverse effects on the individual related to the use of the developed/deployed AI system/technique.
- Demonstrating compliance with the principles of data minimization and privacy by design and by default when processing personal data, as well as the data protection principles of lawfulness, transparency, and fairness of the data processing.
- Transparency regarding the potential impact on the individual, society, or the environment, including by way of an evaluation of the potential negative individual, societal, and/or environmental impacts along with the measures to be set in place to mitigate any potential adverse effect.
- An ethics risk assessment and risk mitigation measures covering the development, deployment, and post-deployment phases of the AI.
- Designing the AI in a way that enables human oversight (human-in-the-loop, human-on-the-loop, human-in-command), traceability, and auditability.
- Whenever possible, an explanation provided to participants or users on how decisions are taken by the developed/used AI, along with the logic behind it.

13.4 Research Misuse

Finally, another core data-related ethics issue briefly worth exploring concerns research misuse. Research misuse concerns the potential for materials, methods, technologies, or knowledge to be misused for unethical purposes. Research misuse occurs when the results (or outcomes) of research are used by others (intentionally or not) for harmful purposes.

In research projects, there is a risk that the resulting data (such as a recipe, code, or description of a process) "could be directly misapplied to pose a significant threat with broad potential consequences to public health and safety, agricultural crops and other plants, animals, the environment, materiel, or national security".[26] Research misuse is frequently associated with concerns about bioterrorism, although the issues associated with it can also cover any kind of result from a research activity that causes harm to others. For example, research involving primary data collection into terrorist activities or other activities that could pose a threat to national security.

While serious concerns about research misuse may more frequently arise in relation to experiments that, for instance, confer resistance to therapeutically useful antibiotics or antiviral agents, or enhance the virulence of a pathogen, or render a

[26] US National Institutes of Health, "Dual-Use Research", available at: https://oir.nih.gov/sourcebook/ethical-conduct/special-research-considerations/dual-use-research.

non-pathogen virulent, it is really about the data underlying these techniques and processes that generate ethics issues and as such require beneficiaries to be cognizant of their moral and legal obligations to the funders and society more broadly.

As the European Commission's guidance note indicates, beneficiaries ought to consider the risks associated with the activities they plan and any unethical ways in which the materials, methods, technologies, and knowledge involved could be used. The guidance note flags several activities that are considered more vulnerable to misuse.[27] These include the development of surveillance technologies that could curtail human rights and civil liberties; the development of social, behavioural, or genetic profiling technologies that could be misused to stigmatise, discriminate against, harass or intimidate people; the development of materials/methods/technologies and knowledge that could harm humans, animals, or the environment if they were released, modified, or enhanced; and in general, the development of materials/methods/technologies and knowledge that could serve purposes other than those intended, and if so, in unethical ways.

Beneficiaries can mitigate the risk of research misuse through various means, including by:[28]

- applying additional safety measures through e.g. safety and security training for staff;
- adjusting the project design, e.g., by using synthetic or dummy data;
- limiting research dissemination, e.g., by publishing only part of the results, regulating export;
- ensuring compliance with any relevant legislation (e.g., national security or counterterrorism legislation);
- appointing an independent ethics advisor or an ethics advisory board; and
- undertaking a risk assessment and explanation as to how they will prevent misuse (including identification of specific measures to be taken).

13.4 Conclusions

As this chapter has demonstrated, applicants and beneficiaries alike who seek EU funds to undertake ground-breaking, innovative research using data (of whatever kind and through whatever methods) must navigate a variety of ethics issues. The EU through its Framework Programmes aims to achieve the highest standards of ethics, accountability, and responsibility. This chapter cannot cover comprehensively the breadth of other kinds of data-related ethics issues that can arise, such as those related to risk of environmental harms or to cybersecurity. Nevertheless, it has been my aim here to demonstrate that as the EU's Framework Programmes intend to support a broad range of innovative activities, and data increasingly feature at the core of those activities, ethics issues and concerns will invariably arise. Beneficiaries

[27] Ethics Self-Assessment Guidance Note, n 11, p. 48.

[28] Ibid., pp. 48-49.

must be attuned to their identification and mitigation, while ethics experts and other stakeholders ought to support beneficiaries in their efforts to make their projects as socially and scientifically valuable—and ethical—as possible.

References

Dove, E.S. 2025. International Health Data Transfers in the United Kingdom – The Legal Framework. *In*: Corrales Compagnucci M. and M. Fenwick [eds.]. International Transfers of Health Data: A Global Perspective. Springer, Singapore.

Dove, E.S. 2018. The EU General Data Protection Regulation: Implications for International Scientific Research in the Digital Era. Journal of Law, Medicine & Ethics. 46(4): 1013–1030.

Dove, E.S. 2020. Regulatory Stewardship of Health Research: Navigating Participant Protection and Research Promotion. Edward Elgar, Cheltenham.

Hasselbalch, G. 2021. Data Ethics of Power: A Human Approach in the Big Data and AI Era. Edward Elgar, Cheltenham.

Lövestam, G. et al. 2025. Fostering scientific integrity and research ethics in a science-for-policy research organization. Research Ethics. 21: 56–75. (doi: https://doi.org/10.1177/17470161241243001).

O'Keefe, K. and D. O Brien. 2023. Data Ethics: Practical Strategies for Implementing Ethical Information Management and Governance, 2nd edn. Kogan Page, London.

Machon, A. 2022. The Privacy Mission: Achieving Ethical Data for Our Lives Online. Wiley, Hoboken.

Martin, K. [eds.]. 2022. Ethics of Data and Analytics: Concepts and Cases. Routledge, London.

O'Keefe, K. and D. O Brien. 2023. Data Ethics: Practical Strategies for Implementing Ethical Information Management and Governance, 2nd edn. Kogan Page, London.

Shukla, S. et al. 2022. Data Ethics and Challenges. Springer, Berlin.

14

Models of Data Sharing and Best Practices

Gabriele Rinck

14.1 Big Data Require a Joint Effort

The endeavour of turning Big Data of increasing complexity into rapid scientific progress relies on a joint effort across the scientific community.

Biomedical research produces a hu ge amount of data with the aim to better understand mechanisms of disease, characterize pathogens, discover new treatments and other crucial interventions. The gained knowledge helps to alleviate the impact disease has on those affected and this benefits society as a whole.

For the enormous task of analyzing and interpreting such a large amount of complex data, it is important for the scientific community to work across the globe – data sharing is the key to this.

14.2 Data Sharing Principles from Bermuda to FAIR

Bermuda Principles

The initial idea of open science and open data goes back to the early days of the Human Genome Project. The International Human Genome Sequencing Consortium consisted of scientists from 20 institutions in six countries (https://www.yourgenome.org/theme/the-human-genome-project/): France, Germany, Japan, China, the UK and the US (https://doe-humangenomeproject.ornl.gov/). Their ethos was to accelerate scientific progress by fostering collaboration and preventing unnecessary duplication of efforts across the consortium. This led to the formulation of the Bermuda Principles (in 1996, refined in 1998) which stressed that sequence assemblies should be deposited immediately into public databases and freely accessible to all researchers (Bentley 1996; Guyer 1998).

European Molecular Biology Laboratory (EMBL), European Bioinformatics Institute (EBI), Wellcome Genome Campus, Hinxton, United Kingdom; Email: rinck@ebi.ac.uk

DOI: 10.1201/9781003475927-17

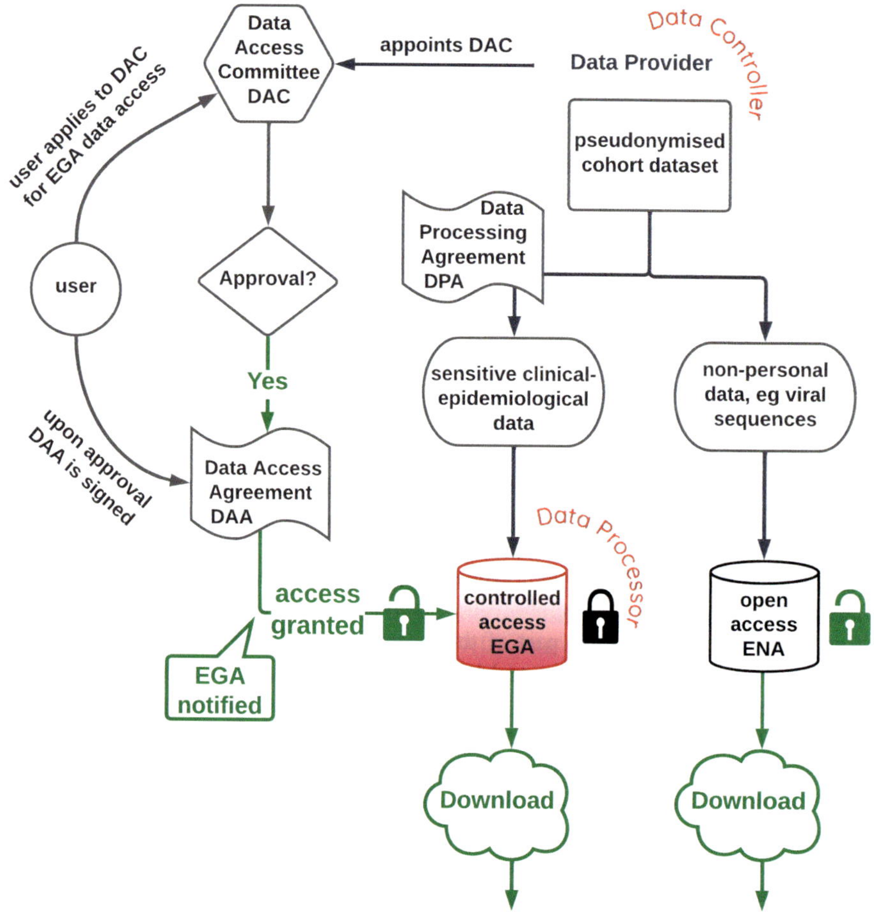

Figure 2: Simplified overview of the data sharing process for cohort datasets using EGA and ENA archives as examples for controlled and open data access, respectively. Non-sensitive data deposited in the open-access ENA repository can be freely downloaded by any user without any restrictions. Pseudonymized sensitive data can be archived in EGA; the organizations hosting EGA act as data processor while the data provider remains the data controller. The data provider establishes a data access committee that will review any data access requests; once a request is approved a Data Access Agreement is signed and EGA is notified. Once the access is granted users can download data for analysis.

Other data types that do not allow reidentification, like for example antibody profiles, could be shared openly. Viral sequences are derived from the pathogen and are not considered personal data.

Each cohort study may have to adapt their data sharing strategy to their project-specific limitations and restrictions. This can be influenced by a number of factors: (1) as mentioned above the data types and information collected can vary from cohort to cohort, (2) the interpretation of GDPR can vary significantly from country to country, (3) the data sharing culture or policies can differ from one institution to the next, (4) the wording of the informed consent may have an impact on the extent of data sharing allowed, (5) anonymization of the dataset might be an option,

(6) the aspect of benefit sharing may also factor into the decision about sharing data, especially if these data were collected in a low or middle income country.

In the spirit of FAIR principles, making data access as open as possible and as closed as necessary, the best option would be to share non-sensitive data via an open access repository while other parts of the data may have limited scope for sharing or require a controlled data access mechanism. Different modalities of sharing are described below.

14.7 Data Access Models

Open access

Ideally, all non-personal data should be deposited in a timely fashion in an open-access data archive; for pathogen sequence data, this could be the European Nucleotide Archive (ENA) (O'Cathail et al. 2025), that is hosted by the European Bioinformatics Institute (EMBL-EBI), which is part of the European Molecular Biology Laboratory (EMBL). ENA is part of the International Nucleotide Sequence Database Collaboration (INSDC) (Arita et al. 2021) and provides a comprehensive record of the world's nucleotide sequencing information, covering raw sequencing data, sequence assembly information and functional annotations. This repository provides open access by means of the ENA browser, search tools, large scale file download and through the API (Application Programming Interface) that allows programmatic retrieval of the data; no authentication is required for the download of the data from ENA and there are no restrictions for the reuse of data. DOI's (Digital Object Identifiers) can be requested for SARS-CoV-2 datasets deposited in ENA and data providers can claim ownership by linking the sequences, datasets or projects to their ORCID (Open Researcher and Contributor Identifier).

Regulated Access

Another widely used repository for sharing SARS-CoV-2 sequences during the pandemic was the GISAID (Global Initiative on Sharing All Influenza Data) database (Khare et al. 2021). In contrast to open-access archives like ENA, GISAID operates under a regulated-access model where all users have to register and agree to GISAID's terms and conditions. GISAID imposes certain restrictions and guidelines on the usage and dissemination of data to address concerns around credit attribution and exploitation (Sheehan et al. 2024). The majority of the data in GISAID are consensus sequences, while ENA has a similar amount of submissions for raw data and consensus sequences.

Controlled Access

In order to comply with data protection regulations, controlled access is an option for sharing pseudonymized human data. One example of a data archive that operates a controlled-access model is the European Genome Phenome Archive (EGA). EGA which is jointly hosted by EMBL-EBI and the Centre for Genomic Regulation (CRG in Barcelona), is a resource for long-term secure archiving of all types of potentially identifiable genetic, phenotypic, and clinical data resulting from biomedical research projects (Freeberg et al. 2022). The data provider remains the data controller, decides about data access requests and can define any data use restrictions. EGA's hosts, EMBL-EBI and CRG act as data processors shown in Figures 2, any researcher wishing to retrieve data requests access from the data controller. Once this request has been approved and a data access agreement is signed between the researcher and the data controller, access is granted and the data can be downloaded from EGA.

The stringent interpretation of GDPR in some countries does not allow data transfer across national borders. The Federated European Genome Phenome Archive (FEGA), launched by EMBL-EBI and CRG in 2022, has emerged as a pivotal project to facilitate the transnational discovery of sensitive human omics data. FEGA consists of a network of national nodes that allows potentially sensitive personal data to be archived in the country of origin; only non-sensitive metadata are shared via the central EGA repository so that users can centrally search for datasets across the entire FEGA system. This federated system is being rolled out across a number of European countries and beyond. At the time of writing, in June 2025, FEGA consists of nodes in eight European countries, and also Canada; discussions on future membership are ongoing with over twenty countries and institutes to expand the network further across the globe. The first datasets have been released in 2024.

Other Federated Data Access Models

DataShield is a secure system that can be used for federated analysis, an approach where the data remain with the data provider (Jones et al. 2012); the data is being analyzed locally with the analyst providing the code for the analysis remotely through a local server. This system is often employed for statistical analyses of clinical-epidemiological data, for example. Any researcher wishing to perform these analyses remotely needs to apply for data access to the data controller, who will assess the access application, similar to the approval process mentioned for the EGA data access.

This approach may be a suitable option for those data providers that, locally or nationally, face particularly stringent restrictions for sharing human data. They will benefit from the fact that the data can remain under their immediate control within their own IT system without the need to transfer the data to external databases.

Chapter 15 describes privacy-preserving AI-based solutions.

Cohort Catalogues

Apart from capturing the information and data from each study participant, it is also important to share study-level data which describe the dataset, the cohort study itself and how the data can be accessed. This enables researchers to identify and locate those cohort datasets that are most suitable for the investigation of a particular research question. A number of cohort catalogues and registries are available that list key study information as well as offer search and filtering functionalities to aid the discovery of the studies. Cohort catalogues like Maelstrom, https://www.maelstrom-research.org/page/catalogue, (Bergeron et al. 2018) or Molgenis, https://

data-catalogue.molgeniscloud.org/catalogue/catalogue/#/cohorts, (Van Der Velde et al. 2019) capture the cohort metadata and which set of clinical-epidemiological variables were collected from the study participants. The BBMRI-ERIC registry (Holub et al. 2016), on the other hand, is more focused on initiatives and studies that include BioBank aspects.

While these catalogues capture a wide range of cohorts across different areas of research, EMBL-EBI's Cohort Browser was specifically designed for infectious disease studies, https://www.pathogensportal.org/cohorts; further details about this Cohort Browser is provided below.

14.8 Data Integration

Deposition Databases

EMBL-EBI hosts numerous deposition databases, which are all specifically tailored to archiving a particular type of a wide range of high-dimensional laboratory data. These data type-specific archives capture the data in an appropriately structured way most suitable for the respective data type and in alignment with FAIR principles. The repositories ENA and EGA have already been mentioned as examples of open access to (pathogen) nucleotide sequences and controlled access to human data, respectively. Other databases hosted by EMBL-EBI can be used to archive further data types such as array data using the ArrayExpress/BioStudies archive (Sarkans et al. 2021), genetic variation data in the European Variation Archive (Cezard et al. 2022), Metabolomics in the MetaboLights archive (Yurekten et al. 2024), structural data on proteins in the Protein Data Bank in Europe (Armstrong et al. 2019), to name just a few. The broad range of EMBL-EBI-hosted archives, which the scientific community has used for many years, provides a solid infrastructure for storing and sharing large biological datasets, ideal for archiving and sharing most data types that make up an infectious disease cohort dataset.

COVID-19 Data Portal

To take advantage of this versatile, established range of archives, at the beginning of the COVID-19 pandemic, the COVID-19 Data Portal (https://www.covid19dataportal.org/) was created as a central point of entry to share and access any SARS-CoV-2-related data from across the EMBL-EBI-hosted repositories and beyond. The COVID-19 Data Portal pulls together all SARS-CoV-2-related data from EMBL-EBI-hosted archives and associated resources to present them in a structured way, together with relevant useful tools. Search and filtering functionalities make it easier for users to find the data of interest, also facilitating searches across all associated resources and listing cross-references to other records (Harrison et al. 2021; Rahman et al. 2024).

The COVID-19 Data Portal alleviates the impediment that data types are often siloed in their respective bespoke database, making it more difficult and time-consuming to locate and link datasets from the same study back together. Even before data deposition in an archive, within a hospital setting, for example, the different data types may be produced, analyzed and possibly even stored in separate departments.

Pathogens Portal

Following the success of the COVID-19 Data Portal and widening the scope towards future pandemic preparedness, a second Data Portal was launched, which covers all other pathogens, following the same concept as the COVID-19 Data Portal. This is the Pathogens Portal (https://www.pathogensportal.org/). The Pathogens Portal includes the Cohort Browser, which, as mentioned above, captures key study information and provides search and filtering functionalities to aid the discovery of the cohort studies. The Pathogens Cohort Browser also indicates which data types have been collected for each study, and for those datasets which have been deposited in one of the EMBL-EBI-hosted repositories, a hyperlink points directly to the dataset in the relevant archive.

Data Linking

While the Cohort Browser provides the link between data types on a dataset level, the EMBL-EBI system can also capture links at a more granular level for each study participant. The BioSamples archive (Courtot et al. 2022) is utilized to create such participant-level links. This archive stores sample information for records that were deposited in archives like ENA, EGA or BioStudies while it can also capture the information about the relationship between records. The dashboard tab in the Cohort Browser provides more detailed information about the comprehensive cohort dataset and shows all samples that are linked on participant level.

By linking different data types on a participant level, a dataset can be analyzed at a greater depth, maximizing the insights and conclusions that can be drawn from a dataset. This will aid the investigation of pathogen-host interactions, understanding

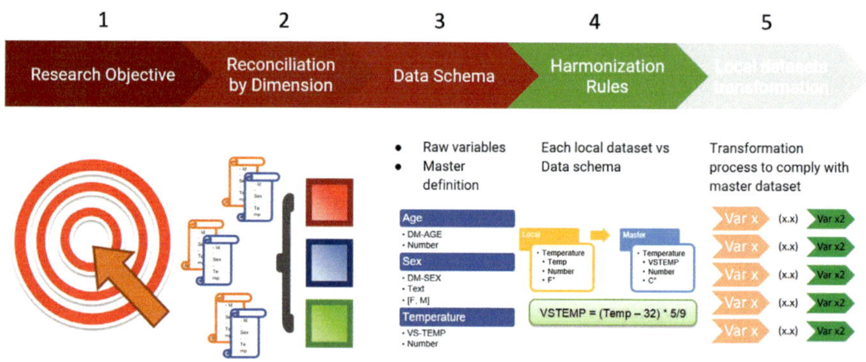

Figure 3: Harmonization process broken down into 5 steps. (1) Defining the Research Objective, (2) Reconciliation of the variables by dimension, (3) Creation of the data schema, (4) Defining the Harmonization Rules, (5) Transformation of the data. (Printed with kind permission from Gustavo Gómez, Universidad Industrial de Santander, Bucaramanga, Colombia. (Gómez et al. 2024))

of viral immune escape or could potentially lead to the identification of a new drug target to combat the pathogen infection.

One example from the 1990s that clearly demonstrated the impact of linking data types resulted in a significant improvement of HIV patient management (Shafer and Vuitton 1999; Shafer 2006). By bringing together HIV sequences, viral phenotyping data from cell culture assays and patient clinical data, such as viral load, CD4+ cell counts and treatment history scientific insights were gained which led to the introduction of a personalized approach to highly active antiretroviral therapy for HIV infections which increased survival rates and the quality of life for those infected.

14.9 Data Harmonization

By performing analyses across multiple cohorts the sample size of the analysis is increased which enhances the statistical power of the study. A prerequisite of including multiple cohorts in the same analysis is the interoperability of datasets across the cohorts, achieved by harmonization of the data. This harmonization entails the use of the same data standards, controlled vocabulary and ontologies, employing the same data dictionary for all cohorts, whenever possible. As this improves the interoperability it contributes to the FAIRification of the dataset.

14.10 Retrospective Harmonization

The retrospective harmonization process can be very time-consuming, sometimes taking several years, involving in-depth discussions with clinical subject matter experts to ensure that the captured variables are aligned across the cohorts; the variable names and descriptions can be interpreted differently from person to person, and different units or formats can further contribute to the variability (Hufstedler et al. 2024; Gómez et al. 2024; Rinaldi et al. 2022).

The retrospective harmonization process compares corresponding variables from the cohorts to be harmonized in order to define the common denominator that can be used across all cohorts. As shown in Figure 3, it consists of the following steps (Gómez et al. 2024): (1) Defining the research questions and study objectives. (2) Variables are clustered by dimension or domains which are collections of data elements with similar characteristics grouped around a topic. (3) A data schema is produced, detailing a Master Data Dictionary. (4) Harmonization rules define how a local dataset can be transformed into the established data schema. (5) The final step is the transformation of the local dataset to comply with the master dataset.

The transformation of a variable information into harmonized information can lead to a loss of information and granularity of the variable data.

The publication by Rinaldi et al. 2022 describes how a harmonization process can utilize a combination of different data standards in order to use the most appropriate data standards for different types of variables (Rinaldi et al. 2022):

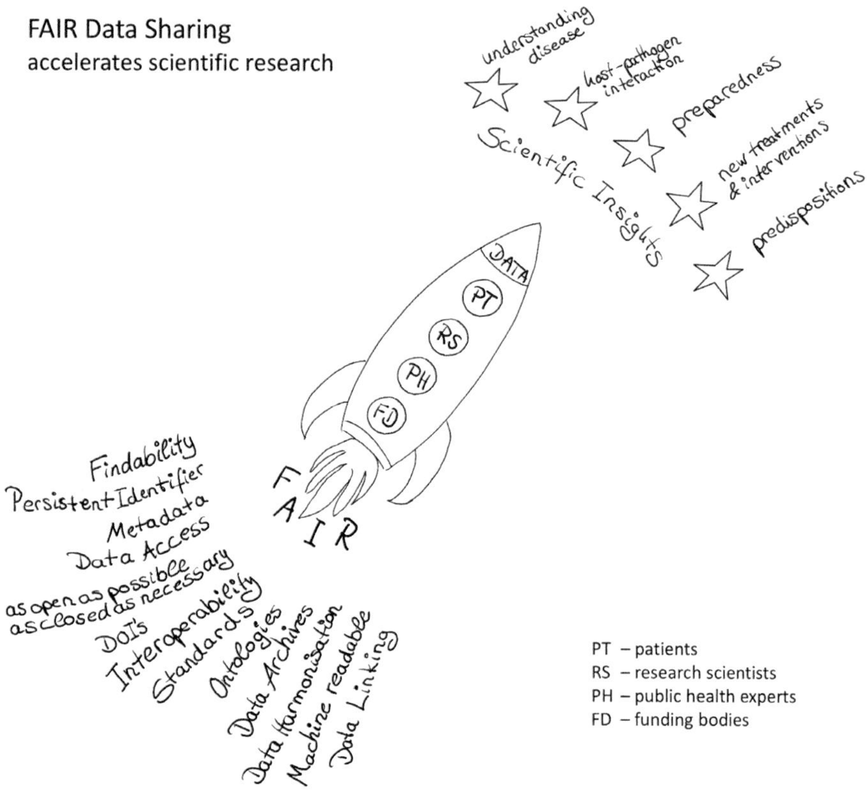

Figure 4: Data Sharing that adheres to FAIR principles enables an acceleration of scientific research. Patients (PT), researchers (RS), public health experts (PH) and funders (FD) are major stakeholders in this process by providing the data, reusing the data and, last but not least, benefiting from the scientific insights gained through FAIR data sharing.

- SNOMED CT best to capture clinical information
- LOINC for laboratory measurements, observations and documents
- ICV-10 (WHO) for information relating to diseases and health conditions
- ATC/DDD Index (WHO) to capture active substances and medicines
- NCIthesaurus describing biomedical concepts including sequencing methodology information

Another overview on key data standards is provided in Hufstedler et al. 2024 and a more extensive list of data standards can be found on the Fairsharing.org website. The Maelstrom team also published guidelines on retrospective harmonization (Fortier et al. 2016).

14.11 Prospective Harmonization

A more efficient way of harmonizing data is a prospective harmonization approach where a common data dictionary is defined from the onset of cohort studies, all using the same standards and ontologies. This would be preferable for any new cohorts that are being set up in order to capture all variables in a consistent format right from the start.

One example of a standardized data dictionary that has been widely used for prospective harmonization during the COVID-19 pandemic was the ISARIC data dictionary (ISARIC Clinical Characterization Group et al. 2022).

14.12 FAIR Principles Key to Propelling Scientific Progress and Pandemic Preparedness

This chapter is aiming to provide an overview of some of the resources and processes that can be employed to establish best data sharing practices in line with FAIR principles. Due to the huge variability of different research studies, there will not be one magic solution that works for all and some researchers may have a preference for different resources which have not been mentioned here. But as long as the life science community continues to strive for the adoption of FAIR principles at the core of their data sharing efforts, it will accelerate the progress of biomedical research. The scientific insights will provide a better understanding of host-pathogen interactions, the mechanisms that lead to disease and affect the disease severity or predisposition, see Figure 4 below. Such new insights will pave the way for a more specific, personalized approach to treatment and patient management and it will put us in a better position to deal with the next public health emergency.

Acknowledgements

I would like to thank the following people for their support and feedback: Nadim Rahman, Gustavo Gómez, Stefan Senger and Guy Cochrane, with additional thanks to Gustavo Gómez for kindly providing Figure 3 on retrospective data harmonization. Financial support for my work on data sharing was provided by the European Union's Horizon 2020 research and innovation program under grant agreements no. 825746 (ReCoDID), 848158 (EarlyCause), 101046203 (BY-COVID) and 101137283 (CONTAGIO).

References

Arita, M., I. Karsch-Mizrachi and G. Cochrane. 2021. The International Nucleotide Sequence Database Collaboration. Nucleic Acids Research 49, no. D1 (January 8): D121–D124.

Armstrong, D.R., J.M. Berrisford, M.J. Conroy, A. Gutmanas, S. Anyango, P. Choudhary, A.R. Clark et al. 2019. PDBe: Improved Findability of Macromolecular Structure Data in the PDB. Nucleic Acids Research (November 6): gkz990.

Bentley, D.R. 1996. Genomic Sequence Information Should Be Released Immediately and Freely in the Public Domain. Science 274, no. 5287 (October 25): 533–534.

Bergeron, J., D. Doiron, Y. Marcon, V. Ferretti and I. Fortier. 2018. /Fostering Population-Based Cohort Data Discovery: The Maelstrom Research Cataloguing Toolkit. Ed. Omid Beiki. PLOS ONE 13, no. 7 (July 24): e0200926.

Birney, E., T.J. Hudson, E.D. Green, C. Gunter, E. Sean, J. Rogers, J.R. Harris and S.D. Ehrlich. 2009. Prepublication Data Sharing. Nature 461, no. 7261 (September): 168–170.

Cezard, T., F. Cunningham, S.E. Hunt, B. Koylass, N. Kumar, G. Saunders, A. Shen et al. 2022. The European Variation Archive: A FAIR Resource of Genomic Variation for All Species. Nucleic Acids Research 50, no. D1 (January 7): D1216–D1220.

Courtot, M., D. Gupta, I. Liyanage, F. Xu and T. Burdett. 2022. BioSamples Database: FAIRer Samples Metadata to Accelerate Research Data Management. Nucleic Acids Research 50, no. D1 (January 7): D1500–D1507.

Fortier, I., P. Raina, E.R. Van Den Heuvel, L.E. Griffith, C. Craig, M. Saliba, D. Doiron et al. 2016. Maelstrom Research Guidelines for Rigorous Retrospective Data Harmonization. International Journal of Epidemiology (June 6): dyw075.

Freeberg, M.A., L.A. Fromont, T. D'Altri, A.F. Romero, J.I. Ciges, A. Jene, G. Kerry et al. 2022. The European Genome-Phenome Archive in 2021. Nucleic Acids Research 50, no. D1 (January 7): D980–D987.

Gómez, G., H. Hufstedler, C. Montenegro Morales, Y. Roell, A. Lozano-Parra, A. Tami et al. 2024. Pooled Cohort Profile: ReCoDID Consortium's Harmonized Acute Febrile Illness Arbovirus Meta-Cohort. JMIR Public Health Surveillance 10 (July 23): e54281.

Guyer, M. 1998. Statement on the Rapid Release of Genomic DNA Sequence. Genome Research 8, no. 5 (May 1): 413–413.

Harrison, P.W., R. Lopez, N. Rahman, S.G. Allen, R. Aslam, N. Buso, C. Cummins et al. 2021. The COVID-19 Data Portal: Accelerating SARS-CoV-2 and COVID-19 Research through Rapid Open Access Data Sharing. Nucleic Acids Research 49, no. W1 (July 2): W619–W623.

Hoeben, A., E.A.J. Joosten, and M.H.J. Van Den Beuken-van Everdingen. 2021. Personalized Medicine: Recent Progress in Cancer Therapy. Cancers 13, no. 2 (January 11): 242.

Holub, P., M. Swertz, R. Reihs, D. Van Enckevort, H. Müller and J.-E. Litton. 2016. BBMRI-ERIC Directory: 515 Biobanks with Over 60 Million Biological Samples. Biopreservation and Biobanking 14, no. 6 (December 1): 559–562.

Hufstedler, H., Y. Roell, A. Peña, A. Krishnan, I. Green, A. Barbosa-Silva, A. Kremer et al. 2024. Navigating Data Standards in Public Health: A Brief Report from a Data-Standards Meeting. Journal of Global Health 14 (April 5): 03024.

ISARIC Clinical Characterization Group, A. Abbas, S.A. Abdukahil, N.N. Abdulkadir, R. Abe, L. Abel, L. Absil et al. 2022. ISARIC-COVID-19 Dataset: A Prospective, Standardized, Global Dataset of Patients Hospitalized with COVID-19. Scientific Data 9, no. 1 (July 30): 454.

Jones, E.M., N.A. Sheehan, N. Masca, S.E. Wallace, M.J. Murtagh and P.R. Burton. 2012. DataSHIELD – Shared Individual-Level Analysis without Sharing the Data: A Biostatistical Perspective. Norsk Epidemiologi 21, no. 2 (April 13): 231–239.

Khare, S., C. Gurry, L. Freitas, M. B Schultz, G. Bach, A. Diallo, N. Akite et al. 2021. GISAID's Role in Pandemic Response. China CDC Weekly 3, no. 49: 1049–1051.

O'Cathail, C., A. Ahamed, J. Burgin, C. Cummins, R. Devaraj, K. Gueye et al. 2025. The European Nucleotide Archive in 2024. Nucleic Acids Research 53, no. D1 (January 6): D49–D55.

Pergl, R., R. Hooft, M. Suchánek, V. Knaisl and J. Slifka. 2019. "Data Stewardship Wizard": A Tool Bringing Together Researchers, Data Stewards, and Data Experts around Data Management Planning. Data Science Journal 18 (December 19): 59.

Rahman, N., C. O'Cathail, A. Zyoud, A. Sokolov, B. Oude Munnink, B. Grüning, C. Cummins et al. 2024. Mobilization and Analyses of Publicly Available SARS-CoV-2 Data for Pandemic Responses. Microbial Genomics 10, no. 2 (February 15). https://www.microbiologyresearch.org/content/journal/mgen/10.1099/mgen.0.001188.

Rinaldi, E., C. Stellmach, N.M.R. Rajkumar, N. Caroccia, C. Dellacasa, M. Giannella, M. Guedes et al. 2022. Harmonization and Standardization of Data for a Pan-European Cohort on SARS- CoV-2 Pandemic. Npj Digital Medicine 5, no. 1 (June 14): 75.

Rocca-Serra, P., W. Gu, V. Ioannidis, T. Abbassi-Daloii, S. Capella-Gutierrez, I. Chandramouliswaran, A. Splendiani et al. 2023. The FAIR Cookbook - the Essential Resource for and by FAIR Doers. Scientific Data 10, no. 1 (May 19): 292.

Sansone, S.-A., P. McQuilton, P. Rocca-Serra, A. Gonzalez-Beltran, M. Izzo, A.L. Lister and M. Thurston. 2019. FAIRsharing as a Community Approach to Standards, Repositories and Policies. Nature Biotechnology 37, no. 4 (April): 358–367.

Sarkans, U., A. Füllgrabe, A. Ali, A. Athar, E. Behrangi, N. Diaz, S. Fexova et al. 2021. From ArrayExpress to BioStudies. Nucleic Acids Research 49, no. D1 (January 8): D1502–D1506.

Shafer, Robert W. 2006. Rationale and Uses of a Public HIV Drug-Resistance Database. The Journal of Infectious Diseases 194, no. s1 (September 15): S51–S58.

Shafer, R.W. and D.A. Vuitton. 1999. Highly Active Antiretroviral Therapy (Haart) for the Treatment of Infection with Human Immunodeficiency Virus Type 1. Biomedicine & Pharmacotherapy 53, no. 2 (March): 73–86.

Sheehan, N., F. Botta and S. Leonelli. 2024. Unrestricted Versus Regulated Open Data Governance: A Bibliometric Comparison of SARS-CoV-2 Nucleotide Sequence Databases. Data Science Journal 23 (May 13): 1–29.

Van Der Velde, K.J., F. Imhann, B. Charbon, C. Pang, D. Van Enckevort, M. Slofstra, R. Barbieri et al. 2019. MOLGENIS Research: Advanced Bioinformatics Data Software for Non-Bioinformaticians. Ed. Jonathan Wren. Bioinformatics 35, no. 6 (March 15): 1076–1078.

Welter, D., N. Juty, P. Rocca-Serra, F. Xu, D. Henderson, W. Gu, J. Strubel et al. 2023. FAIR in Action - a Flexible Framework to Guide FAIRification. Scientific Data 10, no. 1 (May 19): 291.

Wilkinson, M.D., M. Dumontier, Ij.J. Aalbersberg, G. Appleton, M. Axton, A. Baak, N. Blomberg et al. 2016. The FAIR Guiding Principles for Scientific Data Management and Stewardship. Scientific Data 3, no. 1 (March 15): 160018.

Yurekten, O., T. Payne, N. Tejera, F.X. Amaladoss, C. Martin, M. Williams and C. O'Donovan. 2024. MetaboLights: Open Data Repository for Metabolomics. *Nucleic Acids Research* 52, no. D1 (January 5): D640–D646.

15

Privacy-Preserving AI-Based Solutions

Nico Pfeifer

15.1 Introduction

Artificial Intelligence (AI) methods have revolutionized many different scientific disciplines as well as software solutions affecting people world-wide in their daily lives. There are mainly two different areas of AI. One is the classical AI, also called symbolic AI, which led to rule-based systems like the MYCIN system (Shortliffe et al. 1975). In many areas it is however very difficult to create all necessary rules from existing knowledge. Therefore, most of the modern AI approaches are based on Machine Learning (ML) methods that learn certain properties directly from labelled or unlabelled data. In order to learn meaningful insights, the methods usually require large data sets. This can be problematic, if we want to build AI systems in health care, since one entity (e.g., a hospital) usually does not have enough data on its own and, as mentioned in other chapters, there might be limitations on what content can be shared without risking patient privacy.

In this chapter, we define the risk of privacy breach in the context of medical data as the risk that a person's private information could be disclosed. This could happen in different scenarios. The simplest one is that a pseudonymized record can be connected with additional information to re-identify the person. In that scenario the attacker would learn everything shared in the medical record about that person. In another scenario, even if only aggregate information is shared it could be possible for an attacker to find out whether a certain person was part of a study or not. Even that could be problematic, if the study is about a certain disease and the person wants to keep their condition private (Homer et al. 2008).

Taken together, these scenarios imply that keeping data in one central location, as well as sharing data, even in pseudonymized form, poses certain privacy risks.

University of Tübingen, Tübingen, Germany; Email: nico.pfeifer@uni-tuebingen.de

DOI: 10.1201/9781003475927-18

Therefore, researchers have started building systems that are able to query data and train ML models on data that is stored at different locations. These could be, for example, the medical data integration centers established within the German medical informatics initiative (https://www.medizininformatik-initiative.de/en/consortia/data-integration-centres). Nevertheless, one has to keep in mind that these federated data silos and the federated learning (Abdulrahman et al. 2021) on these distributed data instances is not the final answer (Bak et al. 2024), since there are also attacks known to be performed on the learned models to infer something about the underlying data, such as membership or even the data itself (Ferry et al. 2024). Federated learning is one way to perform privacy-preserving Machine Learning (ppML). In general, ppML methods try to protect the privacy of the data during the ML learning process, at least to a certain extent. In federated learning, local models are trained per data location and the parameters of these models are then aggregated and used to update a joint model. The parameters of this model are then sent to the local sites, which can consequently update their local models. The process can encompass many iterations, enabling to develop a joint model, while the data stays at the particular locations. A slightly different scenario is utilized in distributed learning. In this setting a potentially pre-trained model is updated locally. In a slight extension of this scenario, known as the personal health train (Welten et al. 2024), this model can then be updated on further data locations, like a train that visits several stations (the data locations) and uses each station's local data to continuously educate its model further.

Other ppML approaches use cryptographic methods to keep the original data protected. In those approaches, one can usually train the models with similar performance, also called utility, if the data were available in one place. Due to additional computational overhead of the cryptographic methods, one just has to pay in terms of computation time. Another way to perform ppML is by adding noise to the data or the parameter updating process. The most popular method in this direction is Differential Privacy (DP) (Dwork et al. 2006). The main idea in DP is that an attacker is not supposed to be able to distinguish whether the data from one person was part of the data set or not. Instead of having a higher computation time to enable privacy, like with the cryptographic methods, DP methods usually sacrifice utility.

In the following paragraphs, we will describe different platforms, frameworks, and cryptographic approaches that use or extend one or several of the above-described methods.

15.2 DataSHIELD

The concept of DataSHIELD was already introduced in 2010 (Wolfson et al. 2010) but the practical implementation was available only a little bit later (Gaye et al. 2014). The main idea of DataSHIELD is to provide an R package that enables federated learning approaches. In this setup, there is a central Analysis Computer (AC), which sends commands to the Data Computers (DCs) at the data sites. The individual data items are never sent from one computer to another, instead, summary statistics and commands are sent back and forth between DCs and the AC. Since its introduction, DataSHIELD has established a large community of developers as well as users who utilize DataSHIELD for federated learning analyses (e.g., the EUCANN-CONNECT or the MIRACUM consortium).

Kaapana

Kaapana is an open-source toolkit for building medical imaging platforms (Scherer et al. 2020). It offers AI-based workflows mainly based on federated learning scenarios in radiological and radiotherapeutic imaging. Kaapana has been used in several multi-partner collaborations, mainly for radiological image analysis and quantification. This includes a project called RACOON, in which all 38 German university clinics participated.

MedCo

MedCo is a system that offers privacy-preserving machine learning through different cryptographic approaches as well as by providing differential privacy capabilities. It is fully integrated with i2b2, which is used in more than 300 hospitals world-wide to make patient data available to answer research questions.

Privacy-preserving Distributed Algorithms Over the Air (PDA-OTA)

The OHDSI consortium developed a software package that enables multi-institutional data analysis without sharing individual-level patient data by employing distributed learning methods. The main principles were to provide a framework that is communication-efficient, privacy-preserving, heterogeneity-aware, and accurate.

Substra

Substra is an open-source software that allows federated learning for AI and data, hosted by the Linux Foundation. It supports the full ML workflow with training and validation on distributed data sets. Several academic research centres and biopharma companies are using Substra for clinical research, as well as for drug development.

PHT-meDIC

The PHT-meDIC is an implementation of the GO:FAIR concept of a Personal Health Train (PHT). In this implementation, several additional cryptographic approaches are employed to secure the model that is transferred from site to site by the health train for iterative model refinement based on the local data. It allows not just ML approaches but basically any kind of container that performs a certain cleared workflow on the data, enabling, for example, also complex bioinformatics pipelines for data pre-processing (de Arruda Botelho Herre et al. 2022).

SecureML

SecureML offers ppML through cryptographic approaches. It allows to perform several ML approaches by enabling a secure stochastic gradient descent method with secure two-party computation (Mohassel and Zhang 2017).

CECILIA

Cryptographic approaches like SecureML often require much more computational overhead due to the additional communication overhead of the cryptographic protocols. Furthermore, they often only support a limited set of algorithms and learning methods. CECILIA extended SecureML in several aspects and enabled additional SOTA ML approaches (Ünal et al. 2024).

15.3 Conclusion

Many of the above-mentioned approaches enable several ppML techniques. Since most of the approaches were developed in parallel, there is a lot of overlap between the different approaches, but most of them also have unique functionality. This makes it hard to select the appropriate platform or framework, especially since the actual requirements often depend on the specific data analysis task.

Therefore, there are ongoing initiatives that try to unify different approaches or make them interoperable. Within the medical informatics initiative in Germany there is, for example, a project that is building a federated learning platform that will connect Kaapana, PHT-meDIC, and DataSHIELD, as well as add additional functionality for DP and other privacy-preserving data transformations (https://privateaim.de/eng/project.html).

References

Abdulrahman, S., H. Tout, H. Ould-Slimane, A. Mourad, C. Talhi and M. Guizani. 2021. A Survey on Federated Learning: The Journey From Centralized to Distributed On-Site Learning and Beyond, in IEEE Internet of Things Journal 8(7): 5476-5497, 1 April1, 2021, doi: 10.1109/JIOT.2020.3030072.

Bak, Marieke, Madai, Vince I., Celi, Leo Anthony, Kaissis, Georgios A., Cornet, Ronald, Maris, Menno et al. 2024. Federated learning is not a cure-all for data ethics. Nature Machine Intelligence 6: 370–372. doi: 10.1038/s42256-024-00813-x. https://doi.org/10.1038/s42256-024-00813-x.

de Arruda Botelho Herre, M., M. Grafa, P. Placzeka, F. Konig, F. Botte, T. Stickele et al. 2022. Bringing the Algorithms to the Data - Secure Distributed Medical Analytics using the Personal Health Train (PHT-meDIC). https://arxiv.org/pdf/2212.03481.

Dwork, C., F. McSherry, K. Nissim and A. Smith. 2006. Calibrating noise to sensitivity in private data analysis. In: Proceedings of the 3rd Theory of Cryptography Conference, pp. 265–284.

Ferry, J., R. Fukasawa, P. Timothée and V. Thibaut Vidal. 2024. Proceedings of the 41st International Conference on Machine Learning, PMLR 235: 13545–13569, 2024. https://proceedings.mlr.press/v235/ferry24a.html.

Gaye, A., Y. Marcon, J. Isaeva, P. LaFlamme, A. Turner, E.M. Jones EM et al. 2014. DataSHIELD: taking the analysis to the data, not the data to the analysis. Int J Epidemiol. 2014 Dec; 43(6): 1929–44. doi: 10.1093/ije/dyu188. Epub 2014 Sep 26. PMID: 25261970; PMCID: PMC4276062.

Homer, N., S. Szelinger, M. Redman, D. Duggan, W. Tembe, J. Muehling et al. 2008. Resolving individuals contributing trace amounts of DNA to highly complex mixtures using high-density SNP genotyping microarrays. PLoS Genet. 2008 Aug 29; 4(8): e1000167. doi: 10.1371/journal.pgen.1000167. PMID: 18769715; PMCID: PMC2516199.

Kaissis, R. Cornet, M. Maris et al. 2024. Federated learning is not a cure-all for data ethics. Nat. Mach. Intell 6: 370–372. https://doi.org/10.1038/s42256-024-00813-x.

Mohassel, P. and Y. Zhang. 2017. "SecureML: A System for Scalable Privacy-Preserving Machine Learning," in 2017 IEEE Symposium on Security and Privacy (SP), San Jose, CA, USA, 2017, pp. 19–38, doi: 10.1109/SP.2017.12. https://eprint.iacr.org/2017/396.

Scherer, J., M. Nolden, J. Kleesiek, J. Metzger, K. Kades, V. Schneider et al. 2020. Joint Imaging Platform for Federated Clinical Data Analytics. JCO Clin Cancer Inform. 2020 Nov;4:1027-1038. doi: 10.1200/CCI.20.00045. PMID: 33166197; PMCID: PMC7713526.

Shortliffe, E.H., R. Davis, S.G. Axlin, B.G. Buchanan, C.C. Green and S.N. Cohen. 1975. Computer-based consultations in clinical therapeutics: explanation and rule acquisition capabilities of the MYCIN system. Comput Biomed Res. 1975 Aug; 8(4): 303–20. doi: 10.1016/0010-4809(75)90009-9. PMID: 1157471.

Ünal, A.B., N. Pfeifer and M. Akgün. 2024. A privacy-preserving approach for cloud-based protein fold recognition. Patterns (N Y). 2024 Jul. 19;5(9): 101023. doi: 10.1016/j.patter.2024.101023. PMID: 39568647; PMCID: PMC11573750.

Welten, S., M. de Arruda Botelho Herr, L. Hempel, D. Hieber, P. Placzek, M. Graf et al. 2024. A study on interoperability between two Personal Health Train infrastructures in leukodystrophy data analysis. Sci Data. 2024 Jun 22; 11(1): 663. doi: 10.1038/s41597-024-03450-6. PMID: 38909050; PMCID: PMC11193731.

Wolfson, M., S.E. Wallace, N. Masca, G. Rowe, N.A. Sheehan, V. Ferretti et al. 2010. DataSHIELD: resolving a conflict in contemporary bioscience--performing a pooled analysis of individual-level data without sharing the data. Int. J. Epidemiol. 2010 Oct; 39(5): 1372–82. doi: 10.1093/ije/dyq111. Epub 2010 Jul 14. PMID: 20630989; PMCID: PMC2972441.

16

E-health and Older Persons (Between Age Discrimination and Data Protection Issues)

Claudia Morgana Cascione

16.1 Preliminary Remarks

There are a number of perspectives from which the topic of "e-health and older persons" can be approached, given the increasing emphasis in contemporary legal debate on the protection of the elderly on the one hand (Doron 2009, Ayalon and Tesch-Römer 2018, Martin et al. 2015), and on the digitization of health services on the other hand. Digital or electronic health (so-called e-Health), strongly influenced by Information and Communication Technologies (ICT) (Colonna 2021), has firmly become part of individuals' lives, playing an essential and strategic role in the governance of health and care services both at national and European level, also based on the Regulation on a European Health Data Space.[1]

Leaving for a moment in the background all the legal issues related with the global ageing of the population (which is one of the most significant phenomena of 21st century), the aim of the present research is to clarify if and how in contemporary e-health systems, designed to make care and assistance to citizens more efficient, adequate space is left to the special needs of this segment of vulnerable subjects and to ascertain whether adequate protection is provided for the large amount of personal data processed in the digital health systems.

The issue of health data protection in contemporary societies emerges at multiple levels: first, it involves an analysis of the ongoing digital transition in public health

University of Bari; Email: Claudiamorgana.cascione@uniba.it

[1] Regulation (EU) 2025/327 of the European Parliament and of the Council of 11 February 2025 on the European Health Data Space and amending Directive 2011/24/EU and Regulation (EU) 2024/2847 ("European Health Data Space").

DOI: 10.1201/9781003475927-19

systems, especially after the pandemic emergency; second, it implies an examination of the impact of the massive use of increasingly advanced technologies in the health care sector to improve diagnosis, treatment, and monitoring of patients' health conditions.

These phenomena, which in themselves pose problems of effective data protection (Aceto et al. 2018), take on a special significance when considering the position of vulnerable individuals whose data are processed. Since it is not possible to generalize the discourse too much nor to address the issue with regard to all groups of subjects with a 'potential to be injured', the attention will be focused - paradigmatically - on the elderly.

Older persons, more and more numerous also in the light of demographic ageing,[2] have an increasing need for medical assistance, but they are also those who, mainly due to their lack of IT skills, are most at risk of being excluded from the ongoing digitization processes of health services; moreover, as far as data protection is concerned, for aged adults who may be unfamiliar with the ways in which their data are collected, stored, and used, apprehension about the collection of these data may be a significant barrier. Finally, the recent use of AI medical devices in health care, which could be beneficial for personalizing health care for older patients, can exacerbate ageism and discriminatory practices towards older persons (as highlighted by the WHO)[3] and can amplify the problems of protecting health data processed on a large scale.

That said, the main focus of the research will be on three aspects:

- First, the consequences of the digital transition in public health care with regard to senior individuals will be analyzed, in order to ascertain whether the so-called "digital grey divide" (Millward 2003, Gatto and Tak 2008, Gitlow 2014) is amplified in this specific sector, leading to the obvious paradox that those who need care services the most are in fact excluded from them.
- In the second part, the attention will be focused on the problems of privacy and processing of health data of the older persons, analyzing the need to adequate many of the provisions of the GDPR to the peculiar situation of vulnerability represented by old age.
- Finally, the issues posed by the most modern "gerotechnologies" (from wearable devices to medical apps) will be analyzed, highlighting a protection gap in the existing legislation protecting personal data.

16.2 Digital Welfare State and Digital 'Grey Divide'

The digitization of health and care services is an important part of the transition towards the digital welfare state globally, through the implementation, in the delivery of public services, of the use of online platforms, smartphone applications, digital

[2] For updated dates see United Nation Department of Economic and Social Affairs, World Population Prospects 2022, available at https://population.un.org/wpp/.

[3] WHO. Ageism in Artificial Intelligence for Health, Policy brief, 2022, available at: https://iris.who.int/bitstream/handle/10665/351503/9789240040793-eng.pdf?sequence=1.

identities, digitization of information, use of algorithms, big data and various applications of artificial intelligence.

Large investments have been recently made in e-health.

The digitalization of health has been a top priority for the European Union in recent years, as part of its efforts to improve the efficiency and effectiveness of healthcare systems across the Member States.[4] The EU has launched several initiatives and provided grants and loans through the Recovery and Resilience Facility (RRF) to support Member States in advancing their e-health systems.[5]

Health services - as well as many social services in digital welfare state - are increasingly delivered by means of automated procedures, through the collection and processing of large amounts of data and the use of digital technologies employed to predict, identify, monitor, detect and eventually sanction. This, as in other areas involving the public interest, gives rise to clear benefits, improving access to public services and social benefits, standardizing access criteria, speeding up paperwork and ensuring efficiency. Municipalities in various developed democratic countries are experimenting with the use of new technologies and algorithms to help design, implement and manage various aspects of social policies, including care policies.

Nonetheless, as several critical authors have explained, such a transition in social policy is not without risks. The use of new technologies in public services generates a real risk that fundamental personal rights may be violated through them, consequently making the need for certain guarantees to be respected even more stringent (Eubanks 2018).

If, on the one hand, the use of AI tools generates problems of personal data protection, on the other hand, the digitization of social and welfare services poses, upstream, the problem of access to these digital services by the most vulnerable sections of the population. When digital technology is used as an intermediary for rights, the law is often applied automatically regardless of the personal circumstances of the citizens (Ranchordás 2021).

This disproportionately affects vulnerable subjects, who are the most prone to make mistakes in their approach and use of technology. Vulnerable citizens are not only affected by the opacity of automated decision-making, but also risk not claiming the benefits to which they are entitled, making more errors, missing

---

4 European Commission. Communication from the Commission to the European Parliament, the Council, the European Economic and Social Committee and the Committee of the Regions on enabling the digital transformation of health and care in the Digital Single Market; empowering citizens and building a healthier society, COM/2018/233 final.

5 Proof of this is that the Italian Government in the PNRR has provided for an extraordinary allocation for the implementation and dissemination of digital technologies in public health. Already with the "Digital Growth Strategy", the Italian Council of Ministers clarified that "digital innovation of healthcare processes is a fundamental step for improving the cost-quality ratio of healthcare services, limiting waste and inefficiencies, reducing differences between territories, as well as innovating front-end relationships to improve the quality perceived by citizens".

deadlines or being profiled as fraudsters due to their inability to interact with the digital administration (Rachovitsa and Johann 2022).[6]

Therefore, it can be said that the digitization of activities and services has exacerbated personal vulnerabilities and turned them into digital vulnerabilities.

The first obstacle that, in the digitization of public services, must be overcome is to prevent vulnerable citizens from being denied basic rights due to their lack of IT skills or lack of network access. The problem of digital welfare state brings with it the inevitable paradox of disadvantaging those sections of the population that, instead, should be most favoured in the dynamics of social welfare.

With the pervasiveness of digital welfare, new and deeper inequalities emerge, as citizens increasingly need to be autonomous, to use a growing number of online public services and to have sufficient resources to understand the functioning of public platforms.

However, it has been pointed out that millions of citizens in Western countries are unable to make proper use of websites, platforms and services. Digital welfare and automated government decisions are giving rise to situational vulnerabilities.

Governments digitize services having as reference the "liberal legal subject", a fully capable, autonomous and independent adult who can easily use the new digital services. However, this profile is not true for citizens who do not have access to the Internet and digital devices because they live in rural areas or cannot afford them. Nor is it true for those individuals who have poor digital skills because they have intellectual or functional disabilities.

This is more than evident with regard to older persons, who suffer a digital divide, both due to a lack of electronic devices (for a generational reluctance to use new technologies) (Marquié et al. 2002)[7] and a lack of suitable skills, which makes them unfit to interface with electronic portals, request services remotely, and follow automated procedures.

Research on the difficulties of older persons in using new technologies has been conducted for about two decades (Van Dijk 2005, Czaja et al. 2006, Lüders and Gjevjon 2017), in step with technological progress. They have been recently revitalized with the emergence of the digital welfare state, because seniors risk being cut off from receiving digitalized welfare services since they have to dedicate a certain amount of labour and effort to take advantage of the choices available.

[6] The well-known Dutch case SyRI should be mentioned in this perspective District Court of the Hague, *NCJM* et a l. *and FNV v The State of the Netherlands* ('SyRI'), 6 March 2020, ECLI: NL: RBDHA:2020:865. For an analysis in the human rights' perspective see Rachovitsa, A. and Johann, N. 2022. The Human Rights Implications of the Use of AI in the Digital Welfare State: Lessons Learned from the Dutch *SyRI* Case. Human Rights Law Review 22(2): 1–15.

[7] Marquié reports greater fear and anxiety associated with using computers, and in addition, older persons' assessment of their own skills and abilities, with both using and learning to use them, is generally lower than for other age groups.

16.3 Health Digital Services and Older Persons

The problems highlighted are amplified when public health services are digitized.

The number of online services for health care is increasing rapidly in developed countries, especially after the pandemic emergency.

Many countries have implemented measures to support the digital transition in healthcare and enhance health resilience as outlined in the RRF. These measures include e-health services, telemedicine solutions, and improvements in healthcare data management.

As a basic example, it is possible to think to the Electronic Health Record, which is the set of digital health and socio-health data and documents generated by present and past clinical events. With the Electronic Health Record, it is possible to consult and have available online the digital data and documents relating to health care, services and clinical events that make up a patient's medical history. The European Commission, within the framework of the priorities identified in the "Communication on digital transformation of healthcare in the digital single market",[8] has adopted a Recommendation on a European electronic health record exchange format to unlock the flow of health data across borders.[9]

At EU level, there are also several initiatives aimed at supporting the spread of telemedicine and remote care; however, "the deployment of telemedicine to the whole population of a country depends on the country's level of digitalization, including the digital skills of patients and health professionals, as well as the legislation governing the sharing and processing of health data".[10]

Digital solutions for health and care are envisioned to have the potential to increase the wellbeing of millions of citizens and to change radically the way health and care services are provided.

New online services are implemented and adopted to support "self-management", and thus users are expected to take a more competent and active role in maintaining their own well-being and caring for their health and especially long-term illnesses. Some authors have suggested that ICT in healthcare reduces costs, enables diagnosis and improves patient outcomes (Heart and Kalderon 2013). Furthermore, it is proposed that ICT will largely transform healthcare into more proactive, preventive and person-centred care, rather than reactive and hospital-centered care.

Nevertheless, to fully exploit all the benefits suggested by ICT, patients need to use online services and find them beneficial.

In this regard, there have been concerns, first of all, regarding technology's availability and usability, especially with older age groups.

Despite technologies have proven to be beneficial in supporting health assessment and care for older persons (Bujnowska-Fedak and Pirogowicz 2014) (in

---

[8] See *supra* n. 3.

[9] European Commission. Commission Recommendation (EU) 2019/243 of 6 February 2019 on a European Electronic Health Record exchange format.

[10] European Commission. 2018. Market study on telemedicine, available at https://health.ec.europa.eu/document/download/e8937f58-0bbc-4616-b515-08dacef8ae3e_it?filename=2018_provision_marketstudy_telemedicine_en.pdf.

improving health management, increasing independence and safety, decreasing need for institutional care) and in improving health behaviors (Buyl et al. 2020) (increased physical activity and healthy eating) as well as psychological and health outcomes (memory and blood pressure) (Kwan et al. 2020), older adults are the largest individual group that faces challenges in using digital health services for inexperience in the use of technology, poor motivation, financial difficulties, and insufficient technical skills (Fang et al. 2018). In addition, poor health, cognitive decline, the lack of appropriate devices or internet access, and inadequate support and guidance, may hinder older persons' opportunities to benefit from digital health services.

Problems associated with e-health are connected to issues like access, lost privacy or the frivolous use of the devices.

As far as access and use are concerned, there is a concrete risk that older users, who are those most in need of care services, do not profit from them. The difficulties related to the digital care of the elderly became evident during the pandemic, when the part of the population most affected by the virus and by the ensuing blockade was also the one that experienced the greatest difficulties with remote care. The so-called "digital grey divide" has prevented many older people, worldwide, from taking full advantage of health services that have been rapidly digitized, with the obvious paradox that the very people who could have most benefited from remote care have been excluded from it (Cascione 2022a).

The main barriers hindering older adults' access to and use of digital medical technologies have been analyzed and classified into: (i) individual, that can be intrinsic (i.e., related to the physiological ageing process resulting in reduced vision, hearing and memory) and extrinsic (i.e., related to lack of proficiency with computers and technologies); (ii) technological (related to design of e-health programmes that often feature small fonts, small icons and complex functionality); (iii) relational (lack of support or training provided alongside e-health programs; (iv) environmental; (v) organizational (Wilson et al. 2021).

Beyond objective barriers, the spread of digital medicine passes through the perception that it can provide concrete benefits, even greater than traditional medicine.

In a recent study conducted in Finland, which has digitized almost all health services, the perceived benefits of various clusters were examined by interrelating demographic, health, socio-economic position and social participation variables (Heponiemi et al. 2020). It was preliminarily found that perceived benefits in terms of health and efficiency are associated with several factors.

IT skills seem to be the most important factor. In addition, subjects with poor health, financial difficulties, living in rural areas or with low level of education perceived lower levels of economic and health benefits.

Older respondents perceived fewer benefits in each of the dimensions examined, and it has been at least partly explained by their lower access to online services, lower skills in using these services, and lower frequency of use.

This study revealed, with empirical data, that there is a significant risk of digital exclusion among the vulnerable subjects. In an age where services are increasingly provided electronically, this digital divide may predispose people who do not perceive the benefits of online services to be excluded from such services. In order to

ensure that the population, in particular those who are older, in poor health, socially isolated or of low socio-economic status, can benefit equally from eHealth services, it is necessary to provide access, skills and opportunities to use these services. These groups are at the highest risk of digital exclusion and have the greatest need for health and social care services.

The provision of online services has the potential to widen social and health inequalities among the population. Widespread expansion requires rigorous consideration of interventions to address the negative effects that may arise from the provision of health and social care services online and to promote equal opportunities and capabilities among the population.

Targeting education towards vulnerable groups such as the elderly and people with poor health, low levels of education or social isolation can help to ensure that e-health services are accessible and can reach a broad population, improving patients' involvement in their own care (Kaihlanen et al. 2022).

In this perspective, it is remarkable that in Italy in the very recent Legislative Decree on the Elderly,[11] the primary purpose is to "reorganize, simplify, coordinate and make more effective social, health and socio-sanitary care activities for the non-self-sufficient older persons, including through the coordination and reorganization of available resources, and to ensure the economic sustainability and flexibility of long-term care and assistance services for the elderly and the non-self-sufficient elderly" (Article 1). Moreover, Chapter V is dedicated to "Measures on computer literacy and digital facilitation", providing for "activities to train the digital skills of the elderly and to support them in the use of services provided online by public administrations" (...) "in order to promote the computer literacy of the older persons and to ensure their full civil and social participation, including through the use of digital services of public administrations" (Article 19), as well as the activation at educational institutions "of training paths that promote in older people the acquisition of knowledge and skills on the use of digital tools" (Article 20).

16.4 Health Data and Ageing Population

Regarding privacy issues, it is clear that digital health involves the collection and processing of a large amount of personal health data, which requires special protection and warranties. The issue is the basis for the recent proposal for the creation of a European Health Data Space, that calls for a regulation of the primary use of the data and the spin-offs from secondary use.[12]

One of the greatest challenges facing the creation of a European Health Data Space is the ageing of the population: the qualitative and quantitative demographic change poses new frontiers in care and requires adaptation in the processes of digital transformation of health.

---

[11] Legislative Decree No. 29 of March 15, 2024. *Provisions on policies in favor of the elderly, implementing the delegation of authority in Articles 3, 4 and 5 of Law No. 33 of March 23, 2023.* (24G00050), in *GU General Series* No. 65, 18.03.2024.

[12] See *supra* n. 1.

Populations ageing creates in modern societies health-related issues, especially considering the special medical needs of older persons and the expenses of healthcare. The creation of a single Health Data Space can prove to be of utmost benefit for the efficiency of care of the elderly (by teleconsultation, tele-expertise, remote medical monitoring and by the dissemination of medical apps).

Moreover, promoting active ageing is necessary in order to avoid overloading hospitals and care facilities: robots and nursing care devices can play a significant role to offer solutions in this regard. The advancement of wearable technology and the possibilities of using advanced IoT devices to assist older adults with their activity of daily lives and independent living is the subject of many recent research and investigation (Baig et al. 2019). The improvement of *gerontechnologies* is widespread and, globally, there is increasing interest in the use of ICTs for older adults, demonstrated by development initiatives.[13]

Despite the abstract suitability of new devices to facilitate independent living and to improve medical advances in gerontology, there are doubts about their real effectiveness with regard to frail older people, primarily due to "generational" reluctance to use new technologies.

Moreover, devices for the care and assistance of the elderly may collect an incredible amount of personal data and information, many of which of a sensitive nature. For older adults who may be unfamiliar with the ways in which their data are collected, stored, and used, apprehension about the processing of these data may be a significant barrier. So, first of all, there is the need to propose models of 'elder-friendly' devices, especially at the information, data collection and sharing stage.

The spread of information technology-based healthcare solutions for older adults (with expectations for increased efficiency and effectiveness of preventive healthcare and chronic disease management) strongly depends on older adults' willingness to share their personal and health information when using healthcare technologies and on the trust in the procedures for data processing. It also depends on the warranties that the law provides for the protection of older persons' data.

The "sectoral approach" of data protection regulations, focused only on the minors – among vulnerable subjects – creates, upstream, the problem of extending protections to all subjects exposed in the digital ecosystem to potential breaches.[14]

The unstoppable ageing of the world's population, moreover, leads to a further doubt: if the older persons can be classified as an autonomous legal group characterized by the recognition of more or less common traits of vulnerability in its members (Cascione 2022b). This would lead to theorizing a concept of the elderly, intended as a legal category of individuals, united by the progressive decline in physical and mental strength that accompanies ageing (Cascione 2022b).

It is believed that legally recognizing the category of the older persons in the health sector - despite being a harbinger of discrimination - could be useful

---

[13] The Active Assisted Living (AAS) Joint Program, for instance, is a European initiative that has invested over 700 million Euros in improving ICT access to groups such as older adults. Available at: https://www.aal-europe.eu/about/.

[14] See, *infra* § 5.

in "selecting those to (and for) whom to allocate resources, guarantee rights, and arrange for enhanced and specific forms of protection" (Rotolo 2023). Moreover, in certain cases, classifying vulnerable persons can have positive effects by ensuring a higher level of data protection.

This assumption, however, can appear problematic in some respects, given that the older subjects may present different capacities of understanding, different levels of awareness and decision-making capacity, and different degrees and types of weakness. For this reason, claiming that all the old data subjects to whom personal data relate are universally vulnerable may lead to ignoring significant differences among them, depriving those who would need protection the most (Malgieri and Niklas 2020) and unjustifiably discriminating against those who, in spite of their age, retain unaltered their physical and intellectual functions.

The older persons' state of vulnerability must be assessed on a case-by-case basis, on the ground of the person's actual psycho-physical condition, which can only be concretely detected by means of an *ad hoc* and *ex post* investigation.

16.5 The GDPR in an "Elderly-Oriented" Perspective

As is well known, the GDPR and other data protection models do not contain a general approach to vulnerability, but a sectoral one. The subject of specific attention is the child, for whom there is a threshold for digital consent and specific rules on information. However, this approach is also reductive in view of the increased risks arising from the digital dimension (Cuocci 2022). In this regard, the protection of personal data of vulnerable persons is of particular importance, especially in the digital dimension, which could expose them to new vulnerabilities and amplify existing ones.

It is reductive to relate vulnerability only to minor age: there are many cases where people are not able to express their consent consciously (elderly, disabled, illiterate). With reference to personal data, there are different degrees of vulnerability that may vary in intensity depending on the socio-economic context or the degree of disability and that require specific mitigation measures.

The GDPR does not mention vulnerability in its provisions, however, it is possible to interpret the GDPR from a "vulnerable-oriented" perspective. This perspective could lead to a useful solution with regard to older persons, whose vulnerability may emerge both in the evaluation of the capacity to give free consent for the collection of data and in the assessment the risks associated with the processing, being unable to adequately exercise their rights of protection under the law (Irti 2023).

First of all, with regard to transparency and disclosure, Article 12 provides specific guidance with regard to minors; however, the wording of the provision allows the same rule of simplicity and clarity to be applied to other categories of vulnerable who nevertheless deserve "specific protection with regard to their personal data, as they may be equally less aware of the risks, consequences, safeguards provided and their rights in relation to the processing of personal data".[15]

[15] See Recitals 38; 58; 65; 71 GDPR.

In this regard, obtaining informed consent from older adults is confined by many ethical and practical challenges, and therefore, it is much more than obtaining a signature (Altawalbeh et al. 2020). For instance, receiving adequate information, comprehending the information, and then making a voluntary decision are essential components of the process of providing an informed consent. Many Ethics Committee expressly requested to make the information intelligible, clear, considering the possible cognitive degradation of the elderly. There is the need of considering age-related physical and emotional characteristics in the consenting process, and so no "one size consent" can fit all ages.

Moreover, the vulnerability of the data subject (especially if it is represented by senility), is also relevant in the design and organization phase, impacting on the data controller's tasks.

For example, the impact assessment is provided for in Article 35 GDPR, which determines the conditions under which it must be carried out. There is no reference to vulnerability, but there is just one mention in recital 75 about relevant risks to consider when performing a Data Protection Impact Assessment: "where personal data of vulnerable natural persons, in particular of children, are processed". "In particular" means that children are vulnerable subjects, but that also other data subjects might be considered vulnerable. Moreover, the *Guidelines on Impact Assessment* (WP29),[16] require the need for Impacts Assessments for "data relating to vulnerable data subjects" that "results in a significant imbalance of power between the data subjects and the controller because the vulnerable data subjects are unable to consent to or object to their processing or exercise their rights".[17]

Despite the appreciable reference to vulnerability, the guidelines leave many doubts unresolved (Piasecki and Chen 2022, Cuocci 2022). First of all, although they expressly consider as vulnerable the elderly, the disabled, the mentally ill persons, patients, asylum seekers, some of these categories (e.g., the elderly) are so uncertain that they do not provide clear guidance to data controllers. Second, the WP29 prescribes that "a data controller can consider that a processing meeting two criteria would require a DPIA to be carried out"; therefore, the only presence of a vulnerable data subject is not sufficient to make impact assessment mandatory (unless it is the data controller who considers that a processing meeting only one of these criteria requires a DPIA).

---

[16] Art. 29 Data Protection Working Party. Guidelines on Data Protection Impact Assessment (DPIA) and determining whether processing is "likely to result in a high risk" for the purposes of Regulation 2016/679, WP 248 rev.01, adopted on 4 april 2017.

[17] Among the nine criteria listed "in order to provide a more concrete set of processing operations that require a DPIA due to their inherent high risk", the guidelines consider "Data concerning vulnerable data subjects: the processing of this type of data is a criterion because of the increased power imbalance between the data subjects and the data controller, meaning the individuals may be unable to easily consent to, or oppose, the processing of their data, or exercise their rights. Vulnerable data subjects may include children (they can be considered as not able to knowingly and thoughtfully oppose or consent to the processing of their data), employees, more vulnerable segments of the population requiring special protection (mentally ill persons, asylum seekers, or the elderly, patients, etc.), and in any case where an imbalance in the relationship between the position of the data subject and the controller can be identified".

With reference to the specific topic of this essay, it must be considered that the impact assessment prior to the processing of the health data of older persons must be generally carried out for the recurrence of at least two of the criteria set out in the Guidelines: in addition to the vulnerability of the data subject, "sensitive data or data of a highly personal nature" (n.4) are processed; moreover, in e-health, "innovative use or applying new technological or organizational solutions" is concerned (n. 8); lastly data can be processed "on a large scale".

Finally, one of the key principles of the GDPR that should be taken into consideration in a "vulnerable oriented" perspective is the *privacy by design* (Article 25 GDPR), according to which the controller shall, both at the time of the determination of the means for processing and at the time of the processing itself, implement appropriate technical and organizational measures, which are designed to implement data protection principles.

In addition to data protection by design and by default measures, the data controller must implement technical and organizational measures for the exercise of any data subject rights that may be applicable (Articles 15-22 GDPR), with particular reference to the rights of access and rectification of personal data, which are exercized most frequently. In particular, in respect of the condition of vulnerability of the adult data subject, Article 24 requires the adoption by data controllers of particular safeguards when collecting their data, analyzing the level of risk (of the individual's fundamental rights and freedoms): similarly, it can be assumed that it should also be considered the level of vulnerability of the data subject before proceeding to process the data. This implies the need to carry out an assessment of the possible levels of vulnerability of the data subjects in order to implement appropriate safeguards, also in view of the purposes for which the data will be used (Irti 2023).

16.6 Health app and Privacy Issues

These rules assume a particular relevance when considering the widespread of mobile health (so-called *mHealth*), defined by the WHO as "the practice of medicine and public health supported by mobile devices, such as mobile phones, patient monitoring devices, handheld computers and other wireless wearable devices".

Although in recent years the phenomenon of mHealth apps has been deeply studied, many uncertainties remain (Bincoletto 2021), with regard both to their qualification as medical devices and to the requirements deriving from the data protection framework, especially for older persons (Zuraw and Sklar 2020, Guida 2021). Indeed, the gradual spread of the health apps (Raja et al. 2021)[18] has presented again, in a new guise, the pre-existing legal dilemma of data protection in healthcare and, above all, the topic of discrimination arising from particular care and wellness devices that process health-related and biometric data.

[18] Mobile Health includes both applications used in the context of telemedicine (so- called Medical Mobile Health apps) as well as digital applications whose operation, on the other hand, is based on the exclusive interaction between the software and the patient (the simplest Mobile Health apps), who will be acting in the broader capacity of user.

The term 'medical apps' does not correspond to a precise legal definition or organic regulation, as they are taken into account by the European legislator of the 'algorithmic' society in different aspects and for different purposes (Rapisarda 2023).

The relevance of the issue of medical devices in general, which became more acute with the emergence of the AI Act,[19] arose with the Medical Devices Regulation (EU) 2017/745 (so-called MDR). Prior to the AI Act proposal, this regulation already outlined some necessary indications for the qualification of software as medical devices, foreseeing the possibility of marketing them or, simply, disseminating them (if the purpose is the research)[20] within the EU, only if they comply with the stringent EU regulations.[21] Thus, the main criterion for determining whether or not the application pursues one or more of these purposes is the "specific intended use" initially identified by the developer.

The criterion of the "intended use" is also outlined in the EHDS, which states that wellness applications are to be understood as "any device or software intended by the manufacturer to be used by a natural person for the processing of electronic health data for purposes other than health care, such as wellness and the pursuit of healthy lifestyles".[22]

Under the MDR, on the other hand, applications that merely perform the function of a digital repository of data and information or that otherwise perform simple search operations do not constitute medical devices. To be considered 'medical', applications must be able to process, analyze, modify and/or create medical information.

The main challenges arise in relation to those applications where the interaction between the software and the user replaces the therapeutic relationship between the doctor and the patient. Such devices may actually fall within the scope of the MDR, provided that the developer has chosen one of the intended uses specifically outlined in the regulation. The adoption of such a criterion, however, has led to the marketing of applications formally qualified as 'related to the lifestyle and wellbeing'[23] of the patient/user, but basically capable of performing medical-health functions of diagnosis, prevention, monitoring, prognosis or treatment of a disease.

In this regard, "irrespective of whether the app is qualified as a medical device or not, the developer will in any case be required, as data controller, to conform his activity to the rules and principles of the privacy legislation, which substantially reproduces the same approach adopted by the medical devices regulation, in a broader and more general perspective" (Rapisarda 2023). In this way, the complex system

---

[19] See, *infra*, § 6.

[20] On the relationship between AI, personal data and scientific research, see [EHDEN] European Health Data & Evidence Network. 2022. Protecting People while Using their Health Data for Research, a Framework for Understanding Relative Responsibilities and Roles. Available at: https://www.ehden.eu/protecting-people/.

[21] It is necessary that the application in general pursues one of the medical purposes covered by Article 2(1) MDR.

[22] Art. 2(2)(o) EHDS.

[23] See Recital 19 MDR.

of controls, of transparency and traceability of devices, introduced by the MDR,[24] contributes to make safer, also from the data protection point of view,[25] especially those apps intended to be used by patients in total autonomy.

These general considerations take on a special significance with regard to the use that older persons can make of these apps: they are, in many cases, downloadable free of charge, are not subject to control by a doctor and, for this reason, may expose those without sufficient experience with electronic devices to risks and dangers.

The most glaring example is that of wearable devices (Zinzuwadia and Singh 2022, Stanzione 2021, Tirabeni 2022), tools useful for the pursuit of active ageing in one or more areas of the social or personal sphere that allow the older user to freely choose the activity in which to engage according to his or her aspirations and motivations (Irti 2023, McTear et al. 2023, Brunzini et al. 2023, Menassa 2023, Palliotta 2022, Scwhartz et al. 2022).[26]

Such devices are certainly aimed at fulfilling the need to constantly monitor one's own state of health and psycho-physical well-being using methods and devices that are as minimally invasive as possible (Colonna 2021). However, the use of such technologies can generate considerable risks for these individuals.[27] Hence, the need to investigate the functioning, the dynamics and, above all, the safeguards provided for subjects placed in a particular condition of vulnerability (Irti 2023).

The criticality of the use of such devices by vulnerable persons was highlighted, once again, only with specific regard to minors: for instance, the Italian Committee for Bioethics, in its opinion *Mobile health apps: bioethical aspects*,[28] highlighting a lack in the minors of the necessary elements of competence to understand the decisions to be taken, made it clear the need to offer them "simple, concise information, which also includes educational references that encourage awareness of the problems, in suitable language".

Furthermore, for the Bioethics Committee, however, it seems legally questionable that the consent to the processing of data required for the use of these apps can only be given by the minor; since they concern their health, they must necessarily fall under the responsibility of their parents or legal representative.

In its final recommendations, the Committee recommends "monitoring an adequate education of particularly vulnerable groups - children, the elderly, the disabled - in order to ensure non-discrimination and enhance inclusion of the advantages in the use of the new technologies".[29]

---

[24] See Recital 4 MDR.

[25] GPDP Italian Data Protection Authority. 2022. Opinion on the draft legislative decree, containing provisions for the adaptation of national legislation to the provisions of Regulation (EU) 2017/745, relating to medical devices - 26 May 2022 [9782450]. Available at: https://www.garanteprivacy.it/home/docweb/-/docweb-display/docweb/9782450.

[26] The use of electronic devices in the pursuit of Active and Healthy Ageing has also been the subject of a debate related to the classification of so-called wearable devices.

[27] The Lancet Digital Health. 2021. Digital Technologies: A New Determinant of Health. 3(11):e684. Available at: https://www.thelancet.com/journals/landig/article/PIIS2589-7500(21)00238-7/fulltext.

[28] Comitato Nazionale per la Bioetica (Presidenza del Consiglio dei Ministri). 2015. Mobile health apps: bioethical aspects. Available at: https://bioetica.governo.it/media/3214/p121_2015_mobile-health-apps_en.pdf.

[29] Ibidem, 16.

With regard to older users, one may firstly ask whether the guarantees set up to protect minors can also be applied by analogy to them. Their vulnerability may emerge, as for minors, as an expression of a limited capacity to give free consent for the collection of personal data, to understand the information on data processing and the risks connected with the processing and to adequately exercise the rights attributed to them by law.

Of course, except in cases where the older person is subject to guardianship, it should be ruled out that consent can be given by a third party; therefore, the only solution is to enhance the obligations of the data controller.

As expressed in the Transparency Guidelines developed by WP29, "if a data controller is aware that its goods or services are used by (or intended for) (...) vulnerable members of society, including persons with disabilities or persons who may have difficulty understanding the information, the vulnerabilities of such persons should be taken into account by the data controller in assessing how to ensure compliance with its transparency obligations in relation to such persons".[30]

The main objective is to protect the individual and collective health of the users of such applications, helping to strengthen the right to data protection, quality and safety standards of new technologies, through the imposition of specific obligations on the producer/developer and leaving the application of the GDPR unaffected (despite the numerous criticalities), even for those applications that are not considered medical devices.

In such cases, one cannot disregard the identification of possible remedies and specific measures aimed at developing preventive control tools, in full compliance with the principles laid down by the GDPR, which are useful for the development of Mobile Health apps: purpose limitation, data minimization, transparency and attention to information, and compliance with the principle of accountability (Rapisarda 2023).

With specific regard to the privacy by design, it seems crucial to implement clear and effective preventive information with the aim to help the older persons to understand the functioning and characteristics of the applications.

Turning to the instruments of protection introduced by the privacy legislation, it can be observed that, in the absence of *ad hoc* provisions concerning the topic at issue, the rules provided for the processing of personal data will apply and, above all, in the light of what has been observed above, the rules for the processing of data relating to health.

Therefore, regardless of whether the application qualifies as a medical device or not, the developer will still be obliged as data controller to comply with the rules and principles of the data protection legislation. In this regard, it is interesting to note how software manufacturers and developers, in order to escape the more specific obligations required for the creation of 'health' devices only, use the *escamotage* of presenting their digital services as simple wellness and lifestyle monitoring

[30] Article 29 Working Party. 2017. Guidelines on transparency under Regulation 2016/679. Adopted on 29 November 2017 ss last Revised and Adopted on 11 April 2018. WP260 rev.01. https://ec.europa.eu/newsroom/article29/items/622227.

tools. This makes it possible, on the one hand, not to distance themselves from the health market and, on the other, to evade the more specific set of rules and controls prescribed for software qualified as medical devices.

However, the widespread use of these devices entails a number of risks (including self-monitoring and 'do-it-yourself' diagnoses based on algorithms lacking scientific validation), where users download applications on the basis of the liking expressed by users and not on the scientific evaluation of the application.

Given that information on health and wellbeing must be accurate and reliable, since choices and decisions that may adversely affect people's health are based on it, the question remains open as to how to bridge the lack of effective scientific validation of these applications against the traditional parameters of safety and efficacy.

As an example, the design of Health apps intended for the care and wellbeing of older persons (Sethumadhavan et al. 2021), recommends a responsible development of the underlying technologies that takes into account the vulnerabilities of those involved and the sensitive issues related to their privacy (Amram 2023).

The most challenging profiles essentially concern the violation of the processing of particular personal data (especially health data), as well as the possible discrimination suffered by those groups of individuals placed in a specific vulnerable condition.

These considerations raise legal questions regarding the appropriateness of implementing technology in the field of health, if it is not supported by appropriate digital literacy programmes for elderly users to foster their autonomous access to information, and also increase their sense of well- being and satisfaction with their own lives and interpersonal relationships.

Despite the fact that the European legislator tried to introduce a discretionary criterion in order not to excessively restrict the circulation of certain systems considered less dangerous for the health of the elderly, the preventive exclusion of wellness - and lifestyle-related applications could have a negative impact not only on the privacy of the user, but also on his or her own health (Rapisarda 2023).

Despite the rapid pace of technological innovation, the regulatory framework is currently opaque and fragmented.

While it is true that with the MDR regulation, the legislator intended to simplify and rationalize the regulatory framework for medical devices, it is equally true that the advent of the algorithmic society has given rise to complex AI systems that require enhanced quality and safety standards.

Consequently, these provisions will necessarily have to accompany the future AI Act, working towards the promotion of secure, transparent, non-discriminatory systems capable of guaranteeing effective protection of those concerned.

16.7 AI Technologies and Relevant Implications for Older Adults' Data

All the issues concerning discrimination and risks in the processing of personal data of older people are also related to AI systems.

In this regard, the rapid spread of AI (Chang 2023, Wachter and Mittelstadt 2019) in the healthcare sector has led the European institutions to promote and implement a regulatory framework suitable for facilitating the execution of multinational projects necessary for the digital transformation of the EU, thus innovating (Fang Tan et al. 2023) the field of AI-based device production in medicine and the one of personal care and well-being (Alslaity and Orji 2024, Bard 2021, Cai et al. 2023, Chaidi and Drigas 2022).

In such scenarios the "AI Act" highlights the importance of acknowledging that, alongside the numerous benefits brought by the use of AI, this technology also raises concerns about the rights of older people since they are especially vulnerable to manipulation which can compromise their dignity and freedom" (Bobnar et al. 2023).

On this point, machine learning processes endow these technologies with a self-learning capacity that allows them to operate with a high degree of autonomy, despite the persistent 'opacity' of these technologies, with potential detriments to fundamental freedoms and rights of older persons (Chu et al. 2022).

As a result, AI-based devices run the risk of producing discriminatory outcomes, which is even more pressing now that the average age of users is rapidly increasing (La Vattiata 2022).

Thus, discrimination of older people is one of the adverse effects posed by automated decision-making systems and raises many questions on the identification of adequate safeguards for the protection of this particular category of individuals, without interfering with technological innovation.

In this regard, it is relevant to make a brief comparison between the GDPR and the AI Act. Article 22, paragraph 1 of the GDPR on automated decision-making is highly relevant to AI, since automated decisions today are indeed taken through AI-based systems. According to the interpretation suggested above, the Article prohibits any completely automated decisions based on profiling and having legal or significant effects on the data subject (Sartor 2020).

While AI is not explicitly mentioned in the GDPR, the automated decision-making framework serves as a form of indirect control over the use of AI systems, on the basis that AI systems are frequently used to take automated decisions that impact older individuals. Recent studies have highlighted how AI can negatively impact marginalized groups, revealing significant race, gender and age bias in algorithmic systems. Surprisingly, the last category, crucial for social inclusion and equality in ageing societies, has been largely neglected in research on AI bias.

The AI Act, like the GDPR, emphasizes the importance of adequate human oversight of AI systems (especially 'high-risk' ones) to ensure the protection of the fundamental rights and freedoms of this category of individuals.

In other words, AI systems are used to make fully automated decisions, but effective human oversight is needed to protect and ensure equal automated decision-making for the purpose of protecting the rights of individuals, including data protection rights.

In recent years, a wide debate has taken place on the advantages and disadvantages of algorithmic assessments and decisions concerning individuals (Sartor 2020, Kleinberg et al. 2018).

These diverging ideas should not lead to the exclusion of the use of automated decision-making: "in many cases, the best solution consists in integrating human and automated intelligence. The future challenge will consist in finding the best combination between human and automated machines, taking into account the capacities and the limitations of both" (Sartor 2020).

The use of AI systems also raises certain concerns, relating to the protection of older adults' privacy (they may not always understand how the technology works or how their data is being used).

In this context, the first objective should be to provide effective information, so as also to alleviate the concerns of elderly people and to clarify doubts about the types of devices, the data collected, the purpose of processing and the period of storage, and the persons who may have access to this data (Pak and Clark-Shirley 2021).

In this sense, should be provided to older persons - who are more susceptible to misinformation - "clear *instructions* concerning the use, functionalities, and properties of products utilizing AI" (Bobnar et al. 2023).

Essentially, there is a need for such devices to be safe, effective and properly regulated (Natale 2023). To do so, in order to be used in harmony with the new technological tools, data – especially health data – must respect certain fundamental characteristics to guarantee the validity of the results obtained.

On this point, it seems that the data protection framework itself could not be sufficient (Finocchiaro 2019, Nicholson II Price 2021) to regulate AI technologies and may not be adequate to protect older persons. For this reason, a suitable solution would be to reconsider or implement (Van Bekkum and Zuiderveen Borgesius 2022, Wilkinson et al. 2016) traditional control and certification systems (Van Kolfschooten 2023, Vokinger et al. 2022).

The aim is to strengthen the protection of the older adults, safeguarding the reliability of these technologies by means of a specific system of provisions and establishing sufficient control mechanisms to mitigate the negative effects produced by those technologies.

The possible, and often unpredictable, results of the use of artificial intelligence must be investigated. On the one hand, they have the potential to transform even healthcare by deriving new and important insights from the vast amount of health data generated every day, but on the other hand, they create new risks of discrimination against certain groups of people.

In light of these findings, it is imperative that legal practitioners and stakeholders recognize the limitations of existing regulations and work towards a comprehensive and tailored approach to address discrimination and violations of the fundamental rights of older people in AI healthcare systems. "This involves incorporating robust guidelines, promoting transparency, addressing implementation practices and establishing a healthcare-specific legal framework for data governance.

Only through such efforts the potential of AI technology may be harnessed to ensure fair and effective healthcare for all age groups" (Van Kolfschooten 2023).

With specific regard to older persons, it is fundamental to address the design and development phase, in order to promote their dignity and well-being (Bobnar et al. 2023).

References

Aceto, G., Persico, V. and Pescape, A. 2018. The role of information and communication technologies in healthcare: taxonomies, perspectives, and challenges. Journal of Network and Computer Applications 107: 125–154.

Alslaity A. and R. Orji. 2024. Machine learning techniques for emotion detection and sentiment analysis: current state, challenges, and future directions. Behaviour & Information Technology. 43(1): 139–164.

Altawalbeh, S.M., Alkhateeb, F.M. and Attarabeen, O.F. 2020. Ethical issues in consenting older adults: academic researchers and community perspectives. J. Pharm Health Serv. Res. 11(1): 25–32.

Amram, D. 2023. Comparing EU Initiatives on Data: Addressing Risks and Enhancing Harmonisation Opportunities. Opinio Juris in Comparatione. 1: 1–22.

Ayalon, L. and Tesch-Römer, C. [eds.]. 2018. Contemporary Perspectives on Ageism. Springer, Cham.

Baig, M.M., Afifi, S., Gholam Hosseini, H. and Mirza, F. 2019. A systematic review of wearable sensors and iot-based monitoring applications for older adults - a Focus on Ageing Population and Independent Living. J. Med. Syst. 43(8): 233.

Bard, J.S. 2021. Developing legal framework for regulating emotion AI. Boston University Journal of Science and Technology Law. 27(2): 3–43.

Bincoletto, G. 2021. mHealth app per la televisita e il telemonitoraggio. Le nuove frontiere della telemedicina tra disciplina sui dispositivi medici e protezione dei dati personali. BioLaw Journal. 4: 381-407, 386.

Bobnar, L., R. Šaina, B. Vičič, U. Vukotić, N. Zlatanova and L. Veljković. 2023. Introduction. pp. 13, 50, 63. *In*: Tekavčič Veber, M. and M. Kovačič [eds.]. Artificial Intelligence and the Rights of Older Persons. Litteralis, Ljubljana.

Brunzini, A., M. Caragiuli, C. Massera and M. Mandolini. 2023. Healthy ageing: a decision-support algorithm for the patient-specific assignment of ICT Devices and Services. Sensors. 23(4): 1836–1853.

Bujnowska-Fedak, M.M. and I. Pirogowicz. 2014. Support for e-health services among elderly primary care patients. Telemed. J. E Health 20(8): 696–704.

Buyl, R., I. Beogo, M. Fobelets, C. Deletroz, P. Van Landuyt, S. Dequanter, E. Gorus, A. Bourbonnais, A. Giguère, K. Lechasseur and M.P. Gagnon. 2020. E-Health interventions for healthy aging: a systematic review. Syst. Rev. 9(1): 128.

Cai, Y., X. Li and J. Li. 2023. Emotion recognition using different sensors, emotion models, methods and datasets: a comprehensive review. Sensors. 23: 1–33. Available at: chrome-extension://efaidnbmnnnibpcajpcglclefindmkaj/https://www.ncbi.nlm.nih.gov/pmc/articles/PMC10007272/pdf/sensors-23-02455.pdf.

Cascione, C.M. 2022a. The treatment of the elderly in time of Covid-19: attempts at protection or a new form of ageism?" 473–493. *In*: Giannone Codiglione, G. and L. Pierdominici [eds.]. Comparative law in times of emergencies, RomaTre Press, Roma.

Cascione, C.M. 2022b. Il lato grigio del diritto. Invecchiamento della popolazione e tutela degli anziani in prospettiva comparatistica. Giappichelli, Torino.

Chaidi, I. and A. Drigas. 2022. Emotional Intelligence and Learning, and the Role of ICTs. Technium Social Sciences Journal. 35: 56–78.

Chang, D. 2023. AI Regulation for the AI Revolution. Singapore Comparative Law Review. 130–170. With reference to problems arising from AI systems see Castro, D. 2023. The EU's AI Act Is Premature. ITIF. Available at: https://itif.org/publications/2023/12/08/the-eu-ai-act-is-premature/.

Chu, C.H., R. Nyrup, K. Leslie, J. Shi, A. Bianchi, A. Lyn et al. 2022. Digital Ageism: Challenges and Opportunities in Artificial Intelligence for Older Adults. The Gerontologist. 62(7): 947–955.

Colonna, L. 2021. Artificial Intelligence in the Internet of Health Things: Is the Solution to AI Privacy More AI?. Boston University Journal of Science and Technology Law.

Cuocci, V.V. 2022. La protezione dei dati personali dei soggetti vulnerabili nella dimensione digitale. Cacucci, Bari.

Czaja, S.J., N. Charness, A.D. Hertzog C., S.N. Nair, W.A. Rogers and J. Sharit. 2006. Factors predicting the use of technology: findings from the Center for Research and Education on Aging and Technology Enhancement (CREATE). Psychol Aging 21: 333–352.

Doron, I. [ed.]. 2009. Theories on Law and Ageing. The Jurisprudence of Elder Law. Springer, Berlin – Heidelberg.

Eubanks, V. 2018. Automatic Inequalities. How High-Tech Tools Profile, Police and Punish the Poor. St. Martin Press, New York.

Fang Tan, T., A.J. Thirunavukarasu, L. Jin, J. Lim, S. Poh, Z. Ling Teo et al. 2023. Artificial intelligence and digital health in global eye health: opportunities and challenges. The Lancet Global Health. 11: e1432–1443.

Fang, M.L., E. Siden, A. Korol, M. Demestihas, J. Sixsmith and A. Sixsmith 2018. A scoping review exploration of the intended and unintended consequences of eHealth on older people: a health equity impact assessment. Hum Technol. 14(3): 297–323.

Finocchiaro, G. 2019. Artificial intelligence and personal data protection. Giurisprudenza Italiana. 7: 1670–1677.

Gatto, S.L. and S.H. Tak. 2008. Computer, Internet and email users among older adults: benefits and barriers. Educational Gerontology 34: 800–811.

Gitlow, L. 2014. Technology use by older adults and barriers to using technology, physical & occupational therapy. Geriatrics 32(3): 271–280.

Guida, S. 2021. EHealth and mHealth applications between behavioral models, user-centered design and patient protection: examples of best practices. European journal of privacy law & Technologies (EJPLT). 2: 154–183.

Heart, T. and Kalderon, E. 2013. Older adults: are they ready to adopt health-related ICT? Int. J. Med. Inform. 82(11): e209–31.

Heponiemi, T., Jormanainen, V., Leemann, L., Manderbacka, K., Aalto, A.M. and Hyppönen, H. 2020. Digital divide in perceived benefits of online health care and social welfare services: national cross-sectional survey study. J. Med. Internet. Res. 22(7): e17616.

Irti, C. 2023. L'uso delle "tecnologie mobili" applicate alla salute: riflessioni al confine tra la forza del progresso e la vulnerabilità del soggetto anziano. Persona e mercato 1: 32-49, 47.

Kaihlanen, A.M., L. Virtanen, U. Buchert, N. Safarov, P. Valkonen, L. Hietapakka, I. Hörhammer, S. Kujala, A. Kouvonen and T. Heponiemi. 2022. Towards digital health equity - a qualitative study of the challenges experienced by vulnerable groups in using digital health services in the COVID-19 era. BMC Health Serv. Res. 22: 188.

Kleinberg, J., J. Ludwig, S. Mullainathan, and C. R. Sunstein. 2018. Discrimination in the age of algorithm. Journal of Legal Analysis. 10: 113–174.

Kwan, R.Y.C., D. Salihu, P. Hong Lee, M. Tse, D.S. Ki Cheung, I. Roopsawang and K.S. Choi. 2020. The effect of e-health interventions promoting physical activity in older people: a systematic review and meta-analysis. Eur Rev Aging Phys. Act. 17: 7.

La Vattiata, F.C. 2022. AI-based medical devices: the applicable law in the European Union. BioLaw Journal. 4: 411–437.

Lüders, M. and E.R. Gjevjon. 2017. Being old in an always-on culture: older people's perceptions and experiences of online communication. The Information Society 33(2): 64–75.

Malgieri, V.G. and J. Niklas. 2020. Vulnerable data subjects. Computer Law & Security Review. 37: 1015415.

Marquié, J.C., L. Jourdan-Boddaert and N. Huet. 2002. Do older adults underestimate their actual computer knowledge? Behaviour & Information Technology. 21(4): 273–280.

Martin, C., D. Rodriguez-Pinzón and B. Brown. 2015. Human Rights of Older People. Universal and Regional Legal Perspectives. Springer, Dordrecht.

McTear, M., K. Jokinen; M.M. Alam, Q. Saleem, G. Napolitano, F. Szczepaniak et al. 2023. Interaction with a Virtual Coach for Active and Healthy Ageing. Sensors. 23: 2748–2764.

Menassa, M., K. Stronks, F. Khatami, Z.M. Roa Díaz, O. Pano Espinola, M. Gamba et al. 2023. Concepts and definitions of healthy ageing: a systematic review and synthesis of theoretical models. eClinicalMedicine. 56: 1–31.

Millward, P. 2003. The 'grey digital divide': Perception, exclusion and barriers of access to the Internet for older people. First Monday 8(7).

Natale, M. 2023. Quadro normativo vigente e questioni insolute in materia di dispositive medici "intelligenti". 179–190. *In*: Ruffolo, U. and M. Gabbrielli [eds.]. Intelligenza Artificiale, dispositivi medici e diritto. Giappichelli, Torino.

Nicholson Price, W. 2021. Problematic Interactions between AI and Health Privacy. Utah Law Review. 4: 926–927.

Pak, R. and L. Clark-Shirley. 2021. Designing Artificial Intelligence Technologies for Older Adults – Insight Report. World Economic Forum. 6.

Palliotta, A. 2022. Successful, Active and Healthy Ageing: differenze e similarità nell'approccio al tema dell'invecchiamento. Vita e Pensiero. 473–492.

Piasecki, S. and J. Chen. 2022. Complying with the GDPR when vulnerable people use smart devices. International Data Privacy Law. 12(2): 113–131.

Rachovitsa, A. and N. Johann. 2022. The human rights implications of the use of AI in the digital welfare state: lessons learned from the dutch SyRI Case. Human Rights Law Review 22(2): 1–15.

Raja, M., J. Bjerkan, I.G. Kymre, K.T. Galvin and L. Uhrenfeldt. 2021. Telehealth and digital developments in society that persons 75 years and older in european countries have been part of: a scoping review. BMC Health Services Research. 21: 1157–1172.

Ranchordás, S. 2021. Empathy in the Digital Administrative State. University of Groningen Faculty of Law Research Paper Series No 13.

Rapisarda, I. 2023. La privacy sanitaria alla prova del Mobile ecosystem. Il caso delle app mediche. Le nuove leggi civili commentate. 1: 184-213 186, 208, 196.

Rotolo, F. 2023. Il diritto alla cura degli anziani tra vulnerabilità, logiche di mercato e brevettabilità della salute: la necessità di un ripensamento della proprietà intellettuale in chiave sostenibile. Jus Civile. 4: 888–916.

Sartor, G. 2020. The impact of the General Data Protection Regulation (GDPR) on artificial intelligence. STOA -Panel for the Future of Science and Technology. EPRS - European Parliamentary Research Service. 75.

Schwartz, R., A. Vassilev, K.K. Greene, L. Perine, A. Burt and P. Hall Schwartz. 2022. Towards a Standard for Identifying and Managing Bias in Artificial Intelligence. National Institute of Standards and Technology [NIST] Special Publication 1270. 1–77. Available at: https://www.nist.gov/publications/towards-standard-identifying-and-managing-bias-artificial-intelligence.

Sethumadhavan, A., G. Baik, L. D'Ambrosio, A. Fasihi, D. Fried, T. Lustig et al. 2021. Designing Artificial Intelligence Technologies for Older Adults – Insight Report. World Economic Forum. 2–10.

Stanzione, P. 2021. Wearable devices: privacy risks. What happens to the information collected? - Interview with Pasquale Stanzione. Available at: https://www.garanteprivacy.it/home/docweb/-/docweb-display/docweb/9552323.

Tirabeni, L. 2022. I dispositivi indossabili per il benessere. Il Mulino. 3: 119–126.

Van Bekkum, M. and F. Zuiderveen Borgesius. 2022. Using sensitive data to prevent discrimination by artificial intelligence: Does the GDPR Need a New Exception? Computer Law & Security Review. 48: 2–12. Available at: https://www.sciencedirect.com/science/article/pii/S0267364922001133.

Van Dijk, J. 2005. The Deepening Divide: Inequality in the Information Society. Sage, London.

Van Kolfschooten, H. 2023. The AI cycle of health inequity and digital ageism: mitigating biases through the EU regulatory framework on medical devices. Journal of Law and the Biosciences. 10(2): 12.

Vokinger, K.N., T.J. Hwang and A.S. Kesselheim. 2022. Lifecycle regulation and evaluation of artificial intelligence and machine learning-based medical devices. 13–21. *In*: Cohen, I.G., T. Minssen, W.N. Price II, C. Robertson and C. Shachar [eds.]. The Future of Medical Device Regulation: Innovation and Protection. Cambridge University Press, Cambridge.

Wachter, S. and B. Mittelstadt. 2019. A right to reasonable inferences: Re-thinking data protection law in the age of big data and AI. Columbia Business Law Review. 2: 469–620.

Wilkinson, M.D., M. Dumontier, I.J. Aalbersberg, G. Appleton, M. Axton, A. Baak et al. 2016. The FAIR Guiding Principles for scientific data management and stewardship. Scientific Data. 3: 1–9.

Wilson, J., M. Heinsch, D. Betts, D. Booth and F. Kay-Lambkin. 2021. Barriers and facilitators to the use of e-health by older adults: a scoping review. BMC Public Health 21(1): 1556.

Zinzuwadia, A and J.P. Singh. 2022. Wearable devices - Addressing Bias and Inequity. The Lancet - Digital Health. 4: e856–e857. Available at: https://www.thelancet.com/journals/landig/article/PIIS2589-7500(22)00194-7/fulltext.

Zuraw, R. and T. Sklar. 2020. Digital health privacy and age: Quality and safety improvement in long-term-care. Indiana Health Law Review. 17(1): 93 ss.

17

The Role of the Cohort Coordination Board in Supporting Data Sharing and Data Harmonization Efforts in Europe as Part of Pandemic Preparedness

Lorenzo Maria Canziani,[1,*,#] *Ruth Joanna Davis,*[1,#] *Ali Judd,*[2] *Eugenia Rinaldi,*[3] *Giorgia Bincoletto*[4] and *Evelina Tacconelli*[1]

17.1 Introduction

During pandemics, cohort studies can enable rapid exploration of disease epidemiology and effectiveness of interventions. However, differences in research protocols and how data are defined and structured can challenge data harmonization and limit the comparability of study results. Challenges in data sharing due to different data protection laws across legal frameworks also hamper the harnessing of results to inform public health strategy. Support for the implementation of data harmonization efforts and data sharing is crucial to maximise the success of observational research, especially in the context of pandemic preparedness. The Cohort Coordination Board (CCB) was established in April 2022 with the aim of identifying synergies across COVID-19 cohort studies, mapping common tasks, discussing hurdles and solutions, promoting collaboration for optimizing resources, and sharing data for common analyses to reach stronger powered results.

[1] University of Verona, Verona, Italy.
[2] University College London (UCL), London, UK; Fondazione Penta ETS, Padova, Italy.
[3] Charité-Universitaetsmedizin, Berlin, Germany.
[4] University of Trento, Trento, Italy.
# Lorenzo Maria Canziani and Ruth Joanna Davis contributed equally.
* Corresponding author: lorenzomaria.canziani@univr.it

DOI: 10.1201/9781003475927-20

This chapter outlines how the CCB was conceived and how it has evolved into a formal coordination mechanism funded by the European Commission (EC). It describes how the focus of the group has shifted from COVID-19 to the broader topic of pandemic preparedness, with specific research topics being oriented via a bottom-up approach by the core scientific members. It discusses the co-location of the CCB alongside other complementary coordination mechanisms such as the Trial Coordination Board (TCB), within a broader framework supporting a more coordinated approach to research as a cornerstone of pandemic preparedness. Finally, the activities undertaken by the CCB in the area of data sharing and data harmonization are presented, including considerations on the merits of this type of network as a means of support to sustainable health data sharing efforts, with a view to pandemic preparedness.

17.2 Why Are Cohort Studies Important for Pandemic Preparedness?

Cohort studies are a form of longitudinal study, of non-experimental or observational nature which can collect both retrospective (historical) and prospective data. In these studies, researchers select a group of individuals with shared characteristics and follow them over time to measure health-related outcomes. Cohort studies observe events as they naturally occur, offering valuable insights into the complex interplay between risk factors, interventions, and health outcomes. As a cornerstone of epidemiological research, cohort studies can provide a robust framework for understanding disease etiology, describing associations between interventions and health outcomes, and informing evidence-based strategies for disease prevention and control (Ricotta et al. 2023).

The COVID-19 pandemic highlighted the flexibility of cohort studies in pivoting to emerging diseases when an epidemic strikes. Some observational designs are able to quickly adapt research protocols, recruit participants, and collect data in response to evolving public health challenges. Unlike traditional Randomized Controlled Trials (RCTs) and cohort designs, which require extensive planning, simpler observational studies such as case series, perpetual cohort studies, and analyses using routinely collected data such as target trial emulation (Hernán et al. 2022), can be initiated rapidly. This leverages existing infrastructure and networks within health care systems and research institutions, all within a short time frame, across different settings and at a relatively low cost (Ricotta et al. 2023). Moreover, the pandemic highlighted how cohort studies are well-suited for assessing rare outcomes or exposures that may not be feasible to study in experimental settings, thereby expanding the scope of inquiry and enhancing the generalisability of findings (Karlstad et al. 2022).

Indeed, during the COVID-19 pandemic, an unprecedented number of cohort studies were undertaken (Teixeira da Silva et al. 2021). One of the most influential reports was the first cohort of COVID-19 patients in China, which was one of the first studies in the world that described the main characteristics of the acute phase of COVID-19 and attempted to define its main risk factors, as well as clinical adverse events (Zhou et al. 2020). As the pandemic progressed, cohort studies provided important information for vaccination strategies in terms of adverse effects - for example, a nationwide cohort study conducted in Norway showed a rare complication of vaccination, i.e., myocarditis, after vaccination with mRNA-based vaccines (Karlstad et al. 2022).

That being said, the pandemic also highlighted the importance of RCTs alongside observational data. As was the case with RCTs (Bugin and Woodcock 2021), many observational studies initiated during COVID-19 overlapped in focus, were methodologically flawed, and gave misleading results, for example regarding hydroxychloroquine as a treatment for COVID-19 (Alexander et al. 2020). Additionally, several COVID-19 treatments adopted early on in the pandemic (e.g., azithromycin) were confirmed to be ineffective in subsequent large RCTs (Butler et al. 2021; Fiolet et al. 2021).

There is no doubt that there exists a complementarity between RCTs and observational studies. Observational studies have the potential to describe natural history, generate questions for trials and provide estimates for trial feasibility; provide real world comparison arms for some trials, generate data on key patient subgroups who may not be targeted by trials or for whom trials may be unfeasible; follow patients after trial exit to provide long-term safety and effectiveness evidence; and provide insight into issues of trial generalisability (Dickerman et al. 2020; Nickolls et al. 2024).

Despite this complementarity, the pandemic exposed the lack of formal coordination mechanisms or frameworks to enable cohort studies and RCTs to inform each other, and to ensure robust quality of clinical studies and best use of limited resources, representing a missed opportunity for generating even more robust data within an integrated clinical studies "ecosystem".

17.3 The Challenges of Data Sharing and Integration in Clinical Cohort Studies

Despite the advantages that cohort studies can offer to rapidly generate evidence in a pandemic situation as outlined above, the recent COVID-19 pandemic highlighted how differences in data collection methodologies across cohort studies can challenge data harmonization and limit comparability of study results (Tacconelli et al. 2022).

One of the primary challenges in sharing and integrating data from clinical cohort studies lies in the variability of data collection definitions, protocols, and instruments used across different studies. Due to the diverse settings in which cohort studies are conducted, researchers often employ unique data collection methods tailored to their specific research questions, populations, and resources. As a result, data elements, variables, and outcome measures may be inconsistently defined or measured, making it difficult to compare and combine data across studies. This lack of standardization hampers efforts to harmonize data and conduct meta-analyses, leading to heterogeneity and potential bias in the interpretation of study findings (Metelli and Chaimani 2020).

The pandemic also underlined major challenges to sharing of data due to data protection legislation thus representing a missed opportunity to share vital information quickly and efficiently or to inform clinical decision-making and public health strategies, particularly as regards prevention policies (Tacconelli et al. 2022).

The strict regulatory framework surrounding data sharing, coupled with concerns about patient privacy, data protection and confidentiality, has created barriers to the timely exchange and integration of data from clinical cohort studies and trials. Compliance with data protection regulations such as the Regulation 2016/679 - General Data Protection Regulation (GDPR) in the European Union (EU) requires researchers to navigate complex legal provisions and, when necessary, obtain informed consent from participants for data sharing activities. Moreover, concerns about data security, anonymization, pseudoanonymization, and more generally de-identification further complicate efforts to share data concerning health while safeguarding patient privacy as a fundamental right. As a result, researchers may encounter delays or obstacles in accessing and sharing data across institutional or national boundaries, impacting the timely dissemination of findings crucial for informing public health interventions and policies.

Addressing these challenges requires concerted efforts to standardize data collection methods, enhance data sharing infrastructures, and promote collaboration and transparency within the research community. Initiatives such as the development of standardized data dictionaries, set of core variables and outcomes for specific settings, data sharing agreements, Data Management Plans (DMPs), and interoperable data platforms can facilitate the exchange and integration of data across cohort studies. Furthermore, fostering a culture of Open Science, data sharing, and reproducible research practices can promote transparency, accountability, and trust among researchers, funders, and stakeholders (Lindemann and Häberlein 2023). By promoting data harmonization and collaboration, researchers may have greater opportunities to share and integrate data from clinical cohort studies, ultimately enhancing the validity, reliability, and impact of observational research in pandemic situations. Moreover, interdisciplinary work, involving both researchers and legal experts, is vital to define the different requirements in countries involved, by comparing the legal rules, and then to identify appropriate organizational and technical measures to protect individuals' data while promoting sharing for public and scientific interests.

17.4 Rethinking the Clinical Research Framework in Support of Pandemic Preparedness

Now that the emergency phase of the pandemic has come to an end, the focus of research has shifted towards analyzing how we can be better prepared to face a similar pandemic in the future. Specifically, as regards clinical research preparedness and cohort studies, the need to move from retrospective harmonization to prospective adoption of data capture standards, i.e., Findability, Accessibility, Interoperability, and Reusability (FAIR) by design, has been flagged (European Commission et al. 2022). This effort may require involvement of multiple stakeholders, as researchers may not necessarily be the data controllers or have direct influence on the way data are collected or coded. However, by implementing FAIR data principles in the design and conduct of cohort studies, researchers can facilitate seamless data sharing, integration, and reuse across diverse research initiatives and geographical regions.

This proactive approach to data management not only enhances the efficiency and reproducibility of research but also maximizes the value of data for informing public health policies and clinical decision-making.

Moreover, solutions to overcome data sharing challenges such as federated learning are gaining traction, where the data can be analyzed locally at the premises of the data owner without the need for any exchange of data (Abu Attieh et al. 2023). The legal framework is also evolving, particularly with the proposal recently issued by the EC on the European Health Data Space (EHDS) (Directorate-General for Health and Food Safety 2022). The EHDS Regulation aims to promote better exchange and access to different types of electronic health data in the European Union by establishing common standards and practices, infrastructures and a governance framework (Bincoletto 2023).

Finally, the advantages of multicentre perpetual cohort studies, where patients are enrolled on a perpetual basis, according to a core study protocol collecting a set of demographics, clinical characteristics, and outcomes, and "kept warm", ready to be extended when needed, are being put forward as a useful tool for enhancing research efficiency in case of emerging health threats (Hassoun-Kheir et al. 2022). Similarly, multi-country adaptive trials platforms that can support novel diagnostics, vaccines and therapies for infectious diseases in response to an epidemic or pandemic, are being promoted as a solution for rapid trial activation (Adaptive Platform Trials Coalition 2019). Certainly, a more coordinated approach to research is required, where the different instruments are considered not as independent studies but as complementary components of a single framework, with in-built agility to pivot when an epidemic strikes.

17.5 What Is the CCB and How Did It Come About?

The idea to create a Cohort Coordination Board (CCB) (https://cohortcoordinationboard.eu/) came about as a result of the meeting, "*Cohorts united against COVID-19*", organized by the EC in February 2022. With almost 90 participants, the meeting brought together the EU-funded COVID-19 cohort projects, as well as other relevant initiatives, such as the European COVID-19 data platform. From the EC, participants from several Directorates-General (Research and Innovation (DG R&I), Health Emergency Preparedness and Response (HERA)) and Executive Agencies (Health and Digital Executive Agency (HaDEA), Research Executive Agency (REA)) were present, and participants from the European Centre for Disease Prevention and Control (ECDC) and the European Medicines Agency (EMA) also attended. The meeting brought attention to the significant opportunity for constructive collaboration among COVID-19 cohort projects. It also showed the potential for stronger evidence in the field of disease prevention and treatment from the EU-funded cohorts when efforts are joined, for advanced harmonization of data collection across cohorts, using the already-existing enabling tools and initiatives.

The CCB was established to resemble the format of the Trial Coordination Board (TCB) (https://covid19trials.eu/en/tcb), which creates a forum for projects running EU-funded clinical trials with relevance to preparedness.

Specifically, the objectives of the CCB are as follows:

- Provide updates on the status of ongoing cohort studies, including early results and final outcomes;
- Improve understanding of the cohort landscape in the EU and of the role cohorts can play in understanding pandemic preparedness;
- Improve uptake of data standards and coordination in data standards that have been adopted across cohorts and trials;
- Promote the use of cohort data to inform RCTs;
- Avoid overlapping and duplication of efforts (not only in cohort research but also in the creation of tools and infrastructures);
- Combine forces to achieve better results (e.g., larger sample sizes, stronger powered results, and wider dissemination);
- Share approaches to overcome commonly encountered obstacles (e.g., shipment of samples across national borders, lack of common dictionaries, electronic tool to link anonymous patient IDs to multiple samples and Work Packages);
- Organize training activities for best practices within sub-working groups of interest;
- Share documents of relevance for cohorts and with a view to greater harmonization (Data Management Plans, Data Sharing Agreements, Material Sharing Agreements, Informed Consent);
- Make recommendations to the EC for future research in the area of pandemic preparedness;
- Support the CoMeCT Coordination Board by providing input from cohort-based research.

The CCB has been active since the Kick-off Meeting on 29 April 2022, and has expanded both through the addition of new cohorts, and with its inclusion into EU-funded projects. Originally, the CCB was managed within the ORCHESTRA project (https://orchestra-cohort.eu/), then, in September 2022, the activities related to pandemic preparedness were adopted into the VERDI project (https://verdiproject.org/). In November 2023 the Grant Agreement for the CoMeCT project ("*Coordination mechanisms for cohorts and trials*", https://comectproject.org/) was signed, which includes the coordination and expansion of the CCB.

The CoMeCT project officially began on 1 December 2023, and links the running coordination initiatives (TCB and CCB) into one single, coherent coordination mechanism for adaptive platform trials and cohort studies, as well as developing an Outbreak Response Board to liaise with the other two Boards and activate timely action in terms of clinical research pandemic preparedness.

17.6 What has the CCB Done so far to Support Data Sharing and Data Harmonization?

The CCB meets monthly and, to date, 37 meetings have been held. Meetings are presided by the CCB members, which currently stands at 55 representatives from

27 projects and institutions. Membership is open to large scale cohort studies on infections, with relevance to preparedness, and with national, regional or EC funding, as well as to EU-funded projects of relevance for cohort research from other disciplines (e.g., data infrastructures).

At each meeting, the CCB chair welcomes members and external guests to speak on a topic of scientific relevance for the Board. The challenges related to data sharing and data harmonization are recurrent themes among CCB members and discussions within the group have led to several concrete actions in support of data harmonization and data sharing. The staggered start dates of the projects in the CCB means that the older projects are able to share experiences and lessons learnt, and thus offer a head start in rapid project activation for the newer projects.

ORCHESTRA ("*Connecting European cohorts to increase common and effective response to SARS-CoV-2 pandemic*", https://orchestra-cohort.eu/) was one of the first cohort-based research projects to be funded as part of the EC emergency funding allocated to tackle the pandemic. ORCHESTRA includes more than 80 cohorts (COVID-19 and post-COVID-19 syndrome, population based, fragile populations, health care workers) that prospectively enrolled 528,000 individuals. Moreover, the ORCHESTRA retrospective cohorts include 2,350,000 subjects' data (Azzini et al. 2023).

Semantic interoperability was established within ORCHESTRA retrospective and prospective cohorts by mapping all variables to international terminology codes such as SNOMED CT and LOINC. A pool of standardized variables was created that can individually be re-used for new ORCHESTRA studies but also external COVID-19 studies, to facilitate interoperability and merging of data. Following the FAIR principles, the metadata of the study data elements, which include the international codes for different concepts, was published on a public platform. The ORCHESTRA project now counts a total of 3,730 standardized variables (Rinaldi et al. 2022).

As a result of the CCB, ORCHESTRA was able to share the results of the data harmonization efforts with the newer EU-funded projects such as EuCARE (European Cohorts of Patients and Schools to Advance Response to Epidemics - https://eucareresearch.eu/), END-VOC (Ending Covid 19 Variants of Concern through Cohort Studies -https://endvoc.eu/) and REACT (Respiratory Host-Pathogen Interaction - https://react-euproject.eu/). The Data Dictionary of the ORCHESTRA project containing all the variables (questions and answer value sets) coded with international standards was shared with the representatives of EuCARE and END-VOC projects who then adapted their studies accordingly.

The ORCHESTRA standardization experts were called upon to collaborate with the REACT project team concerning the harmonization approach used. REACT focuses on SARS-CoV-2, influenza and Respiratory Syncytial Virus (RSV) infections. The REACT dataset was mapped to the ORCHESTRA variables to check for possible matches. In particular, for variables that covered the same topic but in a different format, the whole list of possibly corresponding ORCHESTRA variables was provided, such as in the case of tobacco and alcohol consumption, treatment, symptoms and comorbidities.

Within the framework of the CCB, the members exchanged their Data Management Plans (DMPs). A DMP is an obligatory document to be produced within the Horizon research programme that describes the data management life cycle for the data to be collected, processed and/or generated in the project and the related procedures, such as improving FAIRness of data and data security. The sources of the data are described through metadata (e.g., type of population of each cohort, samples collected, etc.). By sharing their DMPs, basic organizational tasks can be homogenized, and possible solutions to common problems are shared. Moreover, the metadata of data sources can be explored in order to establish further collaboration.

In addition to facilitating knowledge exchange amongst its members as described above, the CCB has also set up two dedicated working groups on data coordination efforts, namely a *Working Group on data standardization across cohort studies and clinical trials* and a *Working Group on the repurposing of cohorts and the implications for data re-use.*

The CCB has also facilitated the sharing of data and metadata through the data sharing tools that have been developed by the different projects. For example, ORCHESTRA has developed an infrastructure for federated learning, an innovative approach to data analysis in which each dataset is analyzed locally and independently, and only the aggregated results are shared. Federated learning allows for collaborative analysis across multiple datasets without the need to centralize or share sensitive individual-level data. By keeping data local, federated learning preserves the privacy, data protection and confidentiality of participants' personal data, as only summary statistics or model updates are exchanged between sites, i.e., anonymized data. This decentralized approach to analysis not only enhances data security, but also enables researchers to harness the collective power of diverse datasets while mitigating privacy and data protection risks. Through federated learning, ORCHESTRA facilitates global studies on the entire available dataset without compromizing data protection, thereby ensuring the integrity and confidentiality of research outcomes (Rinaldi et al. 2023). The federated learning infrastructure developed by ORCHESTRA has broader implications for collaborative research initiatives such as EuCARE. EuCARE utilizes common variables and data elements defined as in ORCHESTRA. Therefore, ORCHESTRA's federated learning approach enables the seamless integration and analysis of data from both projects without compromizing privacy and data protection. By leveraging federated analysis, researchers involved in EuCARE can collaborate with ORCHESTRA to conduct joint analyses and derive insights from larger and more diverse datasets. This collaborative effort not only maximizes the utility of available data, but also enhances the efficiency and generalizability of research findings. Ultimately, through federated learning, projects can work together to address critical research questions in pandemic preparedness and response while upholding the highest standards of data privacy and security.

In a similar fashion, EuCARE and VERDI, thanks to the connection made through the CCB, have shared their metadata and non-sensitive data with the COVID-19 data portal (https://www.covid19dataportal.org/ – last accessed on 2 May 2024). Despite data sharing restrictions, thematic data repositories with homogenized data can rapidly share valuable information regarding pathogen biology, such as viral genomic sequence and protein structural features, that may lead to development of specific antiviral drugs and vaccines (Harrison et al. 2021).

17.7 Conclusion

The COVID-19 pandemic research response has illustrated the importance of clinical research preparedness and the advantages of coordinating European clinical research initiatives.

Formal coordination mechanisms, such as the TCB and the CCB, can support the longer-term perspective of preparedness for future infection epidemics and pandemics, where the goal is to enable the conduct of perpetual platform trials and of perpetual strategic cohorts with the in-built flexibility to pivot to emerging diseases when an epidemic strikes. Strategic and sustained coordination mechanisms are especially important during inter-pandemic phases for preparedness planning.

The recent formal linking of the CCB with the TCB through the EU-funded project CoMeCT fills an important gap in the research setting, recognizing the need for a structured forum facilitating the exchange of emerging evidence from cohort studies, in a timely fashion, to inform the design of RCTs.

The CCB has identified the areas of data sharing and harmonization as the main barriers for cohort studies to realise their full potential. Since its inception, the CCB has undertaken a number of steps to address these barriers such as the sharing of data dictionaries and of data management plans, as well as the setting up of dedicated working groups to delve into the challenges in more detail. Moreover, CCB members are harnessing the tools developed in their projects (federated learning, data repositories) to share data and enable further analysis.

In the long term, the CCB envisages continued collaboration among cohort-based research projects, a growing membership, and greater alignment with other coordination mechanisms such as the TCB. Thematic areas of investigation will continue to be driven by the scientific experts on the board and therefore it is likely that the challenges of data sharing and harmonization will remain at the forefront of the CCB agenda for the foreseeable future.

Acknowledgements

We would like to express our gratitude to all the CCB members for their commitment and collaboration in the CCB. Sincere thanks also to the European Commission officers for their guidance and support of the CCB activities.

Funding Statement

The ORCHESTRA project is funded by the European Union under project GA n° No 101016167. The CoMeCT project is funded by the European Union under project GA n° 101136531. The VERDI project is funded by the European Union under project GA n° 101045989. Views and opinions expressed are however those of the

18

Health Data in Europe

At the Crossroads of Data Protection and Data Sharing

Giorgio Resta

18.1 The Centrality of Data Protection Within EU Data Law

European data law has traditionally paid more attention to concerns framed in terms of data protection over those of data sharing.

The regulatory framework related to the processing of personal data has gained a culturally and axiologically superordinate position, while a comparatively minor role has been played by other segments of data law, such as that concerning access to data, re-use of public sector information, or intellectual property itself (with the specific offshoots of disciplines on protection of databases and industrial secrecy) (Streinz 2021; Johannes 2022). One could speak, with some simplification, of a unipolar model. The reasons are easy to understand and lie largely in the historical background of legislation on data protection, which has been developed since the 1970s, first within national legal systems and then gradually penetrated by the system of the Council of Europe (Convention 108 of 1981) and the European Union (starting with Directive 95/46/EC) (Rodotà 2021). Fifty years of applying the legislation in question have given it a systematic, organic and complete character that cannot be found in any other statutory scheme on control of information.

In addition, on the level of principles, data protection law undoubtedly stands at the top of the system, in line with a public discourse that has always emphasized the link between these guarantees and post-war constitutionalism, which arose from the rubble of twentieth-century totalitarian experiences (De Hert – Cocito 2024). The choice made by the EU Charter of Fundamental Rights to introduce a specific provision on the right to protection of personal data (Article 8), in addition to the right to respect for private life, now provides clear testimony to this, widening the

University of Roma Tre, Rome, Italy; Email: giorgio.resta@uniroma3.it

DOI: 10.1201/9781003475927-22

transatlantic divide with the US model. Despite the fact that the intellectual roots of privacy and data protection are to be found on American soil, US law still resists the idea of a constitutional right to data protection. As it has been clearly highlighted by Paul Schwartz and Nikolaus Peifer, normative primacy is attributed to the First Amendment and to the principle of free flow of information. "Unlike EU law, U.S. law starts with a principle of free information flow and permits the processing of any personal data unless a law limits this action" (Schwartz and Peifer 2017). By contrast, the European model is based on the assumption that data protection is a fundmantal right and therefore any limitation of such right should rests on a specific legal basis, should pursue a legitimate aim, and should be proportionate. At the statutory level, this idea is embedded in Article 6 of the EU Regulation 2016/679 (GDPR), according to which data processing is prohibited unless a law (or other legitimate basis) allows this action.

All such factors strengthened the position of primacy, both in terms of hierarchy of sources, and in terms of completeness of scientific elaboration, of data protection law with respect to other fields of law designed to regulate circulation of information.

18.2 How the Law is Facing Social Change: The New EU Data Package

The dynamics of information capitalism and the remarkable achievements in the field of AI, are putting the traditional architecture of EU data law under strain. The digitalization of reality and the interconnected networks are driving data collection in all sectors; machine learning is making data processing cost-effective and socially valuable. This development again runs counter to the idea of "data flow limitation" embedded in the basic structures of EU data protection regulation, and namely in the principles of data minimization and purpose limitation, enshrined in Article 5 GDPR. The policy of limiting data collection and data re-use is increasingly perceived as inconsistent with the goals of technological innovation and economic competitiveness, and as a possible obstacle to social progress. As a result, the wisdom of the data protection regime is increasingly being questioned, not only in academic circles but also in the political arena.[1]

All this puts considerable pressure on the legal system, which is called upon to adapt its regulatory models to the changing needs of society, which are expressed in various ways in terms of data reuse. As a result, the normative primacy of data protection law is gradually breaking down. A remarkable – though not uncontroversial – shift within European data law is underway, from a unipolar model to a multipolar model. In the latter, concerns framed in terms of data protection will coexist and carry a comparable weight to those concerning data access and data re-use (Resta 2022). As one author put it, we are witnessing the emergence of a *Datenwirtschaftsrecht,* besides the more traditional *Datenschutzrecht* (Steinrötter 2021; Staudenmeyer 2022).

[1] See lastly the Report by Mario Draghi on *The future of European competitiveness,* Part B, *In depth analysis and recommendations,* 2024. (pages numbers are not included)

The EU Commission's Communication on the "European strategy for data"[2] is a very important document from this point of view, as it not only reflects the above-mentioned concerns, but also acts (as will be better shown below) as a driver for change. The existing approach, confronted with the new technological reality of big data and artificial intelligence, has been shown to be deficient from many points of view, but in particular from two perspectives: the lack of incentives to share data (personal and non-personal) and the ineffectiveness of the protection apparatus in the face of the reality of information capitalism.

Statistical findings show that around 85% of the data collected in Europe is not re-used (Ramge - Mayer Schönberger 2021). This is common to both the personal data sector, which, as the pandemic experience has shown, could be reused more effectively for public interest purposes, such as promoting scientific research and supporting health policies (on this more below); and the non-personal data sector, such as industrial data and data held by public administrations. This represents a *de facto* restraining factor in terms of innovation, given that this – and the reference is not limited to the industrial sector – is today largely data-driven.

On the other hand, even in the regulatory framework in which the preponderant demand is that of control rather than that of sharing, and in particular of personal data, the regulatory mechanism developed by the European legislator has in some respects proved ineffective in the face of the organizational and market practices that have emerged in the context of the digital economy (Wendehorst - Schwamberger and Grinzinger 2020; Gerpott 2020).

Therefore, one of the most frequent criticisms levelled at the European data protection model is that, while the legislation in question has in some cases made re-use of data for public interest purposes more complicated, at the same time it has not succeeded in putting up a strong barrier against commercial parasitism carried out on the basis of resources that are consubstantial to the sphere of personal identity.

18.3 Secondary Uses of Health Data: Legal Constraints

One of the sectors that most clearly illustrates the inherent limitations of the traditional normative infrastructure of data protection, as well as the urgency of a more flexible approach, is the health sector.

The use of digital technologies is rapidly transforming the entire functioning of the healthcare system, making the collection, analysis and sharing of health data a common practice and a widely felt necessity (Freitas 2024). It is widely recognized that safe, efficient and sustainable healthcare systems are strongly dependent on high quality data, which support the delivery of effective care services and are fundamental to advances in research, innovation and public policy-making (Dierks 2023). However, the quantity and quality of health-related data, as well as their availability for further processing, varies widely both internationally and nationally. Indeed, the current data-sharing ecosystem is not keeping pace with the needs of health systems, in part because of a number of legal, technical and administrative

[2] European Commission, 'A European Strategy for Data' (Communication) COM (2020) 66 final.

barriers that make it difficult to collect, access and share such data (especially across borders), thus preventing their full potential from being realized (Saelaert et al. 2023). This was most clearly demonstrated by the Covid-19 pandemic, which highlighted the need to ensure rapid - and possibly cross-border - access to health data for the purposes of patient care and monitoring, research and innovation, and the design and development of health-related public policies.

The GDPR, in particular, has been questioned as a source of potential barriers to the sharing of data that is critical to both the delivery of care (e.g., personalized medicine) and the advancement of biomedical research. This is not straightforward. On the one hand, the Regulation has had positive effects, not only in terms of rights and freedoms, but also from a data quality perspective. On the other hand, both the rights of data subjects and the compliance obligations of data controllers have the potential to limit the amount of personal data that is incomplete, outdated or misleading, and thus to improve data quality overall. In the same vein, it is unquestionable that the GDPR is also about ensuring general and collective interests. Among these interests, scientific research is explicitly recognized as having a special status by Artticle 5 (1) (b) (further processing), Article 9 (2) (j) (exception to the prohibition on processing sensitive data) and 89 (2) (exception to the rights of data subjects). Unlike other regulatory models, and namely the China's Personal Information Protection Law of 2021 (see in this volume Søderlund Kofod – Xiang), the GDPR provides for specific exemptions from data protection obligations when the processing is aimed at scientific research (Wiese Svanberg 2020).

Having said all this, it is undeniable that the GDPR suffers from some shortcomings, some of which are inherent to the legislative choices made, and others related to the inevitable obsolescence of its original architecture. It is worth remembering just three of the GDPR provides for specific derogations to data protection obligations whenever the processing is aimed at scientific research.

(*i*) Legal uncertainty arising from the vagueness or lack of flexibility of the GDPR provisions. I will give two examples. First, the purpose limitation principle and the further use of data for scientific research. Recital 50 states: "the processing of personal data for purposes other than those for which the personal data were initially collected should be allowed only where the processing is compatible with the purposes for which the personal data were initially collected. In such a case, *no legal basis separate from that which allowed the collection of the personal data is required*". It also adds that: "further processing for archiving purposes in the public interest, scientific or historical research purposes or statistical purposes should be considered to be compatible lawful processing operations. The legal basis provided by Union or Member State law for the processing of personal data *may also provide a legal basis for further processing*". However, Articles 5 and 6 do not openly provide for an exception to the general requirement of a legitimate legal basis for the hypothesis of further processing for scientific research. As a result, while some scholars opt for the more liberal interpretation (Dove – Chen 2020), data protection authorities insist on the need for an autonomous legal basis, which could, for example, consist of the consent of the data subject or the legitimate interest of the data controller,[3] even for further processing aimed at scientific research.[4] Second, there is the controversial issue of the permissible scope of consent for biomedical research. It is well known that scientific research is constantly evolving, and it is possible that a certain set of information, of limited use today, may become extremely valuable for new types of studies in the near future. Recital 33 takes this into account, noting that "it is often not possible to fully identify the purpose of personal data processing for scientific research purposes at the time of data collection". It therefore concludes that: "data subjects should be allowed to give their consent *to certain areas of scientific research* when in keeping with recognized ethical standards for scientific research. Data subjects should have the opportunity to give their consent only to certain areas of research or parts of research projects to the extent allowed by the intended purpose". However, the possibility of giving consent for "certain areas of research" tends to clash with the binding nature of Article 4(11) and 6(1)(a), which define consent as a "specific" declaration of will expressed in relation to the processing of personal data for "one or more specific purposes". The stricter interpretation of Articles 4 and 6, first endorsed by Article 29 Working Party[5] and then confirmed by the EDPB,[6] runs counter to the interests of researchers, who need to build their databases - intended to support prospective research activities - on a solid legal basis that could withstand challenges from data protection authorities and law enforcement agencies (Kühling – Schildbach 2024). A debate has therefore emerged on the limits of the acceptability of blanket consent for certain areas of scientific research and the viability of alternative solutions, such as "dynamic consent" (Gefenas et al. 2022; Teare et al. 2022; Comandè 2019).

(*ii*) Regulatory fragmentation. When the GDPR was adopted, some commentators noted that in some sections it had the form of a regulation but the content of a directive. Indeed, the field of scientific research, in particular biomedical research, confirms the validity of this observation (Kühling – Schildbach 2024)). Some important aspects of the regulation of data processing for scientific research are still not harmonized at EU level and show significant differences at national level (Tzortzatou et al. 2021). Two examples will suffice. The first concerns the discretion left to Member States under Article 9 (4) to "maintain or introduce further conditions, including limitations, with regard to the processing of genetic data, biometric data or data concerning health". While some countries

---

3 European Data Protection Board, *Guidelines on Article 6 (1)(f)*, 2024, p. 11.

4 European Data Protection Supervisor, *A preliminary opinion on data protection and scientific research,* 6-1-2020, 22-23.

5 Article 29 Working Party, *Guidelines on consent under Regulation 2016/679,* 2018, 28: "Recital 33 does not disapply the obligations with regard to the requirement of specific consent. This means that, in principle, scientific research projects can only include personal data on the basis of consent if they have a well-described purpose"; see at national level Italian Data Protection Authority, Provv. 30-6-2022, [doc. web n. 9791886].

6 EDPB, *Guidelines on consent under Regulation 2016/679,* n. 5/2020, 33-34.

(especially in Northern Europe) have not made use of this clause, others have made use of the derogation and have made the processing of data for biomedical research subject to a number of additional requirements (Dove – Chen, 2020; Tzortzatou et al. 2021; Comandè 2019).[7] The second relates to Article 89 (2) GDPR and the possibility given to Member States to limit the rights of data subjects referred to in Articles 15, 16, 18 and 21 with regard to processing for scientific purposes. Some countries have taken advantage of this exception. For example, Norway introduced restrictions on the right of access, the right of rectification and the right to restrict processing (Kjersti Befring 2021), Italy on the right of rectification,[8] Denmark and Latvia on all the rights referred to in Articles 15, 16, 18 and 21 (Hartlev 2021); in contrast, several other European countries refrained from doing so, leaving data processing for scientific research under the general provisions of the GDPR (Tzortzatou et al. 2021). This results in a fragmented regulatory framework, which not only creates barriers and additional costs for digital health product manufacturers and healthcare providers to enter markets in other Member States, but also acts as a barrier to studies with large patient cohorts conducted on a transnational scale.

(*iii*) Excessive restrictions on cross-border exchange and transfer of health data (Kuner 2020). The GDPR aims to support the free flow of data within the European single market. At the same time, it places significant restrictions on the transfer of personal data to third countries, on the assumption that the level of protection guaranteed in Europe should also be equivalent in the country of destination. The three possible legal bases for data transfer - formal adequacy decisions (Article 45), appropriate safeguards (such as standard contractual clauses under Article 46) or exemptions for specific situations (Article 49) - are not well suited for transnational cooperation in research projects. During the Covid-19 pandemic, such restrictions were immediately felt by researchers seeking to combine their efforts to understand the morphology of the virus, the genetic basis of severe or critical diseases, and the efficacy of vaccines. On the one hand, even the transfer of data to international organizations such as the WHO (which operates some of the most important databases in the field and plays a crucial role in countering pandemic risks) proved to be difficult, as it is subject to the same substantive and procedural requirements set out in Articles 44-49 GDPR. On the other hand, the exception relating to transfers for important reasons of public interest (such as public health) can only be relied on in exceptional circumstances, because, as the EDPB made clear, the exception can only be used if the transfers are "occasional" and "not repetitive".[9]

---

[7] See, for instance, the Irish Health Research Regulations 2018, which privilege explicit consent for data processing in the health research context; Art. 110 Italian Data Protection Code, requiring a previous impact assessment according to Arts. 35 and 36 GDPR.

[8] Article 110 (2) Italian Data Protection Code.

[9] EDPB, Guidelines 2/2018 on derogations of Article 49 under Regulation 2016/679, 4-5.

18.4 The Way Forward

In its above-mentioned Communication on a European Strategy for Data, the European Commission took into account the legal and technical barriers that still prevent an adequate level of data sharing in Europe. It then set out the objective of creating a Single European Data Space, a true single market open to data from all over the world, where data can flow within the EU and across sectors, in full compliance with data protection rules, through common standards and governance mechanisms for its access, use and re-use.[10] With regard to the health sector, the European Commission highlighted the importance of health data as a "key enabler for the digital transformation of healthcare" and health as an area where the EU can benefit from the data revolution, recognizing the shortcomings of the current health data exchange ecosystem, which is characterized by a deficit in the exchange and accessibility of different types of data, hampering both quality of care and data-driven innovation.

To this end, both the Data Governance Act and the European Health Data Space Regulation work together to create a radically new normative (and technical) infrastructure that will greatly enhance the sharing of health data for secondary purposes. The Data Governance Act (EU Regulation 2022/868) introduces a new horizontal legal framework to support data sharing and re-use in all sectors; the European Health Data Space complements the Data Governance Act in the field of digital health.

18.5 The Data Governance Act: Data Altruism and Data Donation

Among the many aspects of the DGA that deserve attention, two are closely related to the collection and use of health data for health care delivery, innovation and scientific research. The first is the recognition of data altruism organizations. The second is the introduction of a Europe-wide altruism data form.

Article 15 exempts "recognized data altruism organizations" as well as "other not-for-profit entities insofar as their activities consist of seeking to collect data for objectives of general interest, made available by natural or legal persons on the basis of data altruism" from the (quite cumbersome) procedural and substantive regulation of intermediation services set forth by Article 12 DGA. The intention is to carve out a set of favourable rules for collective entities that, operating on a non-profit basis, seek to stimulate collection of personal and non-personal data and their use for purposes of general interest (von Hagen and Völzmann 2022). Data altruism is defined by Article 2(1)(16) as: "the voluntary sharing of data on the basis of the consent of data subjects to process personal data pertaining to them, or permissions of data holders to allow the use of their non-personal data without seeking or receiving a reward that goes beyond compensation related to the costs that they incur

[10] Communication from the Commission to the European Parliament, the Council, the European Economic and Social Committee and the Committee of the Regions, *A European strategy for data* COM/2020/66 final, paragraphs 1 and 3.

where they make their data available for objectives of general interest as provided for in national law, where applicable, such as healthcare, combating climate change, improving mobility, facilitating the development, production and dissemination of official statistics, improving the provision of public services, public policy making or scientific research purposes in the general interest". The pandemic experience brought to the fore the importance of data altruism in Europe, while also showing some of the limitations of the EU data protection regulation. In Germany, the Robert Koch Institut promoted an application (*Corona-Datenspende*) aimed at enabling donation by citizens of personal data collected by smart devices, such as smart watches, for epidemiological and pandemic response purposes (Kühling and Schildbach 2020). Similarly, several organizations, such as Google or Facebook, have made aggregated population mobility data available to governments to identify clusters and track the spread of the virus. Generalizing this approach, the DGA assigns a specific role to non-profit organizations engaged in collection and subsequent allocation of data for purposes of general interest, providing a detailed regulatory framework to ensure the transparency and independence of these organizations, avoiding creation of misaligned incentives. The first requirement is to register with the public register of altruistic data organizations established by Article 19 DGA. This is subject to fulfilment of a series of requirements specified in Article 18 (including the profit-making purpose, the functionally separate structure for the activities in question, and compliance with the code set out in Article 22). Secondly, Article 20 requires maintenance of up-to-date registers, while Article 21 sets out a series of obligations and principles to protect data controllers, including the obligation to provide information (mainly relating to objectives of general interest), a prohibition on using data for objectives other than those of general interest for which the data subject or data holder allows processing, technical assistance in the phase of providing or withdrawing consent, and security in data storage. The underlying intention is to bolster public confidence in the trustworthiness of such organizations (Recital 46), thus enabling other forms of nudging that support altruistic motivations (for instance public awareness campaigns) to find a viable routing system.

With regard to the second point, the DGA provides for the introduction of a European altruism data form to enable consent to be collected in a uniform format in all Member States (Article 25) (von Hagen and Völzmann 2022). This idea deserves to be supported, but it immediately raises the already mentioned question of the limits of the admissibility of "broad consent", an issue which the DGA unfortunately leaves unresolved (see Recital 50).

This is where the scope of European intervention comes to a halt, on the one hand because no precise provisions are foreseen either for the relations between the "donor" of data and the altruistic organizations, or for those between such organizations and third parties who are the final recipients of the data, and on the other hand because Article 16 clearly calls on the Member States to adopt targeted policies to support data altruism, including through the adoption of specific organizational or technical measures.

18.6 The European Health Data Space: Secondary Uses of Health Data

On 3 May 2022, the Commission presented a proposal for a Regulation establishing the European Health Data Space (EHDS).[11] On 15 March 2024, the European Parliament and the Council reached a provisional agreement on the amended text. Recently, on 24 April 2024, the European Parliament approved the Regulation, which is currently awaiting publication in the Official Journal of the EU.[12]

The European Health Data Space is a key pillar of the European Data Strategy and is the first piece of legislation to support a common data space in a specific sector. It builds on the existing legal frameworks for data protection: the General Data Protection Regulation, the Medical Devices Regulation 2017/745 and the In Vitro Diagnostics Regulation 2017/746. It also integrates, as already mentioned, the Data Governance Act, which provides rules that are also applicable to the healthcare sector. Undoubtedly, this proposal represents an ambitious initiative, the core of which revolves around two pillars: primary and secondary use of health data.

With regard to the first pillar, the primary use of electronic health data refers to the processing of electronic health data for the purpose of providing direct healthcare (e.g., assessment or restoration of health, provision of medicines) to the natural person to whom the data relate.[13] Among other things, access to and portability of data are greatly improved by the Regulation (Li – Quinn 2024).

On the other hand, as far as the second pillar is concerned, the EHDS represents a first attempt to consolidate a common EU framework for the secondary use of health data, a timely response to the uncertainties and challenges posed by the GDPR for such use of personal data.[14]

In the EHDS Regulation, the secondary use of electronic health data is defined - by Article 2 (2) (e) - as the processing of electronic health data for purposes other than those for which the data were originally collected or generated.[15] Presumably, this definition refers to data originally collected for the provision of healthcare (the primary use) and then designated for another purpose (the secondary use). The Regulation defines a minimum set of data to be made available for secondary use: Article 33 provides an extensive list that seems to include not only data strictly related to health (e.g., electronic health records, research data), but also data only indirectly related to health and care (e.g., electronic data on insurance status, lifestyle, personally generated data such as from wellness apps or other wearable devices, demographic data).[16] The list is flexible enough to adapt to the evolving needs of data users, while remaining limited to data related to health or known to affect health.

---

[11] European Commission. *Proposal for a Regulation of the European Parliament and of the Council on the European Health Data Space*. COM/2022/197 final.

[12] European Commission, *Commission welcomes European Parliament's adoption of the European Health Data Space and regulation on substances of human origin. Press release.* Brussels, 24 April 2024.

[13] Article 2(2)(d) EHDS.

[14] Article 2(2)(e) EHDS.

[15] By contrast, the GDPR envisions secondary uses only through the lenses of the "further processing" of personal data (Becker et al. 2022).

[16] Article 33 (1) EHDS.

Article 34 sets out the different purposes for which processing of electronic health data for secondary use is allowed. These include: reasons of public interest in the field of public or occupational health (e.g., protection against serious cross-border health threats); scientific research contributing to public health, health technology assessment or quality and safety of care; development and innovation activities for products and services; training, testing and evaluation of algorithms (including medical devices), artificial intelligence systems and digital health applications.

As can be seen in particular from Article 34(1)(e), the re-use of health data is not limited to public bodies operating for purposes of collective interest, but is also allowed for the benefit of private bodies operating for profit, provided that the purposes of re-use fall within one of the legitimate purposes listed in the same provision (Gassner 2024).

Secondary use, on the other hand, is prohibited for certain purposes specifically listed in Article 35, including taking decisions that adversely affect individuals or groups on the basis of their health data. In addition, the EHDS establishes - notably in Article 41 - new obligations for "health data holders" (mostly healthcare providers such as hospitals and clinics, as well as research institutions) to make certain categories of electronic health data available for secondary use by "data users" (e.g., biomedical research organizations, AI developers, pharmaceutical manufacturers). Exceptions for micro-enterprises, individual researchers and natural persons are provided for in Article 32a.

It is important to note that in order for the secondary use of data, as provided for by the EHDS, to be legitimate, an appropriate legal basis must be identified that complies with the provisions of the GDPR. Indeed, the EHDS provides a valid exception to the prohibition of processing of health and genetic data under Article 9 GDPR, but leaves untouched the need for the data user to base the data processing on one of the legal bases provided for by Article 6 GDPR. This is far from an optimal solution, as it reinforces the regulatory fragmentation at national level (which the EHDS was theoretically intended to reduce): as several studies have shown, the laws of the different Member States differ significantly with regard to the legal bases for data processing for scientific research (Beckr et al. 2024).

Also in relation to the issue of regulatory fragmentation, it should be noted that Article 33 (5) EHDS, far from establishing a single golden standard for the data access and re-use regime, allows Member States to introduce "stricter measures and additional safeguards at national level aimed at safeguarding the sensitivity and value" of human genetic, epigenomic and genomic data, as well as data from wellness apps and biobanks with associated data. In other words, while the usual mechanism developed by the EHDS is based on an opt-out logic (see Article 35f, discussed below), this provision allows Member States to exceptionally adopt an opt-in model (Kühling and Schildbach 2024). As a result, a dispersed framework similar to the one created by Artcle 9(4) GDPR is to be expected (Beckr et al. 2024).

With regard to the data access mechanism and its governance structure, the Regulation establishes a completely new mechanism revolving around a system of data permissions granted by a health data access body following a request for access to data by the 'data user' (Roos and Maddaloni 2023). This is the most innovative part of the EHDS, as it envisages a shift from fragmented and individual access to health data (for example, by each research group wishing to carry out a project on population genomics) to a centralized system of data permissions. This is a governance structure that has already been experimented with in some European countries, such as Finland (Southerington 2021),[17] - which as a matter of fact – provided the inspiration for the EHDS (Dierks 2023) – more limited extent, France (Bernelin and Eynard 2024),[18] and recently introduced in Germany (Gassner 2023; Werry and Ntanas 2024),[19] but which is alien to many other legal systems and is therefore open to debate (particularly, but which is alien to many other legal systems and is therefore open to debate (particularly with regard to the nature of the health data access body and its relations with similar bodies). According to Articles 45 and 49 EHDS, two main procedures can be followed to obtain access to data.

If a user requests access to data held by a single trusted health data holder located in a single Member State, he or she may submit the request, in accordance with Article 49 EHDS, to the Health Data Access Body, which will forward it to the data holder.[20] After an assessment and a proposal for a decision, if the outcome is positive, the Health Data Access Body will issue the authorization to access the data in a secure and reliable environment and a fee will be paid according to Article 42.

If the user requests data from controllers established in more than one Member State, pursuant to Article 45 (3) EHDS, the user has to submit "a single data access application through the health data access body of the main establishment of the applicant or of one of these data holder or through the services provided by the Commission in the cross-border infrastructure HealthData@EU. The application shall then be automatically forwarded to the authorized participants and to the health data access bodies of the Member States where the health data holders identified in the data access application are established". However, the decision is not uniform, as Article 46 (3a) that each data access authority decides on the data held by national data holders, and this is another weak point of the EHDS (Beckr et al. 2024). In order to coordinate the decision-making process, the same provisions make it clear that "the concerned health data access bodies and authorized participants shall inform each other of their decisions and may take the information into consideration when deciding on granting or refusing access to electronic health data. A data permit issued by one concerned health data access body may benefit from mutual recognition by the other concerned health data access bodies."

Finally, it should be noted that, following criticism of the Commission's original proposal, namely the insufficient protection of the individual's right to self-determination (Denga 2023), the final text of the Regulation provides for a right of opt-out in favour of the individual whose data are to be re-used (de Frutos Lucas – Haugo 2024). Article 35f(1) provides that: "natural persons shall have the right

[17] Article on the Secondary Use of Health and Social Data (522/2019) s 3(17).

[18] Articles 88 and 89, *Décret no 2019-536 du 29 mai 2019 pris pour l'application de la loi no 78-17 du 6 janvier 1978 relative à l'informatique, aux fichiers et aux libertés.*

[19] *Gesetz zur verbesserten Nutzung von Gesundheitsdaten*, 22-3-2024.

[20] As regards the role of data permit authorities and data users from the perspective of data protection law, Becker-Thorogood, 2022.

to opt-out at any time and without stating reasons from the processing of personal electronic health data relating to them for secondary use under this Regulation. The exercise of this right shall be reversible".

18.7 Conclusions

Health data is some of the most sensitive information about any individual, and as such deserves strong protection. At the same time, unhindered access to health data has become crucial for improving the quality of healthcare (in particular with regard to personalized medicine), for AI innovation and for the advancement of biomedical research. Traditional legal frameworks have not proved particularly suitable for facilitating smooth access to health data, while at the same time continuing to protect fundamental rights and freedoms. Data solidarity has somehow been sacrificed to the individualistic paradigm underlying traditional data protection law. From this point of view, the Covid-19 pandemic was a turning point that marked the emergence of a new attitude towards access to data in Europe (Dierks 2023). Not only economic arguments, but also concerns about social progress and the protection of the most vulnerable are behind the need to improve the secondary use of health data. It is no coincidence that legal reforms have been introduced in recent years, both at national and supranational level.

From the former point of view it is sufficient to the remember the German Act on the re-use of health data (*Gesundheitsdatennutzungsgesetz* 2024)[21] and the Italian reform of Article 110 of the Data Protection Code.[22] At the European level, the Data Governance Act and the European Health Data Space are the clearest examples of such a new legislative trend, which deserves to be supported and encouraged. However, many regulatory barriers remain to be overcome. This paper has identified some of them, most of which stem from the difficulty of reconciling the GDPR, which is centred on the idea of data protection, with the new legal framework, which is based on the conflicting logic of data sharing. The controversies around blanket consent and the legal bases for secondary uses of health data, the regulatory fragmentation caused by exception clauses such as Article 33 (5) EHDS, the restrictions on cross-border data flows (see Thorogood in this volume) are some of the most urgent issues that need to be addressed by lawmakers if the enormous societal value of health data is not to be lost (Kühling and Schildbach 2024).

Acknowledgments

The author of this publication is financed by the Italian Ministry of University, under the Call for tender, PRIN 2022, Project "PRIMA", CUP F53D23003300001.

[21] *Gesetz zur verbesserten Nutzung von Gesundheitsdaten*, 22-3-2024.

[22] See decree-law, 2-3-2024, n. 19.

References

Becker, R., A. Thorogood et al. 2022. Applying GDPR roles and responsibilities to scientific data sharing, 12 International Data Privacy Law 207: 213–216.

Becker, R., D. Chokoshvili, G. Comandé, E.S. Dove, A. Hall, C. Mitchell, F. Molnár-Gábor, P. Nicolàs, S. Tervo and A. Thorogood. 2022. Secondary use of personal health data: When Is It "further processing" under the GDPR, and what are the implications for data controllers', in European Journal of Health Law, 1–29.

Beckr, R., D. Chokoshvili and E.S. Dove. 2024. Legal bases for effective secondary use of health and genetica data in the EU: time for new legislative solutions to better harmonize data for cross-border sharing?, in 14 Int'l Data Privacy Law 223(2024), 225.

Bernelin M. – Eynard, J., Numérique, in Cahiers Droit, Sciences & Technologies, 18, 2024, 79-93.

Comandè, G. 2019. Ricerca in sanità e data protection, un puzzle…risolvibile, in Riv. It. Med. Leg., 187.

de Frutos Lucas, J. and H.T. Haugo. 2024. Moving forward with the European health data space: The need to restore trust in European health systems, in Lancet Reg. Health Eur. 2024 Apr. 12.

De Hert, P. and C. Cocito. 2024. The added value of data protection within the framework of digital constitutionalism in Europe. *In*: De Gregorio, G., O. Pollicino and S. Valke (eds.). The Oxford Handbook of Digital Constitutionalism, Oxford, in print.

Denga, M. 2023. Die Nutzungsgovernance im European Health Data Space als Problem eines Immaterialgütermarkts, in EuZW, 2023, 25.

Dierks, C. 2023. European Health Data Space – Anforderungen und Chancen für die pharmazeutische Industrie, in PharmR, 369.

Dove, E.S. and J. Chen. 2020. Should consent for data processing be privileged in health research? A comparative legal analysis, in International Data Privacy Law, 117.

Freitas, A.T. 2024. Data-Driven Approaches in Healthcare: Challenges and Emerging Trends, in H. Sousa Antunes et al., Multidisciplinary Perspectives on Artificial Intelligence and the Law, Cham, 65.

Gassner, U. 2023. Essentialia eines Gesundheitsdatennutzungsgesetzes, in ZRP, 34.

Gassner, U.M. 2024. Sekundärnutzung von Gesundheitsdaten für die kommerzielle Medizinprodukteforschung, in MPR, 95.

Gefenas E. et al. 2022. Controversies between regulations of research ethics and protection of personal data: informed consent at a cross-road, in 25 Medicine, Health Care and Philosophy, 23.

Gerpott, T.J. 2020. Datenschutzerklärungen. Materiell fundierte Einwilligungen nach der DS-GVO. Empirischer Forschungsstand und Verbesserungsfelder, MultiMedia und Recht (MMR), 220, 739.

Hartlev, M. 2021. Balancing of Individual Rights and Research Interests in Danish Biobank Regulation, in S. Slokenberga – Tzortzatu, O. – Reichel J. (eds.). GDPR and Biobanking. Individual Rights, Public Interest and Research Regulation across Europe, Cham, 223.

Johannes, P.C. 2022. Europäisches Datenrecht – ein Spickzettel, Newsdienst der Zeitschrift für Datenschutz (ZD) – Aktuell, 1166.

Kjersti Befring, A. 2021. Norwegian Biobanks: Increased Complexity with GDPR and National Law. *In:* Slokenberga, S., O. Tzortzatu and J. Reichel (eds.). GDPR and Biobanking. Individual Rights, Public Interest and Research Regulation across Europe, Cham, 323.

Kühling, J. and R. Schildbach. 2020. Corona-Apps. Daten- und Grundrechtsschutz in Krisenzeiten Neue Juristische Wochenschrift (NJW), 1545.

Kühling, J. and R. Schildbach. 2024. Datenschutzrechtliche Spielräume für eine forschungsfreundliche digitale Gesundheitsversorgung – von DSGVO, SGB, etc. zur EHDS-VO und zum GDNG, in ZfDR, 12.

Kuner, K. 2020. Data Crossing Borders: Data Sharing and Protection in Times of Coronavirus, in Verfassungsblog.de, 15 April 2020 < https://verfassungsblog.de/data-crossing-borders/>.

Li, W. and P. Quinn. 2024. The European Health Data Space: An expanded right to data portability?, in Computer Law & Security Review 52: 105913.

Ramge, T. and V. Mayer Schönberger. 2021. Fuori i dati! Rompere i monopoli delle informazioni per rilanciare il progresso, Milano, Egea.

Resta, G. 2022. Pubblico, privato e collettivo nel sistema europeo di governo dei dati, in Riv. trim. dir. pubb. 971.

Rodotà, S. 2021. Tecnologie e diritti, 2nd edn, Bologna, Il Mulino.

Roos P. and G.M. Maddaloni. 2023. Regulierter Datenaustausch zur Gesundheitsforschung. Die legislativen Vorhaben für einen Europäischen Gesundheitsdatenraum und ein Gesundheitsdatennutzungsgesetz, in RDi, 225.

Saelaert, M. et al. 2023. Expanding citizen engagement in the secondary use of health data: an opportunity for national health data access bodies to realise the intentions of the European Health Data Space. Arch Public Health 81, 168.

Schwartz, P.M. and K.N. Peifer, 2017. Transatlantic Privacy Law. 106 Georgetown L.J. 115.

Southerington, T. 2021. Access to Biomedical Research Material and the Right to Data Protection in Finland. *In*: Slokenberga, S., O. Tzortzatu and J. Reichel (eds.). GDPR and Biobanking. Individual Rights, Public Interest and Research Regulation across Europe, Cham, 243.

Staudenmayer, D. 2022. Der Verordnungsvorschlag der Europäischen Kommission zum Datengesetz. Auf dem Weg zum Privatrecht der Datenwirtschaft, Europäische Zeitschrift für Wirtschaftsrecht (EuZW), 596.

Steinrötter, B. 2021. Das Datenwirtschaftsrecht" als neues Teilrechtsgebiet im Recht der Daten. Einführende Bemerkungen zur neuen ZD-Beitragsreihe, Newsdienst der Zeitschrift für Datenschutz (ZD), 543.

Streinz, T. 2021. The evolution of european data law. *In*: Craig, P. and G. de Búrca (eds.). The Evolution of EU Law, 3rd edn, Oxford, OUP, 902.

Teare, H.J.A., M. Prictor and J. Kaye. 2021. Reflections on Dynamic Consent in Biomedical Research: The Story So Far, in 29 Eur. J. Human Genetics 649.

Tzortzatou, O. et al. 2021. Biobanking Across Europe Post-GDPR: A Deliberately Fragmented Landscape. *In*: Slokenberga, S., O. Tzortzatu and J. Reichel (eds.). GDPR and Biobanking. Individual Rights, Public Interest and Research Regulation across Europe, Cham, 397.

von Hagen, P. and L. Völzmann. 2022. Datenaltruismus aus datenschutzrechtlicher Perspektive. Wechselwirkung zwischen DGA und DS-GVO, Multimedia und Recht (MMR), 176.

Wendehorst, C., S. Schwamberger and J. Grinzinger. 2020. Datentreuhand – Wie hilfreich sind sachenrechtliche Konzepte? *In*: Pertot, T. (ed.). Rechte an Daten, Tübingen, Mohr Siebeck, 231.

Werry, S. and E. Ntanas. 2024. Sekundärnutzung von Gesundheitsdaten – Quind deinde? Wie EHDS und GDNG die Landschaft der Gesundheitsvorsorge revolutionieren, in MMR, 641.

Wiese Svanberg, C. sub § 89. *In*: Kuner, C. and L. Bygrave et al. (eds.). The EU General Data Protection Regulation, Oxford, OUP, 2020, 1240.

19

Health Data Sharing for Scientific Research

Closing Data "As Necessary" and Opening it "As Possible"

Giorgia Bincoletto

19.1 Introduction[1]

Data use and sharing are indispensable for medical and biomedical research. A clinical trial collects and analyses data from participants in order to achieve primary and secondary objectives and endpoints during various phases. A clinical investigation typically involves information on human subjects to assess the safety or performance of a medical device. Research that uses a biobank combines biological materials with bioinformatic data (Hallinan 2021; Colcelli et al. 2023). Both prospective and retrospective studies require quality health data, and ultimately, all the mentioned categories of scientific studies aim to publish and share their results. The availability of a large volume of health data, i.e., the so called "Big Data", accelerates innovation, knowledge, reproducibility and replicability of the experiments and tests (Madir 2020).

Data sharing may occur during or after research, and it may involve parties in the same collaborative project (e.g., multicentric studies in a so-called "consortium") or even different entities. This exchange or availability of data may take place within

University of Trento, Trento, Italy; Email: giorgia.bincoletto@unitn.it

[1] This text constitutes a re-elaboration of the speech at the EuCARE Brocher Workshop "Data privacy, data property, and data sharing: an interdisciplinary dialogue for post-pandemic transnational research", Brocher Foundation, Hermance (Switzerland), 20 June 2023. An early and preliminary version of was published in Bincoletto, G. 2023. Scientific research processing health data in the European Union: data protection regime vs. open data. Journal of Open Access to Law. 11(2): 1–24.

DOI: 10.1201/9781003475927-23

or between countries, under different legal frameworks, between private companies, and from a private entity to the public sector or vice versa.

The Covid-19 pandemic tragically demonstrated the urgent need to share scientific progress around the world (Besançon et al. 2021). The advancement of the healthcare sector and the capacity to address public health emergencies depends on the accessibility and the sharing of health data. Hence, the Open Science (OS), Open Access (OA), Open Data (OD) and Open Research Data (ORD) movements and policies aim to facilitate the opening up of scientific research, including the collected information and the outcomes (Margoni et al. 2017; Bartling and Friesike 2014). Scholars considered data sharing a moral imperative in biomedical research (Bauchner et al. 2016). It has also been pointed out that access to and dissemination of scientific knowledge has a strong link with democracy (Rodotà 2012; Paseri 2020).

Nevertheless, the public interests inherent in the implementation of transnational and national data sharing must be balanced with the protection of the interests and rights of the human subjects involved in the scientific research. In particular, the individual's right to the protection of personal data plays a key role in regulating and limiting health data sharing. This right is provided by the legal and ethical frameworks at the international and national levels. Data protection is not an absolute right, and it can be balanced with other rights and interests, e.g., with the "right to science" as the "right to share in scientific advancement and its benefits" (Geiger and Jütte 2023).

The challenge of striking the balance between personal data protection and open data sharing is complex. The research and innovation funding program of the European Union (EU) "Horizon Europe" adheres to the principle of research data being "as open as possible, as closed as necessary".[2] The management of research data necessitates a preventive planning of all the steps of the data processing. Consequently, researchers frequently encounter difficulties in comprehending the circumstances under which it is lawful to open and share the collected datasets with others, especially when the study is conducted in more than one country.

From a legal perspective, comparative law and its plural methodologies are fundamental in order to search the similarities and differences of the rules in different legal frameworks and eventually frame the solutions in common perspectives (Sacco 1992; Graziadei 2003; Bussani and Mattei 2012; Samuel 2014; Siens 2018; Zweigert and Kötz 2018; Kishel 2019; Reinmann and Zimmermann 2019; Resta et al. 2020; Monateri 2022; Husa 2023; Ioriatti 2024). Additionally, an interdisciplinary approach that considers the insights from the legal, ethical, medical and technical fields is necessary to translate and implement the applicable rules in concrete solutions for data sharing.

Against this background, this chapter is structured as follows. The next section provides an overview of the rules set for the protection of personal health data in scientific research at the international level to understand whether there are specific policies for data sharing in the context of transnational health research. Section 3 then examines the Open Science and Open Research Data movements promoted at the same global level to foster circulation of data across countries. After that, Section 4 focuses on the EU legal framework, and it presents the key data protection

[2] See further Section 5.

rules that researchers should carefully consider when designing and carrying out their data management activities, including data sharing. Section 5 explores the Open Research Data paradigm in the EU, which has been innovated by the recent legislative initiatives of the European Commission under the "Data Strategy". An analysis of the interplay between the data protection rules and the open data framework is presented in Section 6 to clarify when it necessary to close data and when it is possible to open it or allow access to it. Finally, Section 7 concludes with some remarks.

19.2 International Rules and Policies for Protecting and Sharing Personal Data for Medical and Biomedical Research

On the global level, data protection is included in the ethical principles and human rights declarations. Article 12 of the Universal Declaration of Human Rights and Article 17 of the International Covenant on Civil and Political Rights provide that no one shall be subjected to arbitrary or unlawful interference with his or her privacy. This term can be interpreted to cover both the right to private life and to protect personal data.

Unauthorized and uncorrected access or sharing of personal health data may cause significant economic, psychological and social harms (e.g., employment and social exclusion, stigma, embarrassment and forms of discrimination). Health data may be considered as the most sensitive and intimate personal data. Paragraph 24 of the pioneering Declaration of Helsinki on Ethical Principles for Medical Research Involving Human Subjects of the World Medical Association establishes that every precaution should be taken to safeguard the participants' privacy interests and the confidentiality of their personal information, i.e., their personal data protection. Privacy and data protection are both essential in the medical sector where the professional duty of confidentiality is also directly instantiated in the Hippocratic Oath (Hoffman and Herveg 2021).

As stated by the Convention on Huma Rights and Biomedicine – known as the "Oviedo Convention", which is directly binding on state parties – the individual is protected by the necessity to collect the informed consent to research participation.[3] This consent is the cornerstone of the research ethics requirements (Laurie et al. 2021). The compliance with the ethical principles and obligations set in various conventions, codes of conduct, guidelines, and recommendations, is verified by the competent local research ethics committee, which preventively approves the scientific project involving human participants. Thus, this institution evaluate how privacy and data protection are safeguarded during research and its phases.

Privacy and data protection as fundamental individual rights are protected by international data protection law, which dates back to the early 1980s. It should be firstly mentioned the Organization for Economic Cooperation and Development (OECD)'s Guidelines on the Protection of Privacy and Transborder Flows of Personal Data of 1980, later revised in 2013 in order to create the "Privacy Framework",

[3] Convention for the protection of Human Rights and Dignity of the Human Being with regard to the Application of Biology and Medicine: Convention on Human Rights and Biomedicine (ETS No. 164).

which recognizes the individual's right to self-determination and defines standard principles (i.e., collection limitation, data quality, purpose specification, use limitation, security safeguards, openness, individual participation, accountability).[4] Although the OECD's principles are not legally binding and they have an abstract formulation, they have been highly influential among countries (Bygrave 2014). As an example, the first guidelines guided the European Directive 95/46/EC and the United States legal framework on informational privacy. It has been stated that the OECD set the "core of data protection law", later transposed into regional and national laws (Townend 2021).

In addition to the Guidelines, the OECD adopted sector-specific recommendations. As far as health data sharing is concerned, the OECD's "Recommendation on Health Data Governance" of 2016 emphasized the lack of a coordinated policy framework to guide health data sharing practices.[5] Therefore, the scope of this policy is to set principles to encourage cross-country harmonization. The second principle boosts common policies and procedures that minimize barriers to data sharing and balance public and private interests in medical research. According to the fourth recommendation, national systems should identify and remove "barriers to effective cross-border cooperation in the processing of personal health data for health-related public interest purposes in a manner consistent with protecting privacy and data security".[6] It may be argued that this international framework highly promotes data sharing, but countries should further provide the rules to govern it.

After the pandemic, the OECD adopted the same approach in the "Recommendation of the Council on Enhancing Access to and Sharing of Data" of 2021.[7] Sharing should be considered as a priority, restrictions to cross-border access and exchange should be limited, but privacy and personal data protection should be protected. The researchers should evaluate the applicable law in every legal framework where the study is carried out.

In 2015, the Human Rights Council of the United Nations created the first mandate of the Special Rapporteur on privacy.[8] Since then, many reports have been published on several privacy-related topics. The Report of the United Nations Special Rapporteur on the right to privacy of 2019 – adopted as the "Recommendation on the protection and use of health-related data" – provides influential guidance to health data sharing.[9]

The UN Report first indicates that the processing of health-related data for the purposes of scientific research should be subject to appropriate safeguards provided for by national law, but also by the same policy recommendations, and it should be carried out for a legitimate purpose.[10] The individual should neither be required

---

[4] OECD. Guidelines on the Protection of Privacy and Transborder Flows of Personal Data, in the form of a Recommendation by the Council of the OECD. 1980.

[5] OECD. Recommendation of the Council on Health Data Governance. 2016, Section III, par. 3.

[6] Ibidem, Section IV.

[7] OECD. Recommendation of the Council on Enhancing Access to and Sharing of Data. 2021.

[8] Human Rights Council Resolution 28/16. The right to privacy in the digital age. A/HRC/RES/28/16.

[9] The report was prepared by the Special Rapporteur on the right to privacy, Prof. Joseph A. Cannataci. It has been presented to the 74th session of the UN General Assembly on 5 August 2019.

[10] Chapter IX, par. 21.1.

nor compelled to participate in scientific research without his or her prior consent.[11] Exceptions to this prior consent may be provided by law and the project should then be subject to ethics committee approval. The same independent body, or another appointed institution, should assess the conditions in which data processing is lawful.[12] It should be stressed that the legal basis for the data processing is in fact different from the mentioned consent to research participation. Indeed, consent is not always the legal ground under national data protection law, and it is not frequently even practicable (e.g., for health conditions of the participant, or in the case of deceased data subjects). The UN Report so specifies that a law providing exceptions to the consent requirement should specifically establish safeguards and respect the data minimization principle, meaning the use of data should be limited to the amount that is adequate, relevant, accurate, up to date to the specific research purposes for which the data processing is planned.[13] These purposes should be determined in advance, at the time of data collection. When it is not possible, the researchers should identify certain area of scientific research or progressively communicate the parts of the evolving project to the data subject.[14] As a result, a dynamic consent may be collected in order to inform the participant during the various phases of the research. This form of consent is not broad: it actually requires the on-going engagement of the participant (Steinsbekk et al. 2013).

Moreover, the UN Report recommends the provision of prior, transparent, and comprehensive information to the participant on concrete characteristics of the scientific research, including aspects related to data sharing.[15] In fact, the list of information to be provided should include the identification of the third parties who will obtain access to the collected data, and the planned cross border data transmission. Thus, the existence of data sharing should be transparent and clear to the research participant once he or she receives the information on the envisage data processing. According to the UN recommendations, data sharing should be also based on a legitimate ground.

The UN Report suggests that health data should be anonymized when it is "technically feasible and practicable in relation to the purposes of the scientific research" to avoid identification of the data subjects.[16] If not, pseudonymization is recommended to secure data and still safeguard the rights of the participants. These measures thus apply to data sharing: researchers should protect data with de-identification or cryptographic techniques. When this sharing is operated with publication, the form of data should not enable data subject's identification, unless he or she has given a specific consent (and there is no withdrawal of it), or the applicable law permits such activity "on the condition that this is indispensable for the presentation of scientific research findings and only to the extent that the interest in publishing the data overrides the interests and fundamental rights and freedoms of the data subject".[17]

---

[11] par. 21.2.
[12] par. 21.3
[13] par. 21.6 and par. 4.1(e).
[14] par. 21.9.
[15] par. 21.7.
[16] par. 21.11.
[17] par. 21.14.

Par. 23 of the UN Report is dedicated to the protection to health-related data transfers. Any cross-border transmission of health data should comply with national law, and also ensure "an appropriate level" of data protection.[18] The legal basis for the transnational sharing is usually defined by specific provisions in the national regulations, such as the explicit and informed consent of the data subject, or public interests laid down by the law with other safeguards.

In sum, this UN framework recommends that data sharing should be based on a legitimate ground and communicated to the data subject in a transparent matter. When the sharing is operated cross-countries, the state of destination should provide appropriate protection to personal data. Technical measures, such as anonymization and pseudonymization should be implemented. Anonymization will prevent the future application of data protection law, however, research in the medical and biomedical field frequently requires as much information as possible. It has been pointed out that anonymization breaks the link between information needed for full traceability of medically relevant research, that it is nearly impossible in some contexts as biobanking, and it is not useful in studies containing stages of follow up of participants (Govarts et al. 2022).

The UN framework represents an international baseline of minimum data protection safeguards for health research and data sharing. These recommendations are seminal points of reference for transnational research; however, researchers should still identify the applicable national data protection rules to identify the legal basis for the data sharing and interpret what "appropriate protection" means regarding a different legal framework.

From an international point of view, it should be lastly added the Convention for the Protection of Individuals with regard to Automatic Processing of Personal Data of the Council of Europe (CoE), which was modernized in 2018 as "Convention 108+". This Convention relies on Article 8 of the European Convention on Human Rights (ECHR), which is regionally binding as an international treaty. It included health data within the category of "special" data, thereby necessitating the implementation of robust safeguards to protect it and it recognizes scientific research purposes as an individual basis for the processing (Bernier et al. 2022). This is the only legally binding multilateral agreement in the field of data protection (de Hert and Papakonstantinou 2014). It has been considered as a global instrument able to ensure minimum standard for cross border data flows even outside the CoE's framework (Ukrow 2018).

In 2019, the Committee of Ministers of the Council of Europe adopted the Recommendation "Protection of health-related data".[19] Chapter V is dedicated to scientific research. Its specifications are comparable to that provided by the UN Report analyzed above. The contents can be considered similar, but the CoE's Recommendation is less detailed. First, this Recommendation indicates that the data processing should be subject to appropriate safeguards set by the law, as well as carried out on specific legal grounds.[20] When the basis is not the data subject's

[18] par. 23.1.

[19] Council of Europe. Recommendation Protection of health-related data. CM/Rec(2019)2, 2019.

[20] par. 15.1.

consent, the law should define specific additional safeguards, including technical and organizational measures to respect the principle of data minimization.

Similar to the UN Report, the CoE's Recommendation indicates that prior, transparent, comprehensive information on the envisaged scientific research should be provided to the data subject.[21] By contrast, it does not mention specific information on data sharing, only specifying the necessity to communicate "any relevant conditions governing the use of the data". However, a legal interpretation of the rule could lead to the same result. When it is not possible to determine the purposes of the scientific project, the CoE's Recommendation suggests identifying certain areas of research or parts of the projects to gradually limit the goal of the research.[22] Anonymization and pseudonymization measures are recommended, while safeguarding the fulfilment of the scientific purposes.[23] Limits on the publication form of the data are mentioned to avoid the identification of the data subjects.[24] If the data should be transferred cross-countries, the CoE's Recommendation mandates the implementation of safeguards to ensure an appropriate level of protection in the destination, and the selection of a ground allowing the data transfer.

Comparing the two frameworks, it may be argued that both the CoE and the UN rules promote common rules for data sharing. The key elements of data governance for health data sharing may be summarized as follows: being transparent with the data subjects on the data circulation, finding a lawful basis for the transmission, and ensuring an appropriate level of data protection while sharing take places cross-borders. Notwithstanding this guidance, the abstract level at which international data law regulates health data sharing detracts its ability to function as a practical guide in concrete processing operations. The recommendations are not legally binding. Anyway, as highly promoted policies at global and regional levels, they may be influential to national legal frameworks in limiting barriers to data sharing and defining common requirements.

The lawfulness of collecting and sharing health data for research purposes needs to be evaluated with regard to the involved national law. As anticipated, data protection could occasionally lead towards closure to safeguard fundamental rights. Even so, the following section will present the international trends towards Open Science and Open Research Data, which aim at making data available without financial, technical and (less) legal constraints.

19.3 Open Science and Open Research Data policies at the International Level

Open Science may be described as a set of principles and practices that seek to make scientific research accessible to everyone. OS as an "umbrella term" promotes a new paradigm of conducting science: openness of every phase of the research, including open access for publications and educational resources (Open Access),

[21] par. 15.4.
[22] par. 15.6.
[23] par. 15.9.
[24] par. 15.11.

open methodologies, technologies and sources for data analysis (Open Source), open collection of data (Open Data), even in the research context (Open Research Data). As a result, innovation may be opened for the benefit of the society as a whole (Chesbrough 2003).

To date, there is no legally recognized definition of OS.[25] It has been noted that it is even difficult to determine the beginning of this movement (Caso 2020). This phenomenon has surely philosophical and sociological foundations in the late 16th and 17th centuries (Merton 1942; David 2008; Resnik 2023). It then arose spontaneously in the late 1980s and early 1990s, when the Web was taking its first steps, and few scientific journals opened their access archives. Open Access to scientific publications was the first form of OS, and its implications have been debated by scholars of every field (Margoni et al. 2017). According to Caso, at the core of OS there is the idea that the Internet represents an extraordinary opportunity to improve science with open connections between scientists, as well as between scientists and citizens (Caso 2020). Another concept that is central to OS is considering the knowledge generated by the scientific community as a "common good" (Hesse (ed) 2007).

To concretely achieve OS, three main macro-levels issues have been recently pointed out: the need for open infrastructures and open technologies, the existence of a governance able to ensure access and reliability of the disseminated knowledge, and the application of OS by public institutions under legal rules allowing and promoting it (Paseri 2020).

The OS approach needs to be balanced with the protection of existing rights and freedoms, including data protection (Guarda 2021). Despite its great potentiality for innovation, one of the main challenges of the OS remains the balance between its openness and the exclusivities recognized by Intellectual Property rights (IP), including copyright, patents, and trade secrets (Gold 2021; Willinsky 2022). It may be highlighted that the need for a balance between opening and closing is common to both data protection and IP law.

From a legal perspective at international level, OS has been promoted by several institutions through reports, declarations, and recommendations.[26] Among all, it may be mentioned the OECD's report "Making Open Science a reality" of 2015, which is the first comprehensive analysis of OS.[27] The OECD considered OS as a key means to support science.

During and after the Covid-19 pandemic, the promotion of Open Science has rapidly increased. In 2020, the adoption of OS has been encouraged by the Committee on Economic, Social and Cultural Rights of the United Nations, in the context of the "General comment" on science and economic, social and cultural rights (Articles 15 (1) (b), (2), (3) and (4) of the International Covenant on Economic, Social and

[25] A systematic taxonomy of OS has been provided by the EU-funded project FOSTER. The "Open Science taxonomy" is available at https://www.fosteropenscience.eu/resources.

[26] The first declarations have been directly promoted by the scientific community: the Budapest Open Access Initiatives (2002), the Bethesda Statement on Open Access Publishing (2003), and the Berlin Declaration on Open Access to Knowledge in the Sciences and Humanities (2003).

[27] OECD. 2015. Making Open Science a Reality. OECD Science, Technology and Industry Policy Papers n. 25. OECD Publishing.

Cultural Rights).[28] In this comment, and in particular in the section dedicated to the essential elements of the "right to participate in and to enjoy the benefits of scientific progress and its applications", the Committee stated that "states should promote open science and open source publication of research". It also specifies that "research findings and research data funded by states should be accessible to the public".[29]

This comment represents an important foundation for OS and Open Research Data since it connects the movements with the right to science enshrined in the Article 27 of the UN Declaration, which is complemented by Article 15 of the mentioned Covenant. These provisions may be now considered the basis for the implementation of the future OS policies. While noting on this comment, some scholars even propose the establishment of a new "human right to Open Science", which would ensure the inclusivity, equity, and sustainability of scientific knowledge for the benefit of the society at a global level (Caso and Binda 2020).

In the same year, the United Nations Educational, Scientific and Cultural Organization (UNESCO), the World Health Organization (WHO), the United Nations High Commissioner for Human Rights and the European Laboratory for Particle Physics (CERN) published a share appeal on Open Science by calling on the international community to adopt all the measures that are necessary to make science more accessible and transparent.[30] In 2021, UNESCO then adopted a comprehensive Recommendation on OS, which provided a comprehensive international framework for open science policies and practices.[31]

UNESCO defined OS in very open terms. OS includes "various movements and practices aiming to make multilingual scientific knowledge openly available, accessible and reusable for everyone, to increase scientific collaborations and sharing of information for the benefits of science and society, and to open the processes of scientific knowledge creation, evaluation and communication to societal actors beyond the traditional scientific community".[32] "Open Research Data" (ORD) are added in the UNESCO's Recommendation as a component of open scientific knowledge. The concept of ORD consists of both digital and not digital data, raw and processed information, and the accompanying metadata. According to UNESCO's definition of ORD, data should be openly used, re-used, retained, and redistributed by anyone. Research Data should be available in a timely and user-friendly manner, and in a format that is machine-readable and actionable, in accordance with the so-called "FAIR principles".[33] This notion is broader than the concept of Open Data, which is instead limited to the information generated and controlled by the public-sector (Dalla Corte and Loenen 2022).

[28] United Nations, Committee on Economic, Social and Cultural Rights. General comment No. 25 (2020) on science and economic, social and cultural rights (article 15 (1) (b), (2), (3) and (4) of the International Covenant on Economic, Social and Cultural Rights).

[29] par. 16.

[30] Reference to the appeal is available at https://events.unesco.org/event?id=1522100236&lang=1033.

[31] The UNESCO's Recommendation of 2021 on Open Science is available at https://unesdoc.unesco.org/ark:/48223/pf0000379949.

[32] Ibidem, p. 7.

[33] Ibidem, pp. 9-10.

The acronym "FAIR" stands for Findable, Accessible, Interoperable and Reusable" (Paseri et al. 2021). Data and its connected metadata are findable when they can be discovered, identified, and located easily by manual and machine searchers under a standard, unique and persistent identifier. They are accessible when they are available and retrievable for the searcher who then has authorization and authentication powers to access. Data is interoperable when the format and language representation are semantically and syntactically understandable. Lastly, it is reusable when it is shared with the least restrictive licenses and detailed attributes. These principles were firstly elaborated for the re-use of scholarly data (Wilkinson et al. 2016) and then they became the guidelines to manage data in scientific research at a global level. They are now mentioned in the UNESCO's recommendations as requisites for Open Research Data.

The access to research data should be "as open as possible".[34] No further guidance on the extent of the closure is provided. UNESCO instead specifies that any restriction should be proportionate and justified because of the protection of rights and freedoms, including privacy, data protection and IP. Even when data is openly shared, Resnik pointed out that if the means used to generate the data, such as methods, protocols, materials, are not shared as well, verifying and reproducing the information are not possible (Resnik 2023). In addition, researchers need programs and technologies that are able to analyze the data and metadata and extract the information from them.

When limitations apply to data sharing, UNESCO suggests the implementation of anonymization to still be able to let data circulate. However, anonymization is a hard task because it should be effective, and the re-identification should be impossible (Stalla-Bourdillon and Knight 2017). Determining whether the dataset is really anonymized is complex, since datasets are often and easily de-anonymized using re-identification techniques (Narayanan and Shmatikov 2008). Combining allegedly anonymized datasets with other outside information available online frequently pries out obscured identities. Therefore, researchers lose faith in anonymization (Ohm 2010). Anyway, as already anticipated, medical and biomedical research often need a lot of attributes to achieve their endpoints.

A document specifically dedicated to the open data sharing in the health context has been published by the WHO[35] in 2022. This policy is intended for the researchers funded by the Organization. It recommends the drafting of a data management and sharing plan, which should consider the nature of the data collected in the dataset and define the appropriate data repository and the license under which the data is made available. The open and pioneering "Creative Commons Attribution 4.0 International (CC BY 4.0)" license is the point of reference for the WHO's policy.[36] As regards the limits to data sharing, the WHO mentions the existence of legal or ethical reasons established at national level. Thus, the policy suggests that the risks of data sharing may be mitigated by "controlling whom the data are shared and

[34] par. 8.

[35] World Health Organization. 2022. Sharing and reuse of health-related data for research purposes: WHO policy and implementation guidance.

[36] The text of the license is available at https://creativecommons.org/licenses/by/4.0/deed.en.

under what conditions, without compromizing the scientific usefulness of the shared data".[37] When data is personal, the WHO recommends the identification of a suitable managed access repository or, once again, the implementation of anonymization techniques. The WHO provides a useful and detailed infographic explaining de-identification and the different degrees of identifiability.[38] It may be noted that the data sharing plan is a valuable instrument for researchers to concretely program and execute the policies on OS and ORD.

It may be here concluded that both Open Science and Open Research Data are no longer merely movements supported by the scientific community. They are policies endorsed internationally by institutions and relevant organizations. Nevertheless, the guidance on how to strike a balance between openness and closure remain inevitably demanded to the specific legal frameworks where data sharing occurs. The following sections attempt to provide some advice by focusing on the EU context, and the data protection and open research data regulations.

19.4 Key Data Protection Rules to Manage Personal Health Data in the European Union

Focusing on the EU legal framework, the individual right to data protection is firstly guaranteed by Article 8 of the EU Charter of Fundamental Rights and by Article 16 of the Treaty on the Functioning of the European Union, which provides the basis for EU legislation in this matter. The general requirements applicable to scientific research processing health data, including data sharing activities, are provided by the Regulation 2016/679 (General Data Protection Regulation or "GDPR").[39] Scholars argued that the GDPR sets the global standard for data protection legislation (Kuner et al. 2020).

During the design and execution phases of data management, governance and sharing the researchers should take into account the following key data protection issues: (1) the notions of personal data vs. anonymized data; (2) the presence of data concerning health as a special category of data and the related legal basis to process and share such data in the EU and outside the European Economic Area; (3) the purpose limitation principle in the context of scientific research; (4) and the application of technical and organizational measures to safeguards data in every phase of the processing.

Regarding the notion of personal data, it should be stressed that only the research activity that processes "information relating to an identified or identifiable natural person"[40] and that is carried out by a data controller according to the material and

[37] See p. 9 of the mentioned WHO's Policy.

[38] Ibidem, p. 17.

[39] Regulation (EU) 2016/679 of the European Parliament and of the Council of 27 April 2016 on the protection of natural persons with regard to the processing of personal data and on the free movement of such data, and repealing Directive 95/46/EC (General Data Protection Regulation), O.J. L. 119, 4.5.2016. Moreover, Directive 2002/58/EC contains special rules for the processing of personal data and the protection of privacy in the electronic communications sector ("e-Privacy Directive").

[40] Article 4(1) GDPR.

territorial scopes of the GDPR should comply with the EU general rules.[41] This notion includes pseudonymized data since this data can be attributed to a natural person using additional information, which has been kept separated by the data controller. The European Union Agency for Network & Information Security provides detailed guidance on pseudonymization process and techniques.[42] Adopting a comparative perspective, the difference between the concept of "personal data" and "personal information" in other legal frameworks, such as in the United States, may represent an issue for data sharing (Guarda and Bincoletto 2023).

In the EU, the researcher should define in advance whether the project includes the use of personal data, and then it should follow the requirements of the GDPR. When data is not personal, Regulation (EU) 2018/1807 "on a framework for the free flow of non-personal data in the European Union" applies in place of the General Regulation.[43] A third category of data is not conceived by the law (Comandè 2022) (Graziadei 2021). According to Recital 26 of the GDPR, data is not personal either if it is "anonymous information, namely information which does not relate to an identified or identifiable natural person" *ab origine* or it is "personal data rendered anonymous in such a manner that the data subject is not or no longer identifiable", meaning personal data that is later subjected to anonymization (Purtova 2018).

The researcher may use anonymous data that are already available in an open dataset or may anonymize personal data. In this last case, data collection should follow the GDPR, but after the anonymization process the rules would not apply. However, two main problems arise: the research goals may not be achieved with the use of non-personal data and the researchers should be aware that anonymization must be effective. According to the Opinion 05/2014 on Anonymization Techniques adopted on 10 April 2014 by the Article 29 Data Protection Working Party, the anonymization process should "irreversibly" prevent identification of the data subject. In the case *Patrick Breyer v. Bundesrepublik Deutschland* (C-582/14), the Court of Justice of the European Union (CJEU) stated the re-identification should be "practically impossible on account of the fact that it requires a disproportionate effort in terms of time, cost and man-power, so that the risk of identification appears in reality to be insignificant".[44] Anonymization should be considered as a dynamic concept that evolves with technological evolution (Podda Palmirani 2021). When it is possible to re-obtain personal data or if the dataset is mixed, data should be treated as personal data, and the GDPR applies.[45]

Secondly, a scientific project in the medical or biomedical filed implies the processing of "data concerning health" which falls amongst special categories of personal data under the GDPR.[46] Data concerning health or "personal health data"

[41] Artt. 2 and 3 GDPR.

[42] European Union Agency for Network & Information Security. 2018. Recommendations on shaping technology according to GDPR provision. An overview on data pseudonymization.

[43] Regulation (EU) 2018/1807 of the European Parliament and of the Council of 14 November 2018 on a framework for the free flow of non-personal data in the European Union, PE/53/2018/REV/1, O.J. L. 303, 28.11.2018.

[44] Patrick Breyer v. Bundesrepublik Deutschland (C- 582/14), par. 46.

[45] Art. 2 and Recital 9 of the Regulation 2018/1807.

[46] Art. 9(1) GDPR.

is a broad category "related to the physical or mental health of a natural person, including the provision of health care services, which reveal information about his or her health status" (Bincoletto 2021).[47] The GDPR generally prohibits the processing of particular categories of personal data, unless one of the specific exceptions provided in Art. 9(2) GDPR applies. A scientific project processing personal health data uses both common and particular data. Thus, the researcher should define lawful legal grounds for both categories.[48]

Since the GDPR leaves space for Member states to further regulate the processing of data concerning health,[49] the researcher should have knowledge of the applicable rules at national level. The Member States' lawful conditions for scientific research may differ (Hansen et al. 2021). When a health project is conducted in multiple countries, the European Data Protection Board (EDPB) recommends using the same legal basis, unless national law provides differently.[50]

As regard the legitimate grounds for carrying out the processing activities, a distinction is necessary for the basis to collect the data and process it for scientific purposes, and the ground to share it, eventually cross-borders.

For the first task, the researcher may choose the data subject's explicit consent,[51] which may be specific, freely given and unambiguous (Chen et al. 2022), or directly the so-called "research exception" when the processing is necessary for scientific purposes, and it is based on Union or Member State law that provides for "suitable and specific measures to safeguard the fundamental rights and the interests of the data subject".[52] The consent is the special legal basis for all the clinical trials conducted in the EU according to Article 28 of the Regulation (EU) No 536/2014.[53] The European Data Protection Supervisor (EDPS) specified that Member State law generally requires prior consent from the participant in a research project processing health data and that the "research exception" is operational only when a specific Union or Member States law has been adopted to include conditions and safeguards.[54]

Adopting a comparative perspective, it may be noted that some Member States requires the consent as the condition for data processing under the "research exception" (e.g., Italy, Ireland, the Netherlands), while others do not (e.g., Germany, France, Estonia). When the identified basis is the consent, it should be emphasized

[47] Art. 4(15) GDPR.

[48] Art. 6 and 9.

[49] Art. 9(4) GDPR.

[50] European Data Protection Board (EDPB). 2021. Document on response to the request from the European Commission for clarification on the consistent application of the GDPR, focusing on health research, p. 5

[51] Art. 9(2)(a) GDPR.

[52] Art. 9(2)(j) GDPR.

[53] Regulation (EU) No 536/2014 of the European Parliament and of the Council of 16 April 2014 on clinical trials on medicinal products for human use, and repealing Directive 2001/20/EC, O.J. L. 158, 27.5.2014. In this context it is relevant the "Opinion 3/2019 concerning the Questions and Answers on the interplay between the Clinical Trials Regulation (CTR) and the General Data Protection regulation (GDPR) (art. 70.1.b))", adopted on 23 January 2019 by the European Data Protection Board.

[54] European Data Protection Supervisor. 2020. Preliminary Opinion on data protection and scientific research, adopted on 6 January 2020, pp. 14, 17.

that data subject also has the right to withdraw it anytime, and forms of dynamic consent are frequently used by researchers to open the uses of the collected data (Mostert et al. 2018). The nature of the consent may be flexible, but not broad.[55] This approach is coherent with the analyzed policy recommendations of the UN and the CoE. In some cases (e.g., post market studies after a clinical trial), where a Union or Member State law provides specific conditions, the research may be carried out under the public health interest exception[56] or under the substantial public interest one[57] (Comandè and Schneider 2022a).[58]

In the case of data sharing, the legal basis varies if the transmission or availability of data is carried out inside the EU or outside the European Economic Area (EEA), meaning to a third country. The main legal basis may even cover the transmission to researchers operating in other Member States (e.g., in the case of the consent), but national rules should be verified. When the activity occurs outside the EEA, the rules of Chapter V of the GDPR applies instead. The protection of data should be appropriate and adequate, which is a requirement similar to the recommendations detailed in the international law analyzed above.

The difference in protection between legal frameworks has been demonstrated by the Court of Justice of the EU in landmark cases.[59] Graziadei pointed out that the CJEU's jurisprudence on international data transfers between the EU and the US is a "wonderful illustration of the increasing demand for the coordination of regulatory regimes at the international level and of how comparative law becomes relevant" (Graziadei 2020). Thus, the researcher should define a specific basis for the data transfer and ensure that the level of protection is not undermined. Various transfer mechanisms are conceivable (e.g., adequacy decision of the Commission, standard contractual clauses), but they cannot be covered in full detail by this chapter. The European Data Protection Board (EDPB) has released the useful "Guidelines 07/2022 on certification as a tool for transfer", the "Guidelines 05/2021 on the Interplay between the application of Article 3 and the provisions on international transfers as per Chapter V of the GDPR" and the "Recommendations 01/2020 on measures

[55] See Section 2. Recital 33 of the GDPR supports such an approach, where it states that "data subjects should be allowed to give their consent to certain areas of scientific research when in keeping with recognized ethical standards for scientific research". The possibility of a flexibility does not imply that a generic consent for "every research purpose" would be lawful, but the researcher can describe a scientific purpose "in more general terms and for specific stages of a research project that are already known to take place at the outset". Such crucial clarification comes from the "Document on response to the request from the European Commission for clarification on the consistent application of the GDPR, focusing on health research" of the EDPB, of 2 February 2021, which provides some guidance in this context.

[56] Art. 9(2)(i) GDPR.

[57] Art. 9(2)(g) GDPR.

[58] As an example of the last exception, the Italian Government has recently proposed a rule to legitimatize scientific research in the medical filed that aims at developing artificial intelligence system without the consent of the data subject. See Disegno di legge "Disposizioni e delega al Governo in materia di intelligenza artificiale", proposed by the Government on 23 April 2024.

[59] Above all see the so-called "Schrems" cases regarding the U.S. legal framework: Maximillian Schrems v. Data Protection Commissioner (C362/14) and Data Protection Commissioner v. Facebook Ireland Ltd, Maximillian Schrems (C-311/18).

that supplement transfer tools to ensure compliance with the EU level of protection of personal data". Later, the Commission adopted an implementing decision on standard contractual clauses for the transfers to third countries.[60] However, as argued by Zeno Zencovich, it is difficult to find more than few countries, among members of the UN, that pass the criterion of the "adequacy of the level of protection" (Zeno Zencovich 2022). Anyway, the data subject should be informed about the transfers.[61] This last requirement is also comparable to the UN's recommendations.

The third issue is related to the data purpose limitation principle, which is considered as a "cornerstone of data protection and a prerequisite for most other fundamental requirements" (De Terwangne 2020). It is one of the general principles of the GDPR (Article 5). Under this principle, personal data must be processed for specified and legitimate purposes and any further processing is lawful only for purposes that are compatible with the original ones. The data processing for a scientific purpose is *a priori* compatible with a primary purpose (the so-called "presumption of compatibility").[62] This legislative choice enhances secondary use of personal data for scientific research, but such possibility is not to be taken for granted: a specific, explicit, and legitimate secondary purpose is still necessary for the secondary use of personal data to be lawful. Scientific research purpose may correspond to the primary use of the collected data (i.e., data is directly collected for the specific study), or it may correspond to the secondary use (i.e., data is further processed after being initially collected for another purpose), by the controller or, possibly, a different entity. In each case, data processing activities, including storing, accessing and sharing, must be lawful (i.e., lawfulness principle) and comply with the all the other requirements of the GDPR.

After all, the fourth issue is the application of technical and organizational measures to safeguards data in every phase of the research, including data sharing. Personal data should be adequate, relevant, and limited to what is necessary in relation to the purpose of the activities (principle of data minimization).[63] This necessity requirement refers to the quantity and the quality of personal data collected (De Terwangne 2020). Achieving data minimization in a scientific research context is problematic, as well as limiting the period of data storage (storage limitation principle).[64] Under the preferential regime established in Article 89 of the GDPR, some obligations of the data controller are limited, but appropriate safeguards should always be put in place (Ducato 2020; Schneider 2020).

The researchers should follow the integrity, confidentiality and accountability principles. Carrying out a Data Protection Impact Assessment (DPIA) on a case-by-case basis is the key to map the processing, its risks and verify the measures and

[60] Commission Implementing Decision (EU) 2021/914 of 4 June 2021 on standard contractual clauses for the transfer of personal data to third countries pursuant to Regulation (EU) 2016/679 of the European Parliament and of the Council (Text with EEA relevance), C/2021/3972, O.J. L. 199, 7.6.2021.

[61] Artt. 13 and 14 GDPR.

[62] Art. 5(b) GDPR.

[63] Art. 5(1)(c) GDPR.

[64] Art. 5(1)(e) GDPR.

safeguards in place to minimize them (Mantelero 2019).[65] State-of-the-art security (e.g., pseudonymization, encryption, authentication and authorization mechanisms) and data protection by design technical and organizational measures are indispensable to protect the personal data at rest, in use and in transit.[66] In addition to the technical measures, specific documentation describing the concrete characteristics of the processing activities is advisable: the research project (in accordance with ethical standards), the information notice on the study, the privacy policy, confidentiality agreements with internal staff, data sharing/transfer agreements with external entities. The WHO's model of data protection plan may also be useful. When multiple entities are involved in the project, the allocation of duties and implementation of technical and organizational measures should be defined in advance.

After the analysis of the data protection issues at EU level, it may be advised to the researchers to map and define: (i) all the applicable data privacy rules at EU and national levels, including sector-specific regulations applying to health data sharing; (ii) the categories of data (Personal data? Which categories? Is it possible, effective, and useful anonymise?); (iii) data flows and activities within the processing, including sharing (Where the data are collected, used, and shared? Inside or outside the EEA?); (iv) the legal grounds for every activity and the limits of the purpose of the processing (What is the legal basis for data collection? For the secondary use? For data sharing? Is sharing compatible with the initial purpose?).

If all these complex aspects are taken into account by the researchers, personal health data will be protected. Data protection law seems then to pose a barrier to use and share information. However, Comandè and Schneider noted that the EU legal framework "does not hamper but rather encourages data-driven research" (Comandè and Schneider 2022a). More specifically, a balance with data sharing can be struck. As anticipated, access to data is necessary to innovation and knowledge in the health sector. Pursuing this goal, the following section analyses the current EU framework on Open Research Data and the relevant new EU data law, focusing on its relevance for the the medical and biomedical research.

19.5 Opening Research Health Data in the European Union: from the Open Data Directive to the European Health Data Space

Since the early 2000s, the Open Data aspires to reduce economic, technological, and legal barriers to accessing to information. The term Open Data refers to data that is freely usable and modifiable for everyone and for every purpose. The EU policies have been firstly focused on releasing public sector information (Borgesius et al. 2015; Richter 2018). Then, the EU started to promote free sharing and free re-use of the data collected by scientific research. According to Article 179 TFEU, scientific knowledge should circulate freely.

[65] Artt. 35 e 36 GDPR.

[66] Artt. 32 and 25 GDPR.

In 2009, the Open Access Infrastructure for Research in Europe (OpenAIRE) project was launched to create an infrastructure able to share information between research data repositories.[67] The European Open Science Cloud (EOSC) was later promoted by European institutions for the data exchange across borders and scientific disciplines.[68] This notable initiative beneffited researchers, universities and research centres at the European level. Other examples of European databases and initiatives that facilitate data re-use in the medical sector are the European Cancer Information System, the European Platform on Rare Disease Registration, the Data analysis and Real World Interrogation Network (DARWIN EU).[69]

The definition of "research data" is originally provided by the Directive (EU) 2019/1024 on Open Data and the re-use of public sector information - "Open Data Directive" (ODD), which includes scientific research as a relevant sector.[70] The reference to this particular sector has been considered as a "change of direction" of the EU approach on Open Data (Caso 2022). Under this framework, research data can be defined as "documents in a digital form, other than scientific publications, which are collected or produced in the course of scientific research activities and are used as evidence in the research process, or are commonly accepted in the research community as necessary to validate research findings and results".[71]

The Open Data Directive mandates the EU Member states and bodies that fund and conduct public scientific research to adopt national policies for data sharing and re-use of research data. According to Article 10(1) Member states should support the availability of research data by "adopting national policies and relevant actions aiming at making publicly funded research data openly available ("open access policies"), following the principle of "open by default" and compatible with the FAIR principles". To achieve this goal, Member States should take into account "concerns relating to intellectual property rights, personal data protection and confidentiality, security and legitimate commercial interests" in accordance with the principle of "as open as possible, as closed as necessary". Here is included the principle mentioned at the beginning of this chapter. National policies will be directed to research performing organizations and research funding organizations.

It has been argued that the provisions of the ODD enhance the role of research data in the data economy and the Digital Single Market (Arisi 2022). Moreover, Article 10 pushes forward openness "by design and by default" and, when possible, data is made available under open licenses (e.g., Creative Commons 4.0) and should follow the FAIR principles. Pre-existing limits to openness remain, including IP rights (of third parties involved in the research), patents, trademarks, registered designs, *sui generis* rights on the databases, privacy and data protection, confidentiality, national security

---

[67] https://www.openaire.eu/.

[68] https://open-science-cloud.ec.europa.eu/.

[69] Information are available at https://ecis.jrc.ec.europa.eu/, https://eu-rd-platform.jrc.ec.europa.eu/_en and https://www.ema.europa.eu/en/about-us/how-we-work/big-data/data-analysis-real-world-interrogation-network-darwin-eu#interaction-with-the-european-health-data-space-section.

[70] Directive (EU) 2019/1024 of the European Parliament and of the Council of 20 June 2019 on open data and the re-use of public sector information, O.J. L. 172/56, 26.6.2019.

[71] Art. 2 ODD.

and legitimate commercial interests (e.g., trade secrets). Once again data protection law works as limiting open data.[72] This legislative choice considerably restricts the use and re-use of research data. The Open Data Directive explicitly states that it abides to the GDPR,[73] while it also remains limited to the public sector information and it requires national implementation, which may hamper harmonization.

In the context of research publicly funded by the EU, current rules promote openness both for science and access. The Regulation (EU) 2021/695 establishing Horizon Europe defines Open Science in the EU as "an approach to the scientific process based on open cooperative work, tools and diffusing knowledge", and Open Access as "online access, provided free of charge to the end user, to research outputs".[74] Article 14 of this Regulation specifies that Open Science implies both the "access to scientific publications" and the "access to research data, including those underlying scientific publications, in accordance with the principle "as open as possible, as closed as necessary". The general principle for research data is openness, but exceptions are possible. These exceptions regard the "legitimate interests" of the researchers that are beneficiaries, meaning "commercial exploitation and any other constraints, such as data protection rules, privacy, confidentiality, trade secrets, Union competitive interests, security rules or intellectual property rights".[75] Limits to openness are undoubtedly based on existing data protection and intellectual property law, but also on other commercial interests. This is coherent with what is illustrated for ORD at international level and for the ODD.

New regulations proposed and adopted under the "Data Strategy" of the European Commission of 2020 aim at opening data sharing and access for research purposes.[76] The free flow of data in the Digital Single Market was identified as a priority. Resta specified that the new European "data law" should be studied in connection with the creation of new infrastructures at EU level and is crucial in terms of both industrial and social policies of the Union (Resta 2022).

Firstly, the Regulation (EU) 2022/868 or "Data Governance Act" (DGA) adds conditions for the re-use of certain categories of data held by public sector bodies, including personal data, except for the categories of data regulated by the Open Data Directive, which fall outside the scope of the DGA.[77] It aspires to make available data in accordance with the principles of "open by design and by default" and of

[72] See also Art. 1(2)(h) of the ODD that excludes from application other documents. Member states are not mandated to adopt rules related to documents which access is limited at national law level to protect personal data.

[73] Art. 1(4) ODD.

[74] Art. 2(5) and (4), Regulation (EU) 2021/695 of the European Parliament and of the Council of 28 April 2021 establishing Horizon Europe - the Framework Programme for Research and Innovation, laying down its rules for participation and dissemination, and repealing Regulations (EU) No 1290/2013 and (EU) No 1291/2013, O.J. L. 170, 12.5.2021.

[75] Art. 39 Regulation 2021/695.

[76] Communication from the Commission to the European Parliament, the Council, the European Economic and Social Committee and the Committee of the Regions. 2020 A European strategy for data. COM/2020/66 final.

[77] Regulation (EU) 2022/868 of the European Parliament and of the Council of 30 May 2022 on European data governance and amending Regulation (EU) 2018/1724 (Data Governance Act), O.J. L. 152, 3.6.2022.

"as open as possible, as closed as necessary" of the ODD.[78] In this sense, the DGA complements the ODD. The DGA broadly defines data sharing as "the provision of data by a data subject or a data holder to a data user for the purpose of the joint or individual use of such data, based on voluntary agreements or Union or national law, directly or through an intermediary, for example under open or commercial licences subject to a fee or free of charge".[79]

As regards the interplay between the DGA and scientific research, the DGA introduces the new concept of "data altruism", which means: "the voluntary sharing of data on the basis of the consent of data subjects to process personal data pertaining to them, or permissions of data holders to allow the use of their non-personal data without seeking or receiving a reward that goes beyond compensation related to the costs that they incur where they make their data available for objectives of general interest as provided for in national law".[80] Scientific research purposes in general interest are examples of objectives. Data sharing relies on a very complex infrastructure and governance framework based on "data altruism organizations", intermediaries and cooperatives (Poletti 2023; Comandè and Schneider 2022b). Data will be made available by not-for-profit organizations, which should be trustworthy and will play a crucial role in the data economy (Carovano and Fink 2023). Data transfers to countries outside the EU should be prevented but they are allowed if based on an international agreement or other specific instruments.[81]

Under the definition of data altruism, research should pursue a general interest. This may be the starting point for data sharing. However, the new concept is very broad, and the researcher may find interpretation difficult. Future EU or national law, or new policies released by competent public entities, who grant data access permits, should delimitate the contours of general and private interests in the research context. The lack of a clear notion of "general interest" in the DGA may lead to interpretation issues and thus legal uncertainly (Baloup et al. 2021). Moreover, the use of the term "general" may lead to confusion in implementation, since the GDPR instead refers to "public interest" (Shabani and Yilmaz 2022). Questions remain regarding whether privately funded research that pursues a public interest is included or not under the objectives of data altruism. In the medical and biomedical context, the French approach for the Plateforme des données de santé ("Health Data Space") and the Finnish framework for the secondary use of health and social data prove important benchmarks (Balint and Buki 2022).

Scholars already noted that the notion of data altruism seems opening the consent of data subject to a "broad consensus for research purposes" (Comandè Schneider 2022a). Nonetheless, in the event of conflict between the DGA and the GDPR, the data protection framework should prevail.[82] When applicable, the necessity of an explicit consent for processing data concerning health for research purposes, therefore, remains. Indeed, Article 25 of the Data Governance Act provides that the

[78] Recitals 9 and 16 DGA.
[79] Art. 2(10) DGA.
[80] Art. 2(16) DGA.
[81] Art. 31 DGA.
[82] Recital 4 of the Data Governance Act.

data altruism consent form should ensure that the data subject is able to give consent to and withdraw consent from a specific data processing. The Commission will adopt a European form to facilitate the collection based on data altruism. In their opinion on the DGA proposal, the EDPB and the EDPS stressed that the GDPR should still apply when the data subject has given consent to the data altruism organization.[83] As a result, the limits of the interpretation of the notion of consent in the scientific context, both for international and EU law, seems still applicable.

In the interplay between the GDPR and the DGA there may also be a lack of semantic consistency. Although the DGA scope is limited to public sector data, the definitions of Article 2 of "data holder" (i.e., who has the right to grant access to or to share certain personal data or non-personal data) and "data user" (i.e., who has lawful access to certain personal or non-personal data and has the right to use that data for commercial or non-commercial purposes) should be coordinated with the data processing roles and principles of the GDPR (e.g., of data subject, data controller, data processor and recipient). This inconsistency might lead to difficulties of practical application: the actors on data sharing and their roles under data protection law should always be clear. It has been noted that the language of the DGA seems more aligned to a theorical framework considering data as an object in the market rather than the traditional approach of considering data as an aspect of personality (Bravo 2021).

The Regulation 2023/2854 - "Data Act" (DA), which will apply from 12 September 2025, is another central initiative to regulate the data economy.[84] It seeks to enhance the interoperability of data and sharing mechanisms and services in the EU.[85] The Data Act concerns several categories of data sharing, including business-to-consumer and business-to-business, business-to-government/public sector body exchanges. Nevertheless, scholars already argued that in the framework there is a lack of a "sound analysis of data as a public asset potentially able to provide the foundations to build an Open Data economy" (Casolari et al. 2023). Provisions able to standardize data sharing agreements are missing. Data should not be opened by default. However, the regulation introduces the concept of "open interoperability" specification, which means "technical specification in the field of information and communication technologies which is performance oriented towards achieving interoperability between data processing services".[86] As regards data sharing for scientific research, the Data Act regulates how a public sector body, the Commission, the European Central Bank or another Union body can request data to be made available to not-for-profit research organizations or individuals carrying out research activities.[87]

83 European Data Protection Board and European Data Protection Supervisor. 2021. The EDPB-EDPS Joint Opinion 03/2021 on the Proposal for a regulation of the European Parliament and of the Council on European data governance (Data Governance Act), par. 161.

84 Regulation (EU) 2023/2854 of the European Parliament and of the Council of 13 December 2023 on harmonized rules on fair access to and use of data and amending Regulation (EU) 2017/2394 and Directive (EU) 2020/1828 (Data Act). O.J. L. 2023/2854. 22.12.2023.

85 Recital 5 DA.

86 Art. 2(41) DA.

87 Art. 21.

With the Regulation on the European Health Data Space (EHDS) 2025/327 the EU created a health specific ecosystem.[88] This is the first regulation of a domain-specific common European space (Wenkai and Quinn 2024; Bincoletto 2022). The EHDS aim at supporting both individuals to have more control over their data and the general public, while fostering healthcare delivery, research, innovation and policy making. The EHDS Regulation provides rules, common standards and practices, infrastructures, and a governance framework on health data, both personal and not personal, for their use and re-use, including for research purposes.

The EHDS Regulation includes provisions applicable to the data sharing inside the EU. A "health data holder" can make data already collected for a primary purpose available for secondary research purposes related to healthcare.[89] Article 53 provides a specification on these purposes. The research should contribute to "public health or health technology assessment" or it should ensure "high levels of quality and safety of health care, of medicinal products or of medical devices, with the aim of benefitting the end-users, such as patients, health professionals and health administrators". This includes both the development and innovation activities for products or services and the training, testing, and evaluating of algorithms (e.g., in medical devices, *in vitro* diagnostic medical devices, AI systems and digital health applications). Health data access bodies appointed at national level oversee the access to data.[90] These entities receive data application requests and, when various conditions are met, they will grant data permits, meaning administrate decisions issued to a specific "health data user". The same bodies will be responsible for several obligations, including informing the data subjects on the legal basis under which the data access is granted, the measures implemented to protect their rights, the arrangements to exercise data protection rights and the results or outcomes of the projects for which the health data were processed.

Similarly the recommendations presented in the context of international data protection law, the EHDS Regulation requires the implementation of the data minimization and purpose limitation principles. By default, health data access bodies will grant access to the data in an anonymized format. Only if the health data user has sufficiently demonstrated that anonymized data is not sufficient for the specific purpose, the access body will provide data in a pseudonymized format. The FAIR principles are quoted in the recitals of the regulation where it is also specified that the new data space should be coordinated with the EOSC.

During the negotiations between the EU Parliament and the Council and in the "EDPB-EDPS Joint Opinion 03/2022 on the Proposal for a Regulation on the European Health Data Space"[91] the issue of identifying the legal basis for the secondary use of health data for research purposes emerged. Before the adoption

---

88 Regulation (EU) 2025/327 of the European Parliament and of the Council of 11 Feburary 2025 on the European Health Data Space and amending Directive 2011/24/EU and Regulation (EU) 2024/2847. O.J. L. 2025/327, 5.3.2025.

89 Chapter IV EHDS Regulation.

90 Art. 55 EHDS Regulation.

91 European Data Protection Board and European Data Protection Supervisor. 2022. EDPB-EDPS Joint Opinion 03/2022 on the Proposal for a Regulation on the European Health Data Space.

of the Regulation, scholars presumed that the ground was the compliance with a legal obligation or the existence of a public interest in combination with the GDPR exceptions and bases (Shabani and Yilmaz 2022). The draft report of the European Parliament on the Proposal of 10 February 2023, specified that allowing a data processing without the prior consent of the data subject needed a "significant shift in data protection law" as it would have created "an important precedent for further legal acts on secondary data use".[92]

The Parliament then proposed the established of a right to opt-out for some or all of the purposes of secondary use in order to ensure the right to object provided under Article 21(6) of the GDPR. As a result, the EHDS Regulation includes a new "right to opt-out from the processing of personal electronic health data for secondary use" (Art. 71), which should be further implemented by national law. Even when the opt-out has been exercized by the data subject (and reasons are not required), Member states law may establish mechanisms to make data available for "scientific research for important reasons of public interest". The EHDS Regulation even mentions the concept of "data altruism" but mandates its future implementation in this context to additional criteria to be established for being coherent with the Data Governance Act. Furthemore, the national implementation of the right of opt-out by Member states has the potential to result in additional fragmentation of the discipline. This might result in a loss of clarity for researchers and necessitate further comparative research.

It may be argued that the EHDS Regulation will open health data (both personal and non-personal), but it does not deploy the notions of Open Data or Open Research Data. In fact, data will be opened to a specific secondary use on the basis of an administrative authorization and the scope of the sharing goes beyond the data held by the public sector, covering the private one also. Recital 17 uses the wording of "selective sharing", which is not "open sharing". Additionally, under the EHDS, data sharing does not operate outside the EU.

Given these preliminary conclusions and looking at the detailed set of rules on EU data protection law and open research, the next session will discuss the interplay between these two frameworks to investigate the way forward health data sharing, even in transnational scientific research.

19.6 The Way Forward A Health Data Sharing that is "As Open As Possible, As Closed As Necessary"

Zeno Zencovich, while investigating the existence of a data protection(ism) in the EU, pointed out that the pervasive data protection law indirectly creates a "data localization principle" (Zeno Zencovich 2022). Data generated in the EU remains within it. An unconditional free and open flow of personal health data for scientific research purposes, appears not compatible with the rules of the GDPR. The legislation on Open Data and Open Research Data is without prejudice to data protection requirements. This safeguard exists for several reasons.

[92] Draft Report on the proposal for a regulation of the European Parliament and of the Council on the European Health Data Space of the Committee on the Environment, Public Health and Food Safety Committee on Civil Liberties, Justice and Home Affairs at the European Parliament. 2023.

Firstly, a strict application of the principle of purpose limitation contrasts with the openness of the uses in an Open Research Data context. This represents a challenge since purpose limitation is a cornerstone in data protection law (Suman, Pierce 2018). At the same time, as already argued, the processing for scientific purposes is considered *a priori* compatible with a primary purpose by the GDPR and this mitigates the application of the principle. The purpose of processing should be specific, explicit, and legitimate. Open sharing has instead broader purposes.

Secondly, the application of the lawfulness principle, which requires the existence of a specific legal ground for every activity, seems not compatible with the different approach of opening data for every use and everywhere. While a full open data approach stresses the ability to use and re-use without conditions (i.e., unlimited openness), the GDPR requires an exception to process data for research purposes, which is frequently the data subject's consent. The need for additional and specific mechanisms when the sharing is operated outside the EEA is even more challenging. Therefore, personal data may be further processed for a secondary research purpose only on a lawful legal basis. A new consent may be necessary before releasing data as ORD (Guarda 2020). When the legal ground is the consent of the data subject, the possibility of the withdrawal, which is always granted by data protection law, is not conceivable in the Open framework, where data are shared online with uncountable users. However, in the research context the withdrawal often does not operate retrospectively in order to protect the goal of the processing and allow the data already collected.

Thirdly, the limitation of the collection and processing grounded on the data minimization and storage limitation principles may be not compatible with the needs of the use and re-use of as much information as possible. The OS and ORD recommendations and policies mention these principles, but their implementation is inevitably mitigated in a research context.

Fourthly, the compliance with the accountability principle, which requires an effective control and responsibility over data that in the open scenario disappears or at least is weakened. Releasing opened data should imply that information can be re-used for any purpose and multiple times. Thus, the data controller has less control over the collected data. The data controller would not be able to implement organizational and technical measures if data is disseminated. Clear conditions for sharing open research data may alleviate this issue.

These incisive words can be shared: "it is not a question of choosing between privacy and data sharing, but of determining which privacy (and more generally also security, transparency and trustworthiness) requirements are necessary to ensure an efficient degree of data sharing in medical research" (Ienca 2023). Thus, it seems possible to achieve two solutions of compromise: anonymization of personal health data for research purposes to exclude the application of the data protection law, or opening data only for specific research purposes authorized time-to-time under terms and conditions but in a way that is FAIR: a selective sharing. Using the Open Access and Open Data frameworks as a reference,[93] the second solution is not full

[93] https://www.eurogct.org/research-pathways/public-involvement-and-data/data-sharing-open-data.

open access of research data (e.g., green or gold OA for scientific publications) but controlled access to FAIR data with as few restrictions as possible.

When it proves feasible as well as adequate to the scientific research purpose, data could be processed in anonymized form to avoid the application of data protection law. The data controller anonymizes the data and then releases the dataset in an open repository (e.g., Open Science Framework – OSF, Zenodo, Science Data Bank).[94] Even the DGA and the EHDS Regulation encourage the application of anonymization techniques to share datasets. International recommendations suggest the same approach. Nevertheless, when choosing anonymization, there are perplexities related to the techniques applicable, since the data controller assumes the responsibility of their effectiveness in making de-identification irreversible. The processing of data rendered anonymized and opened could still present residual risks.

In addition, research often cannot resort to anonymized data. In the medical context, the more information the research has on a single subject, the more accurate the treatment responses and patterns analyzed will be, and the more useful the analysis will prove for replication, so pseudonymization is more common than anonymization. Besides, from a comparative law perspective, the concept of anonymization may differ among countries. As an example, the US Health Insurance Portability and Accountability Act (HIPAA) considers the removal of eighteen identifiers sufficient.[95] The EU law has a higher standard. So, choosing the techniques for the transnational research context depends on where the source of the data is.

A second option is creating the conditions to open a specific dataset for a particular secondary scientific research purpose while still protecting personal data. There is no "one-size-fits-all" solution that is valid for every dataset: data sharing should be managed adopting a data protection by design approach (Bincoletto 2021) and following the FAIR principles. Data sharing agreements between the former data controller and the research entity may be envisaged, in the context of a research that has a public interest especially. Data management plans[96] on the data lifecycle and data protection impact assessments[97] on the risks are also necessary. Transparency on the data sharing in the privacy policies is recommended both by the international and the EU law analyzed above.

The notion of data altruism and the creation of the EHDS have great potentiality in this context, but the new rules should be efficiently coordinated with the GDPR, and the provisions already introduced in the DGA and ultimately all the EU data laws. It has been argued that the altruism mechanism requires prior definition of its interplay with the consent of the GDPR and the notion of research purposes in the general interest (Shabani 2022).

---

[94] See also the Registry of Research Data Repositories, available at https://www.re3data.org/.

[95] Health Insurance Portability and Accountability Act of 1996 (HIPAA), 110 Stat. 1936 (1996); 45 USC § 1320d-2(b).

[96] Useful models are provided by the EU under the Horizon Europe program at https://webgate.ec.europa.eu/funding-tenders-opportunities/display/OM/Online+Manual-

[97] Detailed guidance is provided by the French supervisory authority, Commission Nationale de l'Informatique et des Libertés, CNIL. See upfront Privacy Impact Assessment (PIA). Templates. 2018 and the open-source software for carrying out the DPIA called "PIA".

More clarity on the meaning of data altruism (Resta 2022) and also on the new "opt-out" option in the research sector is therefore required. It may also be suggested that the altruistic purposes detailed in the consent form may be limited to specific areas of medical or biomedical research that pursue public interests, or that consent may be given to a series of scientific projects that have a similar general purpose.

Creating or using technical intermediaries and infrastructures able to contact the data subject to ask for an explicit consent is another possible solution, which needs a robust oversight by designated public entities. Whether the legal ground for data sharing is a legal obligation or a general or public interest according to a provision of the EHDS Regulation or to a Member state law, further safeguards and organizational and technical measures should protect personal health data. National law may even create new conditions applicable to the scientific research that uses personal health data, as already provided in the past (Slokenberga 2022). The conditions for data sharing for scientific research projects "in the general interest" should be precise and all the possible inconsistencies between "data frameworks" should be avoided.

Vogel pointed out the limits of the GDPR as a solid foundation for the new data legislation of the EU (Vogel 2022). This scholar argued that it is increasingly more difficult to ensure a full congruency in data law. The need for adaptation of the GDPR, and eventually modification of its rules, seems conceivable to ensure that the protection of data is not merely on paper but in practice.

Besides, not only data protection law closes data when it is necessary. IP law is relevant. The absence of open infrastructures and interoperability issues impinge upon access and sharing.

Lastly, looking at the EU "data regulations" many authorities and institutions oversees data sharing: the supervisory authorities under the GDPR, the EDPB, the EDPS, national bodies under the ODD, the European Data Innovation Board of the DGA (with also representatives from the previous authorities), the European Health Data Space Board. Member states law will further regulate data sharing where the new data laws have left space to adapt the application of the rules at a national level. They all should promote consistent application of the data sharing rules and policies.

19.7 Conclusions

This chapter attempts to show the tensions between data protection law and the policies of opening science and research data. Health data sharing for scientific research shall comply with the data protection framework to safeguard fundamental rights and freedoms of the data subjects but opening data is indispensable for scientific progress, can benefit society in terms of an improved public health, and it is highly promoted at global level with policies and recommendations.

In the EU, the GDPR establishes high standards over data protection. The implementation of purpose limitation principle (pre-defining the purpose of the processing), the lawfulness requirements (having a lawful legal ground for every activity, including sharing) and accountability (overall protection and implementation of appropriate and effective safeguards, possibly by design and by default) is complex, even before data sharing. Understanding whether it is possible to open the data covers both the case where data are shared within the EU as well as to a third country. The key elements of data governance for health data sharing emerged by the analysis of international and EU requirements and policies may be summarized as follows: being transparent with the data subjects on the data circulation, finding a lawful basis for the transmission, ensuring an appropriate level of data protection while sharing take places cross-borders, applying the FAIR principles and improving interoperability.

Emerging new strategies and legal rules suggest that openness of research data will increase in the near future. The notions of data altruism of the DGA will play a pivotal role. One of the main challenges will be its alignment with the GDPR and the relevant horizontal and sectoral EU and national legislation. The creation of the EHDS may boost data access and sharing in the health domain, and it will have a great impact on scientific research within the EU. The possibility that national law further regulates data concerning health and the processing for scientific research purposes may also be obstacles for harmonization and for the practical development of the EHDS.

Regulating accessibility and openness requires a reasonable approach that takes into account the existing legal requirements. Besides anonymization, selective sharing allows data management practices that open the collected personal health data for specific scientific projects to entities different from the data controller or holder. Currently, *de lege data* (without amendments to the GDPR), personal data does not seem to be opened by default, but it may only be accessible as much as possible.

Transnational data sharing is not specifically promoted by the new EU data law. The creation of new multicentric safe hubs for specific medical and biomedical purposes may be a solution. The UN and CoE policies, as well as the OS and ORD movements at international level, may contribute to this end.

Finally, it is of paramount importance to reiterate the necessity of coordination between all legislation and implementation policies, between the guidelines of the relevant authorities and harmonization between countries. A comparison of the national legal frameworks will be inevitable.

Acknowledgements

The author of this publication is financed by the European Union – Next Generation EU, under the Call for tender PRIN 2022, Project "Clinical trial data between privatization of knowledge and Open Science (acronym: CLIPKOS)" – (2022K4HBFA) – CUP E53D23006760006.

References

Arisi, M. 2022. Open Knowledge. Access and Re-use of Research Data in the European Union Open Data Directive and the Implementation in Italy. Italian Law Journal 1: 33–73.

Balint, F. and B. Buki 2022. Three Ways of Secure Data Reusability in Europe: German Research Data Centres, Finnish Findata and the French Secure Access Data Centre. ELTE Law Journal 1: 81–108.

Baloup, J. et al. 2021. White Paper on the Data Governance Act, available at: SSRN: https://ssrn.com/abstract=3872703.

Bartling, S. and S. Friesike (eds.). 2014. Opening Science - The Evolving Guide on How the Internet is Changing Research, Collaboration and Scholarly Publishing. Springer, Cham.

Bauchner, H. et al. 2016. Data Sharing: An Ethical and Scientific Imperative. JAMA. 315(12): 1238–1240.

Bernier, A., F. Molnar-Gabor, M. Knoppers. 2022. The international data governance landscape. Journal of Law and the Biosciences 9(1): 1–45.

Berti, S.A. and R. Pierce. 2018. Challenges for Citizen Science and the EU Open Science Agenda under the GDPR. European Data Protection Law Review 4(4): 284–295.

Besançon, L. et al. 2021. Open science saves lives: lessons from the COVID-19 pandemic. BMC Medical Research Methodology 1: 1–18.

Bincoletto, G. 2021. Data Protection by Design in the E-Health Care Sector. Nomos Verlagsgesellschaft mbH & Co. KG, Baden-Baden.

Bincoletto, G. 2022. The EDPB-EDPS Joint Opinion on the Commission Proposal for a Regulation on the European Health Data Space: Key Issues to be Considered in the Legislative Process. European Data Protection Law Review 3: 398–404.

Borgesius, F.Z., J. Gray, M. Van Eechoud. 2015. Open Data, Privacy, and Fair Information Principles: Towards a Balancing Framework. Berkeley Technology Law Journal 30(3): 2073–2132.

Bravo, F. 2021. Intermediazione di dati personali e servizi di data sharing dal GDPR al Data Governance Act. Contratto e impresa Europa 1: 199–256.

Bussani, M. and U. Mattei (eds.). 2012. The Cambridge Companion to Comparative Law. Cambridge University Press, Cambridge.

Bygrave, L.A. 2014. Data privacy law: an international perspective. Oxford University Press, Oxford.

Carovano, G. and M. Finck. 2023. Regulating data intermediaries: The impact of the Data Governance Act on the EU's data economy. Computer Law & Security Review 50: 105830.

Caso, R. 2020. La rivoluzione incompiuta. La scienza aperta tra diritto d'autore e proprietà intellettuale. Ledizioni, Milano.

Caso, R. 2022. Open Data, ricerca scientifica e privatizzazione della conoscenza. Diritto dell'informazione e dell'informatica. 4/5: 815–836.

Caso, R., and F. Binda. 2020. Il diritto umano alla scienza aperta. pp. 44–52. *In*: Perrone, G. and M. Perduca (eds.). Così san tutt3 - Diritto alla Scienza, istruzioni per l'uso. Fandango Libri, Roma.

Casolari, F., L. Floridi and C. Buttaboni. 2023. The EU Data Act in Context: A Legal Assessment. International Journal Of Law And Information Technology 31(4): 399–412.

Chen, J., S. Dove and E. Bhakuni. 2022. Explicit consent and alternative data protection processing grounds for health research. pp. 474–502. *In*: Perrone, G. and M. Perduca (eds.). Research Handbook on EU Data Protection Law. Edward Elgar Publishing, Cheltenham.

Chesbrough, H. 2003. Open Innovation: The New Imperative for Creating and Profiting from Technology. Harvard Business School Press, Boston.

Colcelli, V. and R. Cippitani, R. 2023. Circulation of personal data and non-personal data within the European Research Area for research and health purposes. Journal of Open Access to Law, 11(2).

Colcelli, V., R. Cippitani, C. Brochhausen-Delius, R. Arnold (eds.). 2023. GDPR Requirements for Biobanking Activities Across Europe. Springer International Publishing, Cham.

Comandè, G. 2022 (eds.). Elgar Encyclopedia of Law and Data Science, Edward Elgar Publishing, Cheltenham.

Comandè, G. and G. Schneider. 2022a. Differential Data Protection Regimes in Data-Driven Research: Why the GDPR is More Research-Friendly Than You Think. German Law Journal 23(4): 559–596.

Comandè, G. and Schneider, G. 2022b. It's time: Leveraging the GDPR to shift the balance towards research-friendly EU data spaces, Common Market Law Review 59(3): 739–776.

Dalla Corte, L. and B. van Loenen. 2022. Open Data and Public Sector Information. pp. 241–253. *In* G. Comandè (eds.). Encyclopedia of Law and Data Science. Edward Elgar Publishing, Cheltenham.

David, P.A. 2008. The Historical Origins of "Open Science": An Essay on Patronage, Reputation and Common Agency Contracting in the Scientific Revolution. Capitalism and Society. 3(2), available at SSRN: https://ssrn.com/abstract=2209188/

de Hert, P. and V. Papakonstantinou. 2014. The Council of Europe Data Protection Convention reform: Analysis of the new text and critical comment on its global ambition. Computer Law & Security Review 30(4): 633–642

De Terwangne, C. 2020. Chapter II Principles (Articles 5–11). Article 5. Principles relating to processing of personal data. pp. 309–397. *In*: Kuner, C. et al. (eds.). The EU General Data Protection Regulation (GDPR): A Commentary. Oxford University Press, Oxford.
Ducato, R. 2020. Data protection, scientific research, and the role of information. Computer Law & Security Review 37: 105412.
Geiger, C. and B.J. Jütte. 2023. Conceptualizing a "Right to Research" and its Implications for Copyright Law: An International and European Perspective. American University International Law Review 38(1): 1–86.
Giglia, E. 2023. Open? The only way forward for science. Journal of Open Access to Law 11(1).
Gold, E.R. 2021. The fall of the innovation empire and its possible rise through open science. Research Policy 50(5): 104226.
Govarts, E. et al. 2022. Position paper on management of personal data in environment and health research in Europe. Environment International 165: 107334.
Graziadei, M. 2020. The European Court of Justice at Work: Comparative Law on Stage and Behind the Scenes. J. Civ. L. Stud. 13: 1–31.
Graziadei, M. 2021. Commento art. 2 GDPR. pp. 129–143. *In:* D'Orazio, R., G. Finocchiaro, O. Pollicino and G. Resta (eds.). Codice della privacy e data protection, Giuffré, Milano.
Graziadei, M. 2003. The Functionalist Heritage. pp. 100–131. *In:* Legrand, P. and R. Munday (eds.). Comparative Legal Studies: Traditions and Transitions. Cambridge University Press, Cambridge.
Guarda, P. 2020. Free data?: open science in the age of personal data protection. pp. 391–410. *In*: Rooksby, H.J. (ed.). Research Handbook on Intellectual Property and Technology Transfer. Edward Elgar Publishing, Cheltenham.
Guarda, P. 2021. Il regime giuridico dei dati della ricerca scientifica, Editoriale Scientifica, Napoli.
Guarda, P. and G. Bincoletto. 2023. Diritto comparato della privacy e della protezione dei dati personali. Ledizioni, Milano.
Hallinan, D. 2021. Protecting Genetic Privacy in Biobanking through Data Protection Law. Oxford University Press, Oxford.
Hansen, J. et al. 2021. Assessment of the EU Member States' rules on health data in the light of GDPR, Publication Office of the European Union, Luxembourg.
Hansson, M.G. 2021. Striking a Balance Between Personalised Genetics and Privacy Protection from the Perspective of GDPR. pp. 31–42. *In*: Slokenberga, S. et al. (eds.). GDPR and Biobanking. Springer, Cham.
Hesse, C. (ed.). 2007. Understanding Knowledge as a Commons. From Theory to Practice. The MIT Press, Cambridge (MA).
Hoffman, S. and J. Herveg. 2021. Privacy and Integry of medical information. *In*: Orentlicher, D. and T.K. Hervey (eds.). The Oxford Handbook of Comparative Health Law. Oxford University Press. Oxford.
Husa, J. 2023. A New Introduction to Comparative Law. Hart Publishing, Oxford.
Ienca, M. 2023. Medical data sharing and privacy: a false dichotomy? Swiss Med Wkly. 153: 40019.
Ioriatti, E. 2024. Comparative Law Method and the Legal Formants as Catalysts of Normative Realities. *In*: Siliquinicinelli, L. et al. (ed.). The Grand Strategy of Comparative Law: Themes, Methods, Developments. Routledge, Milton Park.
Kischel, U. 2019. Comparative Law. Oxford University Press, Oxford.
Kiseleva, A. and de Hert, P. 2021. Creating European Health Data Space: obstacles in four key legal area. European Pharmaceutical Law Review 5(1): 21–36.
Kuner, C., L.A. Bygrave and C. Docksey (eds.). 2020. The EU General Data Protection Regulation (GDPR): a commentary. Oxford University Press, Oxford.
Laurie, G. et al. (eds.). 2021. The Cambridge Handbook of Health Research Regulation. Cambridge University Press, Cambridge.
Madir, J. (ed.). 2020. Healthtech. Law and Regulation. Edward Elgar Publishing, Cheltenham.
Mantelero, A. 2019. Il nuovo approccio della valutazione del rischio nella sicurezza dei dati. Valutazione d'impatto e consultazione preventiva (Artt. 32–39). pp. 287–330. *In*: Finocchiaro, G. (ed.). Il nuovo Regolamento europeo sulla privacy e protezione dei dati personali. Zanichelli, Bologna.
Margoni, T., R. Caso, R. Ducato and P. Guarda. 2017. Open Access, Open Science, Open Society. pp. 75–86. *In*: Loizides, F. and B. Schmidt (eds.). Positioning and Power in Academic Publishing: Players, Agents and Agendas. Springer, Cham.

Merton, R. 1942. Science and Technology in a Democratic Order. Journal of Legal and Polotical Sociology. 1.
Monateri, P.G. 2022. Advanced Introduction to Comparative Legal Methods. Edward Elgar Publishing, Camberley Surrey.
Mostert, M. et al. 2018. From Privacy to Data Protection in the EU: Implications for Big Data Health Research. European Journal of Health Law 25: 43–55.
Narayanan, A. and V. Shmatikov. 2008. Robust De-anonymization of Large Sparse Datasets. pp. 111–125. *In*: The Institute of Electrical and Electronics Engineers (ed.). 2008 IEEE Symposium on Security and Privacy. IEEE Computer Society, Los Alamitos.
Ohm, P. 2010. Broken Promises of Privacy: Responding to the Surprising Failure of Anonymization. UCLA Law Review 57: 1701–1777.
Paseri, L. 2020. Access to Scientific Information and Knowledge: A Matter of Democracy. pp. 71–82. *In*: Feletig, P., A. Loreggia, A. Resca and S. Quintarelli (eds.). Proceedings of the First International Forum on Digital and Democracy. Towards A Sustainable Evolution 2020. CEUR Workshop Proceedings, Venezia.
Paseri, L., S. Varrette and P. Bouvry. 2021. Protection of Personal Data in High Performance Computing Platform for Scientific Research Purposes. pp. 123–142. *In*: Gruschka, N., L.F.C. Antunes, K. Rannenberg and P. Drogkaris (eds.). Privacy Technologies and Policy. APF 2021. Springer, Cham.
Podda, E. and M. Palmirani. 2021. Inferring the Meaning of Non-personal, Anonymized, and Anonymous Data. pp. 269–282. *In*: Rodrìguez-Doncel, V., M. Palmirani, M. Araszkiewicz, P. Casanovas, U. Pagallo and G. Sartor (eds.). AI Approaches to the Complexity of Legal Systems XI-XII. Springer, Cham.
Poletti, D. 2022. Gli intermediari dei dati. European Journal of Privacy Law and Technologies 1: 45–56.
Purtova, N. 2018. The law of everything. Broad concept of personal data and future of EU data protection law. Law, Innovation and Technology 1: 40–81.
Reinmann, M. and R. Zimmermann. 2019. The Oxford Handbook of Comparative Law. Oxford University Press, Oxford.
Resnik, D. 2023. Openness in Scientific Research: A Historical andPhilosophical Perspective. Journal of Open Access to Law 11(1).
Resta G., A. Somma and V. Zeno-Zencovich (eds.). 2020. Comparare. Una riflessione tra le discipline. Mimesis Edizioni, Milano-Udine.
Resta, G. 2022. Pubblico, privato e collettivo nel sistema europeo di governo dei dati. Rivista trimestrale di diritto pubblico 4: 971–996.
Richter, H. 2018. Open science and public sector information. Journal of Intellectual Property, Information Technology and Electronic Commerce Law 9(1): 51–74.
Rodotà, S. 2012. Il diritto di avere diritti. Laterza & Figli Spa, Roma-Bari.
Sacco, R. 1992. Che cos'è il diritto comparato? Giuffrè, Milano.
Samuel, G. 2014. An Introduction to Comparative Law Theory and Method. Hart Publishing, Oxford and Portland.
Schneider, G. 2020. Health Data Pools under European Policy and Data Protection Law: Research as a New Efficiency Defence?. Journal of Intellectual Property, Information Technology and Electronic Commerce Law 11(1): 49–67.
Shabani, M. 2022. Will the European Health Data Space change data sharing rules? Science 375(6587): 1357–1359.
Shabani, M. and S. Yilmaz. 2022. Lawfulness in secondary use of health data. Interplay between three regulatory frameworks of GDPR, DGA & EHDS. Technology and Regulation. 128–134.
Siens, M. 2018. Comparative Law. Oxford University Press, Oxford.
Slokenberga, S. 2022. Scientific research regime 2.0? Transformations of the research regime and the protection of the data subject that the proposed EHDS regulation promises to bring along. Technology and Regulation. 135–147.
Slokenberga, S., O. Tzortzatou, J. Reichel (eds.). 2021. GDPR and Biobanking: Individual Rights, Public Interest and Research Regulation across Europe. Springer International Publishing, Cham.
Soete, L. and J.-C. Burgelman. 2023. Reconciling Open Science withTechnological Sovereignty: Can the European Union doit?. Journal of Open Access to Law 11(1).

Stalla-Bourdillon, S. and A. Knight. 2017. Anonymous data v. personal data - a false debate: an EU perspective on anonymization, pseudonymization and personal data. Wisconsin International Law Journal 34(2): 284–322.

Steinsbekk, S.K., K.B. Myskja and B. Solberg. 2013. Broad consent versus dynamic consent in biobank research: Is passive participation an ethical problem. European Journal of Human Genetics 21: 897–902.

Suman, A.B and R. Pierce. 2018. Challenges for citizen science and the EU open science agenda under the GDPR. European Data Protection Law Review (EDPL), 4(3): 284–295.

Towned, D. 2021. Privacy. *In*: Laurie, E. et al. (eds.). The Cambridge Handbook of Health Research Regulation. Cambridge University Press, Cambridge.

Ukrow, J. 2018. Data protection without frontiers: On the relationship between EU GDPR and amended CoE Convention 108. European Data Protection Law Review 4(2): 239–247.

Vogel, Y.A. 2022. Stretching the Limit, the Functioning of the GDPR's Notion of Consent in the Context of Data Intermediary Services. European Data Protection Law Review 8(2): 238–249.

Wenkai, L. and P. Quinn. 2024. The European Health Data Space: An expanded right to data portability?. Computer Law & Security Review 52: 105913

Wilkinson, M.D. et al. 2016. The FAIR Guiding Principles for scientific data management and stewardship, available at: https://pubmed.ncbi.nlm.nih.gov/26978244/.

Willinsky, J. 2022. Copyright's broken promise: how to restore the law's ability to promote the progress of science. MIT Press, Cambridge (MA).

Zeno Zencovich, V. 2022. Data Protection[ism]. MEDIA LAWS 2: 1–9.

Zweigert, K. and H. Kötz. 1998. Introduction to comparative law. Clarendon Press, Oxford.

20

Resisting Transparency

The Paradox of the EU Vaccines Strategy

Deirdre Curtin and *Saverio Della Corte**

20.1 Introduction

Key decisions during the pandemic were often made behind closed doors by a "black box" of scientists, civil servants, and politicians. This led to calls for greater transparency and scrutiny (O'Malley et al. 2009). Transparency is understood as a fundamental prerequisite for public accountability and as an essential building block of the legitimacy of executive decision-making. In fact, without transparency in the relevant data, models, information, processes, actors, and decision-making, it will be extremely challenging, not to say impossible, for any oversight body to hold institutions to account. A focus on transparency is not only about its quality as a normative value but also important as a way of providing a more detailed understanding of government and governance processes. According to Pozen, transparency should not be understood as a coherent normative goal at all but rather as a complex social phenomenon capable of explaining the empirical impact on the distribution of powers based on the specific law (Koivisto 2016) in its institutional, historical, political, and cultural context (Pozen 2020).

Transparency turned out to be a core issue in the implementation of the Vaccines Strategy (the Strategy),[1] which represented a flagship initiative within the EU's public health response to the COVID-19 pandemic. Launched by the European Commission on 17 June 2020, the Strategy was a European plan aimed at ensuring safe and effective COVID-19 vaccines for the European Union (EU) and securing timely and equitable access to them. For this purpose, the Strategy was structured around two main pillars: 1. first, ensuring the *safety* of vaccines, through

European University Institute, Florence, Italy; Email: deirdre.curtin@eui.eu
* Corresponding author: saverio.dellacorte@eui.eu

[1] European Commission. 2020. EU strategy for Covid-19 vaccines. COM 245 final.

DOI: 10.1201/9781003475927-24

an accelerated regulatory procedure carried out by the European Medicines Agency (EMA), aimed to fast-track vaccine authorizations while maintaining high standards of quality, safety, and efficacy; 2. second, ensuring the *security* of vaccines, through a joint purchase mechanism, in which the European Commission was delegated by the Member States to negotiate with vaccine manufacturers on their behalf to pool their economic leverage and obtain better contractual conditions.[2]

The acute public interest in COVID-19 vaccines made it compelling to ensure that these objectives were achieved in a visible and transparent manner so that public trust and accountability would be enhanced. However, the dualism of the Strategy, based on the two pillars of *safety* and *security*,[3] posed serious challenges to the full implementation of transparency in its governance. The first pillar of *safety* represents a traditional objective of EU market governance based on Articles 168 (4) and 114 TFEU (Greer 2014; de Ruijter 2019; Delhomme and Hervey 2023) and, therefore, could rely on a consolidated legal and institutional EU framework based on the EMA with its long-established transparency policy. By contrast, the objective of the second pillar – *security* –aligns with the field of health policy, which is a national prerogative with EU law having only a weak complementary competence.[4] As a result, this second pillar was not implemented through the traditional channels of EU public authority but relied on a crisis-driven framework resulting from a temporary delegation of powers based on the emergency clause of Article 122 (1) TFEU[5] (Purnhagen et al. 2020; Hervey and de Ruijter 2020).

In our contribution, we argue that this dualism in the functioning of the Strategy brought about what we term a "transparency paradox". Transparency in the sense of visibility was explicitly engaged with only in the context of the authorization procedures within the first pillar, where it led to a virtuous inter-institutional dialogue between the EMA and EU oversight institutions, such as the European Ombudsman (EO) and the Court of Justice (CJEU). Conversely, and this is the nub of the paradox, transparency as a fundamental public value was largely ignored where it mattered most. This was within the second pillar of the Strategy, which entailed highly political and redistributive activities, such as the negotiation, development, and purchase of vaccines, with significant expenditure of public money and considerable executive discretion. Unveiling this paradox is in our view particularly compelling vis-à-vis the

---

2 Ibidem.

3 I draw on an established distinction between security and safety developed in the context of food law. Security can be considered as physical and economic access to resources, whereas safety is related to the fact that a certain product is not harmful for human health. On this distinction in food law, see FAO. 1996. Rome Declaration on World Food Security and World Food Summit Plan of Action: World Food Summit Rome, Italy, available at https://digitallibrary.un.org/record/195568.

4 See Article 168 (7) TFEU: "Union action shall respect the responsibilities of the Member States for the definition of their health policy and for the organization and delivery of health services and medical care. […]".

5 Article 122 (1) TFEU: "Without prejudice to any other procedures provided for in the Treaties, the Council, on a proposal from the Commission, may decide, in a spirit of solidarity between Member States, upon the measures appropriate to the economic situation, in particular if severe difficulties arise in the supply of certain products, notably in the area of energy."

Strategy's progressive institutionalization in the framework of the "European Health Union"[6] (Alemanno 2020), the ambitious policy project proposed and implemented by the Commission to make the European Union more resilient in the face of future health crises.

The rest of our contribution is structured as follows. We start by taking stock of the characteristics of the Strategy, with the aim of highlighting its functional, legal, and institutional dualism based on the two pillars of safety and security (Paragraph II). In the paragraphs that follow, we analyze the two pillars of the Strategy in order to unveil the transparency paradox. We first analyze how the principle of transparency was implemented throughout the pandemic in the first pillar, i.e., the COVID-19 vaccine authorization procedures conducted by the EMA (Paragraph III); in Paragraph IV, we carry out a similar analysis of the functioning of the joint procurement mechanism at the basis of the second pillar and its subsequent institutionalization in the aftermath of the pandemic. Finally, in our conclusions (Paragraph V), we carve out our understanding of the "transparency paradox" based on the previous analysis and explain its main components and ways forward.

20.2 The Dual Governance of the Vaccines Strategy

20.2.1 Providing for Safety*: Adapting Market Regulation to Crisis-Management*

In general, the authorization of a pharmaceutical product by the EMA is a lengthy process including several phases of scientific evaluation, with the result that it can take years before it is launched in the market.[7] The urgent need for a COVID-19 vaccine during the pandemic required that this same authorization process would take much less time. The first pillar of the EU Vaccines Strategy included, therefore, the not-easy task of conjugating the fast-track procedure with the necessary prerequisite of vaccine *safety*. In fact, doing so did not require any relevant legal innovation, but only adapting the flexibilities in the existing EU pharmaceutical regulation.[8] Pharmaceutical safety has historically been a central object of market governance based on Articles 114 and 168 (4) TFEU[9] (Greer 2014; Micklitz 2015). Originally introduced in the 1990s[10] (Hofmann 2008), the current EU regime for pharmaceutical

[6] European Commission. 2024. The European Health Union: acting together for people's health. COM(2024) 206 final.

[7] European Medicines Agency (EMA). Marketing authorization, available at https://www.ema.europa.eu/en/human-regulatory-overview/marketing-authorization.

[8] See European Commission. 2020. EU strategy for Covid-19 vaccines. COM 245 final: "Adapting the EU's regulatory framework to the current urgency and making use of existing regulatory flexibility to accelerate the development, authorization and availability of vaccines while maintaining the standards for vaccine quality, safety and efficacy."

[9] Pharmaceutical regulation can be ascribed to the broader area of risk regulation in the EU market governance.

[10] A centralized approach based on a European agency was introduced to overcome the risk that a regulatory system fully based on mutual recognition would lead to a race to the bottom in the quality and transparency of risk assessment.

authorization envisages a centralized procedure (Regulation (EC) n. 726/2004)[11] based on the scientific assessment of the EMA, applying to pharmaceutical products derived from biotechnology, orphan medicinal products, as well as medicines with certain active substances,[12] which couples with a decentralized procedure for other conventional medicines, based on the mutual recognition of national authorizations (Klamert 2018).

COVID-19 vaccines fell under the scope of the centralized authorization procedure according to Regulation (EC) n. 726/2004, as they are "medicinal products for human use containing a new active substance [...], for which the therapeutic indication is the treatment of [...] viral diseases".[13]

Unlike other pharmaceuticals, however, the rollout of COVID-19 vaccines required a fast-track process that would not be possible under the ordinary authorization procedures. For this reason, the Commission[14] resorted to the so-called Conditional Marketing Authorization (CMA) regulated by Commission Regulation (EC) n. 507/2006.[15] The CMA allows medicines to be approved based on less comprehensive data than normally required, provided that those are sufficient to demonstrate that the benefits outweigh the risks, primarily when addressing unmet medical needs or during public health emergencies.[16] The conditional authorization is valid for one year and can be renewed while additional data is gathered to confirm its benefits and risks. This way, the CMA balances the need for timely access to critical medicines with ongoing monitoring and data collection to ensure safety and efficacy.

The EMA played a central role in the market authorization of COVID-19 vaccines as it is the core institutional actor in the administrative procedure for the centralized authorization of pharmaceuticals. The centralized procedure revolves around the scientific assessment of the EMA, which is subsequently taken into consideration by the Commission for its final decision. Although the formal procedure attributes the decision-making power to the Commission, the EMA can be defined as the *de facto* decision-maker (Hofmann et al. 2011) because the Commission's decisions strongly rely on the EMA's scientific expertise. This makes the EMA a "quasi-regulatory" agency, i.e., an agency whose recommendations are not binding for the Commission but carry considerable weight for being technically and scientifically authoritative (Craig 2018).

To expedite the authorization under the CMA, the EMA conducted its scientific assessment through a rolling review procedure, whereby data are reviewed as soon

---

11 European Parliament and the Council. 2004. Regulation (EC) No 726/2004 of the European Parliament and of the Council of 31 March 2004 laying down Community procedures for the authorization and supervision of medicinal products for human and veterinary use and establishing a European Medicines Agency. OJ L136/1.

12 Ibidem, Article 3.

13 See the Point 1.3 Annex and Art. 3 of Regulation (EC) n. 726/2004.

14 European Commission. 2020. EU strategy for Covid-19 vaccines. COM 245 final.

15 European Commission. 2006. Regulation (EC) n. 507/2006 of 29 March 2006 of the conditional marketing authorization for medicinal products for human use falling within the scope of Regulation (EC) No 726/2004 of the European Parliament and of the Council. OJ L 92/6.

16 Ibidem, Art. 2.

as they are submitted without waiting for the full data package to be available.[17] This assessment on a rolling basis was conducted by two of the EMA's scientific committees – the Committee for Medicinal Products for Human Use (CHMP) and the Pharmacovigilance Risk Assessment Committee (PRAC) – under prior approval by the COVID-19 EMA pandemic Task Force (COVID-ETF), a newly-established body composed of EMA and national experts that supplemented the existing Committees during the pandemic. As a part of the EMA's emergency governance,[18] the COVID-ETF was designed to facilitate the coordination between the Commission and the Member States in the development, authorization and pharmacovigilance of COVID-19 vaccines and accelerating the authorization procedures. The COVID-19 ETF's tasks were subsequently taken over and institutionalized with the introduction of the Emergency Task Force (ETF)[19] established by Articles 15-20 of Regulation 123/2022[20] to offer scientific guidance on the development of treatments, vaccines, and clinical trial protocols (Brooks et al. 2022).

20.2.2 *Providing for Security: A Form of Temporary European Health Policy*

While the first pillar of the Strategy was based on an adaptation to the crisis response of well-established bodies and procedures of EU law, the second pillar was implemented through a crisis-driven mechanism with innovative characteristics. Unlike *vaccine safety*, the objective of *vaccine security* is a prerogative of national health policies, with no direct EU competencies. For this reason, the second pillar was based on Article 122(1) TFEU, which allows for the temporary conferral of exceptional powers to the EU in case of an emergency.[21] In April 2020, the Emergency Support Instrument (ESI) Regulation, an EU budget financial tool

---

17 European Medicines Agency (EMA). COVID-19 guidance: evaluation and marketing authorization. Available at: https://www.ema.europa.eu/en/human-regulatory-overview/public-health-threats/coronavirus-disease-covid-19/covid-19-public-health-emergency-international-concern-2020-23/guidance-medicine-developers-and-other-stakeholders-covid-19/covid-19-guidance-evaluation-marketing-authorization.

18 European Medicines Agency (EMA). EMA's governance during the COVID-19 public health emergency. Available at: https://www.ema.europa.eu/en/human-regulatory-overview/public-health-threats/coronavirus-disease-covid-19/covid-19-public-health-emergency-international-concern-2020-23/emas-governance-during-covid-19-public-health-emergency.

19 European Medicines Agency (EMA). Emergency Task Force (ETF). https://www.ema.europa.eu/en/committees/working-parties-other-groups/emergency-task-force-etf.

20 European Parliament and Council. 2022. Regulation (EU) 2022/123 of the European Parliament and of the Council of 25 January 2022 on a reinforced role for the European Medicines Agency in crisis preparedness and management for medicinal products and medical devices. OJ L 20/1.

21 Article 122 (1) TFEU: "1. Without prejudice to any other procedures provided for in the Treaties, the Council, on a proposal from the Commission, may decide, in a spirit of solidarity between Member States, upon the measures appropriate to the economic situation, in particular if severe difficulties arise in the supply of certain products, notably in the area of energy."

based on Article 122 TFEU, was amended[22] to introduce the possibility for the Commission to coordinate a centralized procurement procedure based on a previous intergovernmental agreement.[23] According to the intergovernmental agreement of June 2020,[24] the Commission was delegated to negotiate with the pharmaceutical companies the "Advance Purchase Agreements" (APAs), complex contractual arrangements involving the Commission, the Member States, and the pharmaceutical companies.

The APAs covered the negotiation, development, and final purchase of vaccine doses. Initially, the Commission would fund the R&D activities of certain pharmaceutical companies to develop and test the COVID-19 vaccines through the ESI funds.[25] This initial disbursement of European funds was meant to de-risk the essential investments and increase the speed and scale of vaccine manufacturing. In exchange for that, the pharmaceutical companies would ensure the Member States a right – or, depending on the contract, an obligation – to acquire a pre-determined stock of vaccine doses once authorized by the EMA.[26] The Commission would allocate the available stocks to the Member States in proportion to their population as soon as the manufacturers delivered them. Although the Commission funded the initial de-risking via EU money, the Member States remained responsible for the final purchase through their national budgets.[27]

Although the APA framework is not the first joint procurement scheme on a European level in the area of public health[28] (McEvoy and Ferri 2020), it nevertheless represented a unique governance experiment due to the high degree of centralization of decision making power conferred on the Commission, which

---

22 European Council. 2016. Council Regulation (EU) 2016/369 of 15 March 2016 on the provision of emergency support within the Union OJ [2016] L 70/1. The centralized procurement procedure of the Vaccines Strategy was included in the ESI Regulation through Council Regulation 2022/521 in April 2020 which amended part of the Council Regulation (EU) 2016/369 (Council Regulation (EU) 2020/521 of 14 April 2020 activating the emergency support under Regulation (EU) 2016/369 and amending its provisions taking into account the COVID-19 outbreak OJ [2020] L 117/3).

23 Article 4 (5) (b) ESI Regulation.

24 The intergovernmental agreement was subsequently ratified with a Commission Decision. See Commission Decision of 18.6.2020 approving the agreement with Member States on procuring COVID-19 vaccines on behalf of the Member States and related procedures C (2020) 4192 final.

25 European Commission. 2020. Annex to the Commission Decision on approving the agreement with Member States on procuring Covid-19 vaccines on behalf of the Member States and related procedures C (2020) 4192 final.

26 See Articles 3-4 C (2020) 4192 final ANNEX.

27 COM (2020) 245 final: "Once any of the vaccines supported proves successful, Member States will be able to acquire that vaccine directly from the producer on the basis of the conditions laid down in the APA. Allocation of access to vaccine doses between Member States would be according to a population-based distribution key."

28 A joint procurement mechanism was already envisaged by the so-called Health Threat Decision, adopted in the aftermath of the 2009 H1N1 crisis, (Decision No 1082/2013/EU of the European Parliament and of the Council of 22 October 2013 on serious cross-border threats to health and repealing Decision No 2119/98/EC [2013] OJ L 293/1). Similarly, joint procurement provisions were included also in the RescEU decision, which established legal measures for the stockpiling and allocation of supplies in times of emergency (Decision n. 2019/420 of the European Parliament and of the Council of 13 March 2019 amending Decision n. 1313/2013/EU on a Union Civil Protection Mechanism [2019] OJ L 77I).

would not normally be permitted given the limited competence of Article 168 TFEU (Della Corte 2023). Negotiating on behalf of the Member States meant that the Commission was empowered to choose the commercial partners it preferred, decide contractual conditions regarding prices, timing, and delivery of the products, and allocate the available vaccine doses among the Member States. On their side, the Member States were bound to procure vaccines exclusively within the APA framework, as the intergovernmental agreement explicitly envisaged an "obligation not to negotiate separately".[29] As the Member States were bound to purchase the vaccines negotiated under the APAs, the Commission *de facto* steered a huge amount of national funds, which needed to be coupled with the European money spent to de-risk the pharmaceutical companies' R&D efforts.

This exceptional conferral of powers on the EU executive was counterbalanced by an *ad hoc* governance structure, which differed from the consolidated governance in the first pillar of the Strategy. The core decisions regarding the APAs were made by the so-called Steering Group, a board composed of senior national officials co-chaired by the Commission and established within the DG SANTE.[30] Likewise, the negotiation team that signed the APAs comprized national experts from six Member States.[31] The intergovernmental nature of the decision-making process also characterized the subsequent institutionalization of the Strategy in a permanent operational framework (Della Corte 2023). In September 2021, the Commission established the so-called Health Emergency Preparedness and Response Authority (HERA),[32] which took over the task previously attributed to DG SANTE and the Steering Group. The HERA Board reflects the intergovernmental composition of the Steering group, as it comprises one high-level representative from each Member State, appointed by the Commission based on nominations by the relevant national authorities.[33]

The establishment of HERA represented one of the core initiatives in the context of the so-called European Health Union launched by the Commission in the aftermath of the pandemic.[34] Although some would have expected the new Health Union to establish a new public health agency taking the EMA as a model (Bartlett 2020). HERA was designed as an internal DG of the Commission characterized by a peculiar way of functioning.[35] In fact, HERA represents a unique body of EU governance based on a twofold nature. In ordinary times, HERA is a spending hub aimed at funding strategic projects and building up long-term preparedness against potential new public health threats (Anderson et al. 2021). In the event of a public

29 Article 7 C (2020) 4192 final ANNEX.

30 Article 2 C (2020) 4192 final.

31 Ibidem.

32 European Commission. 2021. Introducing HERA, the European Health Emergency Preparedness and Response Authority, the next step towards completing the European Health Union. COM (2021) 576 final; European Commission. 2021. Commission decision of 16.9.2021 establishing the Health Emergency Preparedness and Response Authority. C (2021) 6712 final.

33 Article 6 C (2021) 6712 final.

34 COM (2024) 206 final.

35 C (2021) 6712 final Recital 6 and Article 4.

health emergency, the new Council Regulation n. 2022/2372[36] provides that the Council can activate HERA's "crisis mode" and confer on the Authority additional emergency powers, including the power to sign APAs, such as during the Vaccines Strategy.[37]

20.3 A Revolution in the EMA's Transparency Policy

Having outlined the contours of the dualism found within the governance of the Strategy, we can now assess how transparency was implemented across its two pillars. Starting with the first pillar, we recall first the transparency policy that the EMA pursued since 2010. Although before 2010, the agency had maintained that data and documentation submitted for market authorization of medicinal products were confidential by default, it subsequently initiated efforts to negotiate with pharmaceutical companies to release redacted versions of clinical study reports on specific drugs for public access (Kim 2017).

In 2014, the EO initiated an inquiry to evaluate how the EMA should manage the public disclosure of safety-related documents for medicinal products. She determined that the EMA placed excessive emphasis on the commercial interests of pharmaceutical companies, rather than prioritizing the overarching public health interest.[38] She observed that although exceptions to accessing information exist, such as safeguarding privacy and protecting commercial interests, these exceptions are not applicable when there is a compelling public interest in disclosing the document. In particular, she found "that there is always a public interest in the disclosure of documents containing information on the safety and efficacy of medicines".[39] In response to the EO's recommendations, the EMA later introduced new proactive policies aimed at increasing transparency within the organization based on Regulation n. 1049/2001 and Article 15 TFEU.[40] As a result, the public is now able to request access to data submitted during the authorization process, and reports are available

36 European Council. 2022. Council Regulation (EU) 2022/2372 of 24 October 2022 on a framework of measures for ensuring the supply of crisis-relevant medical countermeasures in the event of a public health emergency at Union level OJ (2022) L 314/64.

37 Article 8 Council Regulation 2022/2372. These additional powers, however, are decided on a case-by-case basis through a Council Regulation, which activates the crisis mode and defines the type of activities, their duration, and their potential extension (Articles 3 and 4 Council Regulation 2022/2372).

38 European Ombudsman, Decision on Own-Initiative Inquiry OI/3/214/FOR Concerning the Partial Refusal of the European Medicines Agency to Give Public Access to Studies Related to the Approval of a Medicinal Product (8 June 2016, Case OI/3/2014/FOR) http://ombudsman.europa.eu/en/decision/en/68107.

39 Ibidem, para 32.

40 See EMA, European Medicines Agency Policy on Access to Documents (Related to Medicinal Products for Human and Veterinary Use) (2010, Policy/0043, EMA/110196/2006); See EMA, European Medicines Agency Policy on Access to Documents (2018, Policy/0043, EMA/729522/2016). The former European Ombudsman had made major recommendations to overhaul transparency procedures at the EMA. See European Ombudsman, Decision of the European Ombudsman Closing His Inquiry into Complaint 2560/2007/BEH against the European Medicines Agency (24 November 2010, 2560/2007/BEH) para 80.

online on the EMA's website (Coombes 2019). In 2015, an additional policy – the so-called Policy n. 0070 – was implemented to allow the publication of clinical data.[41]

Subsequently, in 2017, the EO launched another inquiry into the EMA's activities on her own initiative (Stein 2019).[42] She sought to investigate "pre-submission activities", a practice in which developers of medicinal products obtain scientific advice from the EMA before formally submitting marketing authorization requests of a specific medicinal product. While the EO recognized that this communication could be beneficial for public health outcomes, she cautioned that it was "important to avoid even the perception that the eventual opinions of EMA on medicines were influenced by these earlier interactions".[43] In her conclusion, she determined that the EMA should aim for the highest possible level of separation between individuals offering advice during the initial stages and developers and those who are later involved in evaluating the product for market authorization.[44]

This inter-institutional dialogue between the EMA and the EO also continued during the pandemic given the central role assumed by the EMA in the implementation of the Vaccines Strategy. In July 2020, the EO launched another investigation into the EMA, this time focusing on the COVID-ETF, which, as we saw above, was a task force created by the EMA to support the European Commission and the Member States in the development and regulation of COVID-19 medicines. Concerns were raised about the independence of the vaccine and treatment evaluations, as some members of the task force were also responsible for preparing draft evaluation reports on COVID-19 treatments for the EMA.[45] The EO essentially asked the EMA to confirm that it applies the same principles previously established for its pre-submission activities to the work of the COVID-ETF. She also urged the EMA to ensure transparency in all COVID-related activities, including the prompt publication of clinical data for the relevant products.[46] The EO has long been concerned that there should be a clear separation between those offering scientific advice to medicine developers and those later responsible for evaluating the marketing authorization

---

41 EMA, The European Medicines Agency Policy on Publication of Clinical Data for Medicinal Products for Human Use (2014, Policy/0070, EMA/240810/2013) www.ema.europa.eu/docs/en_GB/document_library/Other/2014/10/WC500174796.pdf.

42 It is worth noting that the current EO's practice of initiating inquiries has prompted criticism in some quarters.

43 European Ombudsman. 2019. Decision in Strategic Inquiry OI/7/2017/KR on How the European Medicines Agency Engages with Medicine Developers in the Period Leading Up to Applications for Authorizations to Market New Medicines in the EU (17 July 2019, OI/2017/KR) www.ombudsman.europa.eu/en/decision/en/116683.

44 Ibid. para 32. See the EMA's response to the inquiry: EMA. 2019. EMA Takes Note of the European Ombudsman's Decision on Pre-submission Activities' (19 July 2019) http://ema.europa.eu/en/news/ema-takes-note-european-ombudsmans-decision-pre-submission-activities.

45 European Ombudsman. 2019. Letter to the European Medicines Agency (EMA) Concerning the Transparency and Independence of the Work of the EMA in Supporting the Development and Evaluation of COVID-19 Medicines (29 July 2020, Case SI/5/2020/DDJ) available at http://ombudsman.europa.eu/en/correspondence/en/130852.

46 Ibidem.

application for the same product.[47] The EMA indeed implemented measures to ensure a clear separation between these two roles, and these revised measures also extend to COVID-19 medicines. Additionally, the EMA clarified that these protocols apply to the newly formed COVID-ETF, ensuring that no member plays a leading or prominent role during the pre-submission phase.[48]

During the pandemic and in response to both the EO inquiry and mounting public pressure,[49] the EMA adopted what it called exceptional measures to enhance the transparency of its regulatory activities regarding COVID-19 treatments and vaccines, whether already approved or under evaluation.[50] The EMA achieved this by reducing its typical publication timelines and releasing information it would not normally disclose for other medicines. This included making public the list of medicines that had received scientific advice or guidance from the COVID-ETF,[51] which is information that would typically remain unpublished. In response to the EO's inquiry, the EMA confirmed that measures ensuring independence in pre-submission activities were applied to the development of COVID-19 treatments. Additionally, it stated that it was implementing "exceptional measures with regard to medicines for COVID-19, speeding up standard publication timelines and providing more information than is usually the case".[52]

In general terms, the EO's method of regularly initiating its own investigations and providing specific recommendations has proven highly valuable for the EMA in advancing public access to its documents throughout various stages of the regulatory process.[53] This progressive inter-institutional dialogue is still ongoing. In June 2023, the EO started a strategic own initiative enquiry on the EMA's transparency policy

47 European Ombudsman. 2017. Strategic Inquiry into Pre-submission Activities Organised by the European Medicines Agency (Case OI/7/2017/KR, opened 17 July 2017), available at http://ombudsman.europa.eu/en/case/en/49999.

48 EMA. 2020. Mandate, Objectives and Rules of Procedure of the COVID-19 EMA Pandemic Task Force (COVID-ETF) (31 March 2020, EMA/166423/2020), available at: http://ema.europa.eu/en/documents/other/mandate-objectives-rules-procedure-covid-19-ema-pandemic-task-force-covid-etf_en.pdf.

49 See, e.g., an open letter from 15 October 2020 and EMA's answer to that letter: http://ema.europa.eu/documents/other/reply-open-letter-concerning-transparency-evaluation-vaccinescovid-19_en.pdf. The EMA publishes many responses to open letters on its website (http://ema.europa.eu/en/news-events/publications/open-letters).

50 EMA. 2020. Transparency: Exceptional Measures for COVID-19 Medicines, available at: www.ema.europa. eu/en/human-regulatory/overview/public-health-threats/coronavirus-disease-covid-19/treatments-vaccines/transparency-exceptional-measures-covid-19-medicines.

51 EMA. 2020. Treatment and Vaccines for COVID-19, available at http://ema.europa.eu/en/humanregulatory/overview/public-health-threats/coronavirus-disease-covid-19/treatments-vaccines-covid-19.

52 EMA. 2020. Reply to the European Ombudsman's Letter Concerning Transparency and Independence of the Work of the European Medicines Agency in Supporting the Development and Evaluation of COVID-19 Medicines (30 September 2020, EMA/516771/2020, available at http://ema.europa.eu/en/documents/other/reply-european-ombudsmans-letter-concerning-transparency-independence-work-europeanmedicines-agency_en.pdf.

53 See D Curtin, 'Opening Executive Technocratic Bubbles: Gusts of Transparency in a Turbulent Europe', in Francisco Miguel De Abreu Duarte and Francesca Palmiotto, Sovereignty, Technology and Governance after COVID-19 : Legal Challenges in a Post-Pandemic Europe (Hart Publishing 2022).

over clinical trial data (Curtin 2022).[54] The EMA's transparency policy, based on the 2016 Policy n. 0070 for proactive transparency, was suspended in 2018 to cope with the complications deriving from staff relocation after Brexit and increased workload from the COVID-19 pandemic. As the EMA announced the restarting up of its transparency policy, the EO reviewed how the EMA planned to address the backlog of clinical trial data accumulated within the downtime period from 2018 to 2023. She also sought clarification on the handling of data, since the EMA chose to prioritize the disclosure of information regarding "new active substances". The EMA outlined the reasons for this approach and the way stakeholders were involved in the related decisions. The EO encouraged the EMA to share its preliminary plans and next steps promptly to allow input from experts, industry, and civil society.[55]

The EO reviewed on the concrete reopening of the EMA's Policy n. 0070 too, which was relaunched in October 2023 for products containing new active substances, including COVID-19 vaccines. Given the huge backlog of clinical trial data stored during the period 2018–2023, the EMA faced problems in managing effective disclosure and publication. To cope with this backlog, the EMA implemented a chronological queue system for handling access requests and a "5-2 rule" that limited the number of documents applicants could request at once and restricted the number of simultaneous requests per individual. The EO assessed these disclosure techniques in her decision of December 2023.[56] The EO criticized the practice of chronological queueing for not aligning with EU legislation that mandates prompt processing of such requests, and found no legal basis for the arbitrary limitation of access based on the "5-2 rule", thus labelling it as maladministration. However, since the EMA was already working to phase out these temporary practices, the Ombudsman refrained from immediate recommendations and instead requested a progress report from the EMA.

This shift towards increasing transparency and accessibility of clinical trial data in the EU has encountered considerable resistance from the market. Prominent pharmaceutical companies have often opposed this trend, bringing legal actions to challenge the extent of the data that the EMA is permitted to disclose (Kim 2017). Their conviction is that the documents contain commercially sensitive information that the company's consider as their own proprietary data.[57] The courts dismissed their argument that a general assumption of confidentiality applied to such documents. They also concluded that the applicants had not provided sufficient evidence to show that the documentation in question posed a serious risk to their commercial interests, such as revealing business strategies or new scientific discoveries.

---

54 European Ombudsman, How the European Medicines Agency ensures proactive transparency concerning clinical trial data of medicines seeking EU marketing authorization (case SI/3/2023/MIK) https://www.ombudsman.europa.eu/en/opening-summary/en/171508.

55 European Ombudsman, Closing Note on the Strategic Initiative on how the European Medicines Agency ensures proactive transparency concerning clinical trial data of medicines seeking EU marketing authorization (SI/3/2023/MIK) https://www.ombudsman.europa.eu/en/doc/correspondence/en/177918.

56 European Ombudsman, Decision on how the European Medicines Agency (EMA) deals with requests for public access to documents (case 2243/2022/SF) https://www.ombudsman.europa.eu/it/decision/en/179137.

57 General Court. 5 February 2018. Case T-235/15 Pari Pharma v EMA EU:T:2018:65.

In subsequent cases,[58] two pharmaceutical companies attempted to overturn EMA decisions that permitted third-party access to their documentation, which included toxicology and clinical study reports submitted to the EMA in support of applications for the authorization of human and animal medicinal products. The applicants contended that competitors could use this data and should be considered commercially sensitive. However, the court ruled that no general presumption of confidentiality applies to documents submitted in support of a marketing authorization application. On appeal, Advocate General Hogan reached the opposite conclusion, stating that a general presumption of confidentiality did exist. He argued that allowing access to the clinical data submitted with the application would potentially harm the applicant's commercial interests.[59] Following Advocate General Hogan's opinion, the EMA released a press statement indicating that if the Court of Justice supported his opinion, the agency's transparency policies "would be impaired and have to be revised". The EMA emphasized that its existing policy promotes "trust in the EU regulatory network, broadens the scientific knowledge base, fosters the development of medicines and ultimately benefits public health".[60] This case provides an interesting example of an expert agency publicly advocating for its own transparency policies while legal proceedings were still ongoing. The Court of Justice ultimately upheld the General Court's decision, ruling that no automatic presumption of confidentiality exists. EU institutions and agencies retain however the authority to conduct individual assessments of documents to determine if they fall under the exceptions outlined in Article 4 of Regulation 1049/2001. In this instance, the EMA conducted such an assessment of the reports, resulting in a limited amount of redaction.

Overall, the proactive approach taken by the European Ombudsman (EO) through multiple inquiries, combined with the support of the General Court and the Court of Justice on appeal, has significantly advanced the EMA's move toward greater transparency in its reports, clinical trials, and other data, which used to have limited disclosure due to the potential commercial sensitivity of such information. By engaging in constructive dialogue with key accountability bodies – the EO and the courts – the EMA has adjusted its practices to promote more public access whenever possible. As a result, the agency's stakeholders have expanded beyond pharmaceutical companies seeking authorizations or participating in trials to now include the broader public. The public's interest in scientific documentation is much more widely acknowledged these days, including by the EMA itself, compared to a decade ago.

---

58 Case C-175/18 P PTC Therapeutics International v EMA; Case C-178/18 P MSD Animal Health Innovation and Intervet International v EMA.

59 Opinion of Advocate General Hogan. 11 September 2019. Case C-178/18 P MSD Animal Health Innovation and Intervet International v EMA EU:C:2019:710, paras 72 – 80. The concept of a general presumption of confidentiality has been the subject of considerable criticism, with scholars characterizing it as 'detrimental to accountability'. See HCH Hofmann and PL Sandberg, 'An Agenda for Transparency in the EU' (European Law Blog, 23 October 2019) http://europeanlawblog.eu/2019/10/23/an-agenda-for-transparency-in-the-eu.

60 EMA. 4 October 2019. EMA Management Board: Highlights of October 2019 Meeting' available at http://ema.europa.eu/en/news/ema-management-board-highlights-October-2019-meeting.

20.4 The APA framework: the revolution that never happend

Transparency constituted arguably an even more compelling precondition for overall public control and accountability in the APA framework due to the considerable political discretion involved in the negotiation, development, and purchase of vaccines, whereas the EMA's activities were limited by their technical nature and based on scientific authority as a source of legitimation (Kruck and Weiss 2023). In fact, claims for a transparent decision-making process were raised by civil society[61] and institutional bodies, such as the EO (Natsis 2020)[62] and the European Court of Auditors (Bauer-Babef 2024; EULawLive 2024) and recently reached the European Courts.[63] These claims translated into a dialogue between the Commission, the EU institutional accountability forums, and broader civil society actors - a dialogue lending itself to a comparison with the EMA's process of transparency-building as outlined above. While the EMA could rely on its long-established transparency policy, the APAs were based on an *ad-hoc* governance framework based on intergovernmental consensus. Did the Commission take the cue from the complaints of civil society and institutional watchdogs, as the EMA had virtuously done? Did this dialogue contribute to designing a consistent transparency policy? Although the APA framework was created during the public health crisis as a temporary mechanism, these questions remain compelling given the institutionalization of its crisis-driven patterns within HERA.

20.4.1 The transparency pitfalls of the APA framework

20.4.1.1 Disclosure of the members of the joint negotiation team

The first transparency challenge concerned the composition of the Joint Negotiation Team designated to conclude the APAs with the pharmaceutical companies.[64] During the negotiations, neither the identity of the experts nor their country of origin was disclosed to the public. Many argued that secrecy about the Joint Negotiation Team may have covered potential conflicts of interest with some pharmaceutical companies or certain Member States (Peigné 2021; Wegener 2022). A complaint was brought

---

[61] With this document, thirty-nine civil society organizations requested that the European Commission and EU national governments ensured the maximum degree of transparency in the EU's exchanges, negotiations and deals with pharmaceutical companies over COVID-19 vaccines.

[62] See European Ombudsman, Overview of European Ombudsman's Initiative Looking into the COVID-19 Response of the EU Administration (5 January 2021) available at <https://www.ombudsman.europa.eu/fi/news-document/en/136499>.

[63] See ECA. 2022. Special report 19/2022: EU COVID-19 vaccine procurement – Sufficient doses secured after initial challenges, but performance of the process not sufficiently assessed. <https://www.eca.europa.eu/en/publications?did=61899>.

[64] As we saw in Paragraph III, the Joint Negotiation Team was a limited group of national experts from six Member States with production capacities for vaccines, and they were responsible for the negotiation of the APA with the pharmaceutical companies.

to the EO,[65] who decided that there was no maladministration in the Commission's refusal to disclose the names because secrecy in this regard was meant to preserve the independence and privacy of the negotiators.[66] According to the EO, transparency needed to be counterbalanced with the protection of personal data protected by Article 4(1)(b) Regulation (EC) n. 1049/2001 and Regulation (EU) n. 2018/1725.[67] The arguments of the complainant did not establish a "specific purpose in the public interest" in the sense of Article 9 Regulation (EU) n. 2018/1725 capable of prevailing over the private interest in data protection.[68] However, the EO recommended publishing at least some information related to the experts, such as their nationality, which is possible without revealing their specific identities.[69] Following the EO's decision, the Commission responded by disclosing the nationality of the experts involved in the Joint Negotiation Team.[70]

A different interpretation of Article 4(1)(b) Regulation (EC) n. 1049/2001 and Article 9 (1) (b) Regulation (EU) 2018/1725 was given in the recent ruling *Courtois and Others v Commission* of the General Court on 17 July 2024.[71] The applicants sought the annulment under Article 263 TFEU of a Commission Decision, which granted only partial access to the declarations of interest made by the members of the Joint Negotiation Team without disclosing their personal details due to reasons of data protection. The applicants complained that they needed those personal details to verify the presence of any potential conflicts of interest. Unlike the EO, the General Court found that the applicants had adequately demonstrated a specific public interest under Article 9(1)(b) of Regulation 2018/1725 and the Commission failed to adequately counterbalance the public interest in detecting potential conflicts of interest and the protection of the negotiator's right to privacy. The General

65 European Ombudsman. 2021. How the European Commission ensures transparency in relation to 'advanced purchase agreements' negotiated with pharmaceutical companies for vaccines against COVID-19 (29 January 2021) available at < https://www.ombudsman.europa.eu/en/opening-summary/en/137486>.

66 European Ombudsman. 2021. Decision on how the European Commission ensures transparency in relation to the team responsible for negotiating 'advanced purchase agreements' with pharmaceutical companies for vaccines against COVID-19 (case 175/2021/DL) <https://www.ombudsman.europa.eu/en/decision/en/139613>.

67 European Commission. 2018. Regulation (EU) 2018/1725 on the protection of natural persons with regard to the processing of personal data by the Union institutions, bodies, offices and agencies and on the free movement of such data, and repealing Regulation (EC) No 45/2001 and Decision No 1247/2002/EC [2018] L 295/39.

68 European Ombudsman. 2018. Decision on case 175/2021/DL, para. 19. See Article 9 Regulation (EU) n. 2018/1725: "1. […], personal data shall only be transmitted to recipients established in the Union other than Union institutions and bodies if: […] (b) the recipient establishes that it is necessary to have the data transmitted for a specific purpose in the public interest […]."

69 Ibidem, para. 22.

70 European Ombudsman. 2021. Follow-up reply from the European Commission to the European Ombudsman's decision on how the European Commission ensures transparency in relation to experts negotiating 'advanced purchase agreements' with pharmaceutical companies for vaccines against COVID-19 (19 August 2021) available at <https://www.ombudsman.europa.eu/en/doc/correspondence/en/145953>.

71 Case T-761/21 Courtois and Others v Commission ECLI:EU:T:2024:477.

Court went even further in its reasoning, stating that even the simple transparency of the negotiations and the conclusion of the contracts "could, in fact, constitute a specific aim of public interest within the meaning of Article 9(1)(b) of Regulation 2018/1725".[72] As a result, the General Court's judgement broadened the reach of transparency considerably beyond the interpretation conducted three years earlier by the EO on a similar issue and must be considered the authoritative ruling on the matter of the names.

20.4.1.2 Disclosure of the content of the APAs

A second transparency claim was about disclosing the APA contracts. In the implementation of the Strategy, the Commission only published redacted versions of the APAs, with key information blacked out or missing. A European Parliament resolution of 21 October 2021 pointed out that partial disclosure caused "an unnecessary lack of transparency and trust in society" and urged for the full disclosure of the non-redacted versions of the APAs, as well as the names of members of the Negotiation Team.[73] This complaint was made despite the prior EO's decision of March 2021, in which the EO acknowledged the Commission's efforts towards greater transparency around the vaccine negotiations.[74] In particular, the EO considered that the publication of redacted versions of the APAs satisfactorily balanced transparency with the need to protect commercial interests under Article 4 (2) Regulation (EC) n. 1049/2001.[75] The EO also noted that it is justifiable that the Commission could not process requests for greater disclosure in time "given the considerable number of documents covered by the complainant's request and the additional workload faced by the relevant department in the context of the COVID-19 pandemic".[76] Nonetheless, she also urged that transparency requirements be safeguarded in the "ongoing and future vaccine negotiations" "to avoid similar situations" in the future.[77]

The EO's decision was recently revisited by the General Court in *Auken and Others v Commission*,[78] which offered a more granular interpretation of the commercial interest exception under Article 4 (2) Regulation 1049/2001. The decision partially upheld the request of some applicants to annul, pursuant to Article 263 TFEU, the Commission's decision to grant public access only to a redacted version of the APA. According to the applicants, the Commission wrongly applied the exception relating to the protection of commercial interests and infringed Articles 11(2) (*Freedom of*

---

72 Case T-761/21, Para 71.

73 European Parliament. 2021. Resolution 2021/2678 of 21 October 2021 on EU Transparency in the development, purchase and distribution of COVID-19 vaccines (2021) l. K and para. 6-8 available at <https://www.europarl.europa.eu/doceo/document/TA-9-2021-0435_EN.html>.

74 European Ombudsman. 2021. Decision on the European Commission's refusal to give public access to documents concerning the purchase of vaccines against COVID-19 (Joint cases 85/2021/MIG and 86/2021/MIG) < https://www.ombudsman.europa.eu/en/decision/en/141706>.

75 Ibidem, Para. 17,

76 Ibidem, Para 19.

77 Ibidem, Para. 20.

78 The case was published on the same day as Courtois and Others v Commission (17 July 2024). See Case T-689/21 Auken and Others v Commission ECLI:EU:T:2024:476.

expression and information) and 42 (*Right of access to documents*). The General Court upheld the applicants' complaint regarding the Commission's refusal to ensure access to the APAs' provisions on indemnification. The indemnification clauses provided that the Member States shall reimburse the pharmaceutical companies for any damages they may have to pay if their vaccines were found to be defective to cover the risks related to the accelerated vaccine development timeline. The General Court concluded that the Commission failed to prove that broader access to these clauses would genuinely harm the commercial interests of the pharmaceutical companies. A second irregularity detected by the General Court was related to the fact the redacted copies of the APAs did not include the definitions of "wilful misconduct" or "best reasonable efforts" and the provisions on vaccine donations and resales, as it did not explain how granting access to this information would actually and specifically harm those commercial interests.

20.4.1.3 *Disclosure of the text messages of the Commission's President*

Another concern regarding the APA framework was the high degree of informality of the negotiation procedures. In April 2021, the New York Times leaked information regarding the existence of an exchange of text messages between the Commission's President and the CEO of an important pharmaceutical company over the purchase of additional vaccine doses to cope with the shortages that the EU was facing in that timeframe (Stevis-Gridneff 2021). This "personal diplomacy" conducted directly by the Commission's President raised criticisms about the lack of transparency, as the Commission refused to disclose the text messages related to the negotiation. In September 2021, the EO was requested to open an inquiry,[79] which resulted in declaring that the Commission's conduct constituted maladministration.[80]

During the inquiry, the Commission defended its position, stating that the right of public access applies only to existing documents possessed by the institution and that there was no obligation for the institution to "preserve each and every document".[81] The Commission referred to its record-keeping policy, whereby a document must be registered only if it contains important information, whereas a text message or other forms of instant messaging are inherently short-lived documents not containing, in principle, important information.[82]

79 European Ombudsman, The European Commission's refusal of public access to text messages exchanged between the Commission President and the CEO of a pharmaceutical company on the purchase of a COVID-19 vaccine (16 September 2021) <https://www.ombudsman.europa.eu/en/opening-summary/en/146850>.

80 European Ombudsman. 2021. Decision on the European Commission's refusal of public access to text messages exchanged between the Commission President and the CEO of a pharmaceutical company on the purchase of a COVID 19 vaccine (case 1316/2021/MIG) available at <https://www.ombudsman.europa.eu/en/decision/en/158295>.

81 European Ombudsman. 2021. Recommendation on the European Commission's refusal of public access to text messages exchanged between the Commission President and the CEO of a pharmaceutical company on the purchase of a COVID-19 vaccine (case 1316/2021/MIG) available at <https://www.ombudsman.europa.eu/en/recommendation/en/151678#_ftnref8>.

82 Ibidem, Para. 10.

According to the EO, these motivations were unsatisfactory. The case was not about whether text messages should be registered or not,[83] but about a refusal to give public access to documents under Regulation (EC) n. 1049/2001. In fact, it appears that text messages can be considered "documents" within the meaning of the Regulation as long as their content may be "saved, copied or consulted after it has been generated".[84] Article 3 (a) refers to "any content whatever its medium [...] concerning a matter relating to the policies, activities and decisions falling within the institution's sphere of responsibility". This means that neither the medium nor the recording of a document matters, but only its content "and whether or not it relates to the 'policies, activities and decisions' for which the institution is responsible".[85]

For this reason, disclosure of text messages should not be limited to documents that fulfil the Commission's recording criteria. The Commission's search for the text messages was flawed and inadequate because it was limited to registered documents.[86] The EO recommended a new search, "making it clear that the search should not be limited to registered documents or documents that fulfil its recording criteria".[87] Although the Commission acknowledged that text messages are documents within the meaning of Regulation 1049/2001,[88] it did not inform the Ombudsman whether it had conducted the recommended search of non-registered documents with the result that as a matter of practice of text messages were excluded from the scope of Regulation 1049/2001".[89]

In the meantime, the failure to disclose text messages was brought before the criminal courts. On 14 October 2022, the European Public Prosecutor's Office (EPPO) confirmed that it was investigating the Commission's President for the acquisition of COVID-19 vaccines in the European Union,[90] following other investigations of Belgian Prosecutors for the usurpation of functions and titles, the destruction of public documents, conflict of interest, and corruption.[91] More recently, in February 2023, the New York Times managed to bring the so-called "*Pfizergate*" before the Court of Justice, for adjudication[92] alleging that the Commission refused access to

---

83 See, on this, European Ombudsman, How EU institutions, bodies, offices and agencies record text and instant messages sent/received by staff members in their professional capacity (SI/4/2021/MIG) <https://www.ombudsman.europa.eu/en/case/en/59322).

84 Case T-188/12, Breyer v Commission, ECLI:EU:T:2015:124 Para. 42.

85 Ibidem, Para. 16.

86 European Ombudsman, Decision (Case 1316/2021/MIG) Para. 9.

87 European Ombudsman, Recommendation (Case 1316/2021/MIG).

88 European Ombudsman. 2022. Reply of the European Commission to the Recommendation from the European Ombudsman regarding the European Commission's refusal of public access to text messages exchanged between the Commission President and the CEO of a pharmaceutical company on the purchase of a COVID-19 (27 June 2022) available at <https://www.ombudsman.europa.eu/en/doc/correspondence/en/157681>.

89 European Ombudsman. 2021. Decision (Case 1316/2021/MIG) Para. 18.

90 EPPO, Ongoing EPPO investigation into the acquisition of COVID-19 vaccines in the EU (14 October 2022) <https://www.eppo.europa.eu/en/media/news/ongoing-eppo-investigation-acquisition-covid-19-vaccines-eu>

91 EPPO, Investigation into the acquisition of Covid-19 vaccines: clarifications (17 May 2024) <https://www.eppo.europa.eu/en/media/news/investigation-acquisition-covid-19-vaccines-clarifications>

92 Case T-36/23: Action brought on 25 January 2023 — Stevi and The New York Times v Commission [2023] OJ C 112/37.

the text messages on unlawful grounds, as it breached Regulation (EC) N. 1049/2001 and Article 11 of the Charter of Fundamental Rights of the European Union, as well as the principles of good administration and the obligation to state reasons.[93]

20.4.2 HERA's missed revolution

The overall picture of transparency that can be drawn from the above analysis is quite negative. The disclosure of relevant information regarding the negotiation, development, and purchase of vaccines was resisted by the pharmaceutical companies, which could rely on their bargaining power to prevent the Commission from publishing any detail of the contracts which could have weakened their own commercial interests[94] (Mersh 2023). Resistance to disclosure also was rooted in the intergovernmental nature of APA governance, which made the disclosure of some relevant pieces of information, such as the members of the Joint Negotiation Team, a matter of national choice. More generally, the APA framework was characterized by a high level of informality to the detriment of a coherent transparency approach, which could have enhanced accountability and public trust in the Vaccine Strategy.

This state of affairs is in stark contrast with the first pillar of the Strategy, centred around the EMA, which implemented transparency as a key principle of the authorization procedures. This contrast was confirmed in the recent COVI Report,[95] presented by the European Parliament in July 2023. The Report points to the lack of transparency in the APAs and recommends that the Commission disclose "complete, timely and accurate information" on ongoing negotiations, concluded contracts, delivery schedules, terms of liabilities and indemnifications, and manufacturing locations, and provide the negotiators' identities and their declarations of interest.[96] Conversely, the Report deemed the EMA to have provided "clear, transparent, accurate and timely information on the authorization and supervision of COVID-19 vaccines and therapeutics, with unprecedented speed and frequency" with *ad hoc* measures to increase the transparency of its regulatory activities on COVID-19 vaccines and treatments.[97]

Against this backdrop, the recent institutionalization of the Vaccines Strategy through HERA could have represented an opportunity to solve the transparency flaws affecting the pandemic mechanism. The setting up of a permanent legal and institutional framework for pharmaceutical security could draw from the lessons

---

93 Hearing was held in October 2024.

94 "'When we were negotiating those contracts, we were forced to accept conditions, which were not necessarily what we wanted, but which were there. So we had no choice,' he said, adding that the priority was getting vaccines as fast as possible amid the crisis. 'Would those criticising us have accepted that we said: sorry, we don't want to purchase your vaccines because we want to be transparent. (…) Would that be an acceptable answer? Not at all. The priority was to save lives,' he concluded".

95 European Parliament. 2023. Report on the COVID-19 pandemic: lessons learned and recommendations for the future, 26 June 2023 (2022/2076(INI)) available at <https://www.europarl.europa.eu/doceo/document/A-9-2023-0217_EN.html>.

96 Ibidem, Para. 198-200 and 302-315.

97 Ibidem, Para. 319.

of the EMA and the evolution it underwent and start an analogous "transparency revolution" through a virtuous dialogue with the EU accountability forums. Instead, the revolution seems to be missing in action in HERA's emerging legal and operational framework.

HERA's lack of transparency appears structural and systemic. The new authority was established as an internal Commission DG rather than a standalone agency in order to avoid the lengthiness of the ordinary legislative procedure in the pandemic times. This means, however, that HERA is not subject to the transparency requirements envisaged in the Common Approach for the decentralized agencies.[98] This choice also cut off the European Parliament from supervizing HERA's governance. In its Resolution of 21 October 2021,[99] the Parliament claimed that establishing HERA as an internal office of the Commission was "another shortcoming that has undermined transparency and accountability for public spending and decision-making in the area of public health",[100] as it had negative repercussions on the definition of HERA's mandate and the public scrutiny of its activities. The recent COVI Report recommended that HERA be upgraded to a fully-fledged agency as a means of increasing its transparency and democratic control.[101]

The lack of a coherent transparency policy also affects the legal framework regulating the functioning and the internal organization of HERA. The Commission Decision of 19 June 2021 establishing HERA[102] includes no provisions on how transparency ought to be ensured in the operation of its internal governance, such as the HERA Board and the HERA Advisory Forum, whose members still remain undisclosed.[103] Several civil society organizations also complained about the lack of transparency in the interplay between HERA and the pharmaceutical industry grouped in the so-called Joint Industrial Forum and the recently adopted Critical Medicines Alliance.[104] The opacity of HERA's governance and practices must be seen as a critical alarm bell, as HERA acts as a spending and purchasing hub aimed at bolstering the EU's preparedness against future crises.

No transparency arrangements have been included either in Council Regulation (EU) n. 2022/2372, which specifies HERA's emergency powers in case of a public

---

98 Joint Statement on decentralized agencies https://european-union.europa.eu/system/files/2022-06/joint_statement_on_decentralised_agencies_en.pdf.

99 European Parliament, Resolution 2021/2678 of 21 October 2021 (n. 68) para. 11.

100 Ibidem.

101 European Parliament, Report on the COVID-19 pandemic (n. 91) para. 72.

102 C (2021) 6712 final (n. 44).

103 Andrea Renda and others, 'Improving the mission, governance and operations of the EU HERA' (January 2023) 1 CEPS in-depth analysis, < https://www.ceps.eu/ceps-publications/improving-the-mission-governance-and-operations-of-the-eu-hera/>

104 See Hera's mission should be "public health first" (European Economic and Social Committee, 26 January 2022) available at <https://www.eesc.europa.eu/en/news-media/news/heras-mission-should-be-public-health-first> "[...] 'We worry that HERA could become not a health but an industrial policy instrument hidden away in the Commission's structure. We see this as a real risk, as the Commission's proposal fails to guarantee proper public scrutiny. We believe that HERA's operations must clearly show that its primary objective is to improve public health and health security, and that it's not about economic considerations', said Piotr Kolczynski of the Standing Committee of European Doctors (CPME)".

health crisis. Article 5 of this Council Regulation refers to transparency but only as a tool to ensure intergovernmental control over the decisions of the Health Crisis Board ("provide all Member State representatives with equal access to information") rather than fostering public trust and accountability.[105] No reference to transparency was moreover made in relation to the declarations of interest of the Health Crisis Board's members whose disclosure to the public is not ensured by Article 6.

The new legal framework of HERA also seems to replicate the high degree of informality characterizing the APA negotiation procedures. Article 8 of Council Regulation (EU) n. 2022/2372 envisages a procedure of delegation of powers through intergovernmental agreements that resembles that of the APA framework.[106] This institutional choice means that the design of potential future joint procurement mechanisms is left to *ad hoc* intergovernmental agreements with the result that procedural safeguards for transparency and public control will be decided on a case-by-case basis by the participating Member States. No transparency clause was for example included in Article 8 for the disclosure of future public contracts or negotiating teams.

All in all, the "missed revolution" in the institutionalization process of the second pillar of the Vaccines Strategy did not allow the new framework of HERA to overcome the various forms of resistance to transparency that has affected the functioning of the APAs. HERA did not solve the problems relating to the excessive power of pharmaceutical monopolies, which raised the need to protect their commercial interests against the disclosure of relevant information regarding the procurement of vaccines, such as the content of APA contracts. Instead, the opaque governance of the new Authority seems to confirm this secrecy trend and even casts shadows on public funds invested in HERA's activities. Moreover, HERA rules confirm the intergovernmental patterns of the Strategy, with the result of leaving transparency to the outcomes of contingent intergovernmental bargaining.

20.5 Conclusions: Beyond the Transparency Paradox

Our analysis has shown that the dualism affecting the Vaccine Strategy also translated into a split way of coping with the challenges of transparency. The EMA has established an admirable reputation for transparency in its decision-making processes during the COVID-19 pandemic, exemplified by its proactive dissemination of clinical trial data and regulatory outcomes. This virtuous cycle of transparency-building contrasts sharply with the absence of a coherent approach to transparency in the governance first of the APAs and later of HERA. Unravelling the complex institutional landscape of the governance of the Vaccines Strategy allowed us to juxtapose the EMA's achievements with the transparency pitfalls evident in the institutionalization of the APA framework.

[105] In fact, the provision was not even included in the initial proposal (See Article 5 Proposal) but was added subsequently after the Political Agreement of 17 December 2021 with the Council (See Article 5 Political Agreement).

[106] Article 8 (2) Regulation (EU) n. 2022/2372.

The result of this juxtaposition is what we term a "transparency paradox". The paradox lies in the fact that transparency was lacking precisely where it was needed most, i.e., the APA framework. In fact, the APA empowered the Commission to steer the expenditure of an enormous amount of European and national resources and carry out highly discretionary tasks (Brooks and Geyer 2020) related to the negotiation, development, and procurement of vaccines in a very informal setting, with risks of collusion with pharmaceutical monopolies and moral hazards by the Member States. By contrast, the EMA had already undergone a process of progressive disclosure over the course of the previous decade and further enhanced its positive reputation in this regard in the wake of the pandemic.

Clearly, while it is understandable that the initial response to the pandemic crisis required fast and efficient decision-making, even at the expense of some procedural safeguards, well-established bodies such as the EMA could quickly fall back on their consolidated transparency policies. We rather assume that the "transparency paradox" refers to the overall process of inter-institutional dialogue and transparency-building that accompanies the subsequent institutionalization of the crisis response. In this second phase, opacity was not a necessity but a more or less deliberate policy choice - a road that did not need to be taken and with other routes available within easy reach.

Let us explicitly mention two possible paths that could have been followed instead and that would not have involved calling into question the current division of competencies between the EU and its Member States in health policies. First, HERA's governance would have been improved through the establishment of robust financial and administrative capabilities akin to the US Advanced Research Projects Agency (ARPA) rather than a small Commission DG. This approach could have helped diffuse major resistance against transparency, by rebalancing the public/private gap in terms of bargaining power and abandoning forms of intergovernmental collaboration. Second, even keeping the intergovernmental nature of HERA governance, there was no need to delegate the whole legal organization and procedural aspects of joint procurement to *ad hoc* intergovernmental agreements, which perpetuate crisis-driven patterns that hinder a coherent implementation of transparency. Rather, a more structured *ex-ante* legal framework could set the procedural pre-conditions in which intergovernmental consensus can be reached. By implementing these strategies, the EU could move towards a more transparent, accountable, and effective governance model in health policy, ensuring that the lessons learned during the pandemic translate into improved public health responses in the future. The revolution still lurks in the shadows.

References

Alemanno, A. 2020. Towards a European Health Union. Time to Level Up. European Journal of Risk Regulation 11: 721.

Anderson, M., R. Forman and E. Mossialos. 2021. Navigating the Role of the EU Health Emergency Preparedness and Response Authority (HERA) in Europe and Beyond. The Lancet Regional Health. Europe. 9: 100203.

Bartlett, O. 2020. COVID-19, the European Health Union and the CJEU: Lessons from the Case Law on the Banking Union. European Journal of Risk Regulation. 11(4): 781.

Bauer-Babef, C. 2024. General Court slams Commission for lack of transparency over COVID vaccines contracts. Euractiv available at https://www.euractiv.com/section/justice-home-affairs/news/eu-court-of-justice-slams-commission-for-lack-of-transparency-over-covid-vaccines-contracts/.

Brooks, E. and Geyer, R. 2020. The development of EU health policy and the Covid-19 pandemic: trends and implications. Journal of European Integration 42(8): 1057.

Brooks, E., A. de Ruijter, S.L. Greer and S. Rozenblum. 2022. EU Health Policy in the Aftermath of COVID-19: Neofunctionalism and Crisis-Driven Integration. Journal of European Public Policy 30(4): 721.

Coombes, R. 2019. European Drug Regulator Fears Return to Days of Data Secrecy. British Medical Journal 367: l 6133.

Craig, P. 2018. EU Administrative Law. Oxford University Press. New York. 164 –167.

Curtin, D. 2022. Opening Executive Technocratic Bubbles: Gusts of Transparency in a Turbulent Europe. *In*: De Abreu Duarte, F.M. and F. Palmiotto (eds.). Sovereignty, Technology and Governance after COVID-19 : Legal Challenges in a Post-Pandemic Europe. Hart Publishing.

de Ruijter, A. 2019. Eu Health Law & Policy: Expansion of EU Power in Public Health and Health Care. First edition, University Press.

Delhomme, V. and T. Hervey. 2023. The European Union's Response to the Covid-19 Crisis and (the Legitimacy of) the Union's Legal Order. Yearbook of European Law.

Della Corte, FS. 2023. The EU Vaccines Strategy: A Missed Opportunity for EU Public Health? European Journal of Risk Regulation 15(2): 333–345.

EULawLive. 2024. General Court rules Commission failed to provide sufficient access to Covid-19 vaccine purchase agreements. EULawLive available at https://eulawlive.com/general-court-rules-commission-failed-to-provide-sufficient-access-to-covid-19-vaccine-purchase-agreements/.

Greer, S.L. 2014. The Three Faces of European Union Health Policy: Policy, Markets, and Austerity. Policy and Society 33: 13.

Hervey, T. and de Ruijter A. 2020. The Dynamic Potential of European Union Health Law. European Journal of Risk Regulation 11: 726.

Hofmann H., Rowe G.C. and Türk A. 2011. Administrative Law and Policy of the European Union. Oxford University Press. 243 – 244.

Hofmann, H.CH. 2008. Mapping the European Administrative Space. West European Politics 31: 662.

Kim, D. 2017. Transparency Policies of the European Medicines Agency: Has the Paradigm Shifted? Medical Law Review 1; 25(3): 456,465.

Klamert, M. 2018. Public Health Policy. *In*: Herwig, CH. Hofmann, G.C. Rowe and A.H. Türk (eds.). Specialized Administrative Law of the European Union: A Sectoral Review. Oxford University Press. New York.

Koivisto, I. 2016. The Anatomy of Transparency: the Concept and its Multifarious Implications. European University Institute Working Paper, Max Weber Programme MWP 2016/09, available at https://cadmus.eui.eu/bitstream/handle/1814/41166/MWP_2016_09.pdf?sequence=1&isAllowed=y.

Kruck, A. and M. Weiss. 2023. The Regulatory Security State in Europe. Journal of European Public Policy 30(7): 1205, 1221.

McEvoy, E. and D. Ferri. 2020. The Role of the Joint Procurement Agreement during the COVID-19 Pandemic: Assessing Its Usefulness and Discussing Its Potential to Support a European Health Union. European Journal of Risk Regulation. 11:851. 1–13.

Mersh, A.H. 2023. Delsaux: HERA has taken steps to make pandemic preparedness transparent. Euractiv available at <https://www.euractiv.com/section/health-consumers/news/delsaux-hera-has-taken-steps-to-make-pandemic-preparedness-transparent/>.

Micklitz, H.W. 2015. Risk and EU Law. Edwar Elgar Publishing.

Natsis, Y. 2020. Transparency Is a Fundamental Pillar for the Success of the EU Vaccines Strategy. European Public Health Alliance, available at https://epha.org/transparency-is-a-fundamental-pillar-for-the-success-of-the-eu-vaccine-strategy/.

O'Malley, P., J. Rainford and A. Thompson. 2009. Transparency during Public Health Emergencies: From Rhetoric to Reality. Bulletin of the World Health Organization 87(8): 614.

Peigné, M. 2021. EU's 'Secret' Vaccine Negotiators: Who's Dealing with Big Pharma? EUObserver, available at <https://euobserver.com/health-and-society/153412>.

Pozen, D.E. 2020. Seeing Transparency More Clearly, Pub. Admin. Rev. 80: 326.

Purnhagen, K.P., A. De Ruijter, M.L. Flear, T.K. Hervey and A. Herwig. 2020. More Competences than You Knew? The Web of Health Competence for European Union Action in Response to the COVID-19 Outbreak. European Journal of Risk Regulation. 11: 297.

Stein, T. 2019. A Supervisory Agency of Its Own Making"? The Questionable Political Agenda of the Current European Ombudsman. Verfassungsblog.

Stevis-Gridneff, M. 2021. How Europe Sealed a Pfizer Vaccine Deal With Texts and Calls. New York Times available at <https://www.nytimes.com/2021/04/28/world/europe/european-union-pfizer-von-der-leyen-coronavirus-vaccine.html>.

Wegener, E. 2022. EU's Lack of Transparency Around Covid-19 Vaccine Negotiations Is Damaging Public Trust. Transparency International available at <https://ti-health.org/content/european-union-covid-19-vaccine-negotitaitons-contracts-transparency/>.

21

Realizing International Data Transfers for Health Research through Risk-based Regulation and Secure Infrastructure

Adrian Thorogood

21.1 Introduction

International data sharing is an essential component of cross-border collaborations in health research (Allea et al. 2021). It is the lifeblood of international collaboration in areas of genomics, clinical trials, and public health, and can enhance the scale and representativeness of databases, ensure results have broad application, and strengthen the integrity of scientific results. Rare disease research necessitates recruitment of small numbers of patients across multiple centres and countries to achieve necessary sample sizes. Personalized medicine involves stratifying study groups according to granular molecular subtypes or other risk factors, also requiring recruitment across institutions or countries. This stratification can be supported by comparisons of populations with different genetics, lifestyles and environments (Bentzen et al. 2021). In addition to the importance of data sharing to address the COVID-19 pandemic, various international collaborations leverage national healthcare data across countries to study prescription drug use, health system performance, and cancer incidence and survival (Oderkirk 2021). International data sharing with regulators around the world is also a key for monitoring medical product safety (Bentzen et al. 2023).

Open science and data sharing policy highlight how the broad and timely availability of research findings and data with the international scientific community can accelerate scientific progress by improving research integrity, and providing

Terry Fox Research Institute, Vancouver, Canada; Email: athorogood@tfri.ca

DOI: 10.1201/9781003475927-25

opportunities for collaboration and creative reuse of research outputs (Hallinan et al. 2021). Health research networks also rely on (commercial) cloud computing services to efficiently and securely store, remotely access, and process large datasets and access cutting-edge analytics, often involving data transfers to companies with servers located in or legal connections to third countries.

At the same time, health data derived from human patients and research participants are typically rich and sensitive health and genetic data, and so likely to be personal data captured by evolving data protection laws and human subjects research regulations. The EU General Data Protection Regulation (GDPR) applies to the processing of personal data, including pseudonymized (i.e., key-coded) data, which is often necessary to use in health research.[1] The GDPR and other data protection regimes place legal restrictions on cross-border transfers of personal data, to ensure that the export of data does not compromise individual privacy protection. Chapter V of the GDPR prohibits transfers of personal data outside the EU/EEA, unless a legal transfer mechanism is in place to ensure essentially equivalent protection for the data in the third country.

A number of scientific groups and commentators have highlighted in recent years that cross-border transfer restrictions make international collaboration in health research incredibly cumbersome if not downright impossible[2] (Bentzen et al. 2021; Peloquin et al. 2020; Hallinan et al. 2021). This has led to a range of calls and proposals for more clear and research-friendly regulation. Barriers to transfers have led to the pursuit of alternative strategies, including the use of Privacy-Enhancing Technologies (PETs). Uptake of PETs, however, has been limited by uncertainty over their legal effects and the limits they place on the utility of data for collaborative scientific endeavours. Legislative and infrastructure initiatives, like the recently approved European Health Data Space (EHDS), promise to facilitate secure and appropriately governed national and regional data sharing for secondary use. In so doing, however, these systems may introduce additional legal and socio-technical barriers to international data sharing and collaboration.

This paper will proceed as follows. First, it will describe the prominent EU GDPR regime regulating cross-border transfers. Next it will outline the problems of these sorts of regimes for health research and propose risk-based approaches to regulation in the context of health research. The paper will then explore the degree to which PETs and health data access regimes like the EHDS could facilitate or hamper legally-compliant international data sharing.

---

[1] European Union. 2016. Regulation (EU) 2016/679 of the European Parliament and of the Council of 27 April 2016 on the Protection of Natural Persons with Regard to the Processing of Personal Data and on the Free Movement of Such Data, and Repealing Directive 95/46/EC (General Data Protection Regulation).

[2] ALLEA, EASAC and FEAM joint initiative. 2021. International Sharing of Personal Health Data for Research, available at: https://easac.eu/publications/details/international-sharing-of-personal-health-data-for-research/.

21.2 Restrictions on Cross-Border Transfers

The GDPR generally restricts international transfers of personal data except where a comparable level of data protection can be ensured (Article 44). The rationale is to "ensure that international transfers of personal data from the EU/EEA do not deprive data subjects in the EU/EEA of those guarantees" (Knoppers et al. 2023). Commonly mentioned risks of transferring data to a third country are: (1) conflicting national laws; (2) access to data by foreign intelligence agencies and law enforcement; (3) lack of independent supervisory authorities; and (4) legal and practical difficulties for data subjects to enforce obligations and obtain administrative or judicial redress against foreign organizations.[3] Under the GDPR, international transfers of personal data are only exceptionally permitted under a legal transfer mechanism. Transfer mechanisms of particular relevance to health research include adequacy decisions, appropriate safeguards (e.g., standard contractual clauses), and derogations.

21.2.1 Adequacy

The clearest and simplest avenue to international transfers is an adequacy decision by the European Commission, which declares that a third country offers "essentially equivalent" protection to the GDPR, thus allowing transfers to those countries without any specific authorization.[4] Importers only have to worry about complying with their own country's laws to be deemed to comply with the GDPR. Adequacy decisions refer to the equivalency of a third country's entire legal framework, not to a specific processing operation. The European Data Protection Board (EDPB) maintains an Adequacy Referential to guide evaluations of third countries, which covers core concerns and oversight processes.[5] The comparative law approach is functional, focusing on both the aims and effectiveness of the third country's legal framework. Adequate has been interpreted to mean "essentially equivalent" by the Court of Justice of the European Union (CJEU).

Only a little over a dozen countries have been recognized as adequate to date, excluding many research intensive countries like China and Australia.[6] This is because of the high essentially equivalent standard, and practical issues that adequacy decisions are time and resource intensive and potentially influenced by politics (Stoddart et al. 2016). Adequacy decisions are not always all-or-nothing. Canada's adequacy decision applies only to the commercial sector. Japan's adequacy decision requires importers to take additional measures to protect certain types of special category data.[7] The Privacy Shield 2.0 permits transfers between the EU and certain

---

3 Art 45, Rec 104, GDPR.

4 Art. 45 GDPR.

5 European Commission. 2017. Adequacy Referential WP 254 Rev.01 (Adopted on 28 November 2017; as Last Revised and Adopted on 6 February 2018).

6 European Commission. Adequacy Decisions. Text. European Commission - European Commission. Accessed February 10, 2022.

7 European Parliament and Council. 2016. Commission Implementing Decision (EU) 2016/1250 of 12 July 2016 Pursuant to Directive 95/46/EC of the European Parliament and of the Council on the Adequacy of the Protection Provided by the EU-U.S. Privacy Shield.

self-certified commercial organizations in the US. Adequacy decisions can also be conditional and subject to regular review. In early 2024 the European Commission also reviewed past adequacy decisions of 11 countries, which were all maintained.[8] The United Kingdom (UK) obtained an adequacy decision post-Brexit, which appears to be more conditional than past decisions. The UK decision includes a 4-year sunset clause, alongside detailed monitoring and reporting obligations.[9] The UK now maintains its own list of adequate countries, but significant divergence from the EU list would likely risk its own EU adequacy decision. This highlights conceptual challenges if the EU's unilateral adequacy approach starts to scale internationally.

21.2.2 EU-US Privacy Shield 2.0

EU-US transborder flows have long been complicated by concerns over national surveillance and economic protectionism, as well as the fact that the US does not currently have a general data protection framework (Cory et al. 2020). Several agreements have been established over the past decades to facilitate transfers from the EU to the United States. The Safe Harbor was established in 2000 but invalidated by the CJEU in 2015 in the first *Schrems* case.[10] The successor Privacy Shield was similarly invalidated by the Court in *Schrems II*. The reasoning was that US national surveillance laws granted authorization to intelligence agencies and law enforcement to access electronic communications data without limitations of (strict) necessity and proportionality or providing sufficient judicial remedies, in contravention of fundamental rights under the EU Charter to respect for private life, data protection, and effective judicial protection.[11]

The Trans-Atlantic Data Privacy Framework, or Privacy Shield 2.0, has now been implemented for transatlantic data flows and commercial exchanges. The Framework includes strong rules for self-certified US companies to follow a strict set of privacy principles, under the oversight of the US Department of Commerce. The framework is also buttressed by *Executive Order 14086* to enhance safeguards for US signals intelligence established to address the concerns of the EU courts.[12] Binding safeguards are introduced to limit access by US intelligence agencies to what is strictly necessary and proportionate for a list of legitimate intelligence objectives. Additionally, an independent redress mechanism is established including complaints

[8] European Commission. 2024. Report on the First Review of the Functioning of the Adequacy Decisions Adopted Pursuant to Article 25(6) of Directive 95/46/EC | European Commission. 2024.

[9] European Commission. 2020. Commission Implementing Decision (EU) 2020/1308 of 21 September 2020 Determining, for a Limited Period of Time, That the Regulatory Framework Applicable to Central Counterparties in the United Kingdom of Great Britain and Northern Ireland Is Equivalent, in Accordance with Regulation (EU) No 648/2012 of the European Parliament and of the Council (Notified under Document C(2020) 6539) (Text with EEA Relevance).

[10] European Court of Justice. 2015. Maximilian Schrems v. Data Protection Commissioner. ECJ, Case C-362/14.

[11] European Court of Justice. 2020. Data Protection Commissioner v. Facebook Ireland Ltd, Maximillian Schrems. ECLI:EU:C:2020:559. Hereafter "Schrems II".

[12] The White House. 2022. Executive Order 14086 On Enhancing Safeguards For United States Signals Intelligence Activities. The White House.

investigation by a Civil Liberties Protection Officer who can decide on breaches and remediations, a Data Protection Review Court to handle appeals, and annual review by an Oversight Board. The European Commission issued an adequacy decision covering the framework on July 10, 2023, following non-binding opinions from the European Data Protection Board and European Parliament.[13]

There remains a risk that Privacy Shield 2.0 will be declared invalid by the CJEU in a subsequent case. Concerns have already been raised about whether the framework does enough to ensure necessity and proportionality of processing, as well as about the independent and transparency of the redress mechanism (Gerke and Rezaeikhonakdar 2023). At the same time, the European Commission has made significant efforts to ensure the framework survives judicial scrutiny and the CJEU appears to be taking "a more pragmatic and procedural approach to surveillance practice" in recent case law (Tricco 2024).

21.2.3 Appropriate Safeguards (e.g., Standard Contractual Clauses)

In the absence of an adequacy decision, the next alternative is using "appropriate safeguards, and on condition that enforceable data subject rights and effective legal remedies for data subjects are available".[14] These safeguards often require the approval of a tool by a data protection authority as well as a risk-assessment by a controller. Standard contractual clauses (SCCs) v 2.0 are an important type of appropriate safeguard used for health research, approved by the EU Commission.[15] They impose contractual obligations on a data exporter and a third country data importer to follow data protection principles, and provide third party beneficiary rights for data subjects, allowing them to directly enforce their rights against either party. The SCCs can be incorporated into a broader contract, as long as the other clauses do not contradict the standard clauses or undermine protection.[16]

The European Court of Justice in *Schrems II* upheld the validity of the SCCs, but only where the data exporter conducts a case-by-case "Transfer Impact Assessment" (TIA) before proceeding to determine if third country laws or practices undermine the contractual clauses in the specific context of the transfer.[17] A TIA must consider the specific circumstances of the transfer; the laws and practices of the third country – including the rule of law, general framework, and those requiring access by public authorities (which cannot be bound by contractual clauses); and the relevant contractual, technical, and organizational safeguards in place.[18]

[13] European Commission. 2023. Adequacy Decision for the EU-US Data Privacy Framework.

[14] Art. 46 (1) GDPR.

[15] European Commission. 2021. Annex to the Commission Implementing Decision (EU) 2021/914 of 4 June 2021 on Standard Contractual Clauses for the Transfer of Personal Data to Third Countries Pursuant to Regulation (EU) 2016/679 of the European Parliament and of the Council.

[16] European Commission. 2021. Commission Implementing Decision (EU) 2021/914 of 4 June 2021 on Standard Contractual Clauses for the Transfer of Personal Data to Third Countries Pursuant to Regulation (EU) 2016/679 of the European Parliament and of the Council. Para 3.

[17] Schrems II.

[18] European Commission. 2021. Annex to the Commission Implementing Decision (EU) 2021/914 of 4 June 2021 on Standard Contractual Clauses for the Transfer of Personal Data to Third Countries Pursuant to Regulation (EU) 2016/679 of the European Parliament and of the Councill, Clause 14 (b).

SCCs may still be used if a TIA concludes that a third country's laws or practices limit the effectiveness of the clauses for the transfer, if this can be remedied by supplementary measures.[19] These are contractual, organizational or technical measures *beyond* those required by the GDPR's general security and accountability requirements.[20] Supplementary contractual measures could require the data importer to take a lead role in providing information about the law and practices in its country, to report any requests, or to review the legality of any requests and to challenge if there are grounds to do so.[21] Generally only technical measures such as end-to-end encryption and pseudonymization are likely to be effective against access by foreign public authorities.[22] Even technical measures may fail to provide sufficient guarantees, as highlighted by recent high-profile cases against Big Tech (Manancourt and Kayali 2022).

21.2.4 Derogations

Article 49 derogations provide exceptional mechanisms for transfer where there is no adequacy decision or appropriate safeguard available. One derogation relevant for health research is the explicit consent of the data subject to a transfer or set of transfers, who is "informed of the possible risks of such transfers for the data subject due to the absence of an adequacy decision and appropriate safeguards".[23] The general conditions for GDPR consent apply here, namely that consent be "freely given, specific, informed and unambiguous"[24] and may be withdrawn at any time.[25] The EDPB view is that derogations should only be used where necessary and must still ensure the level of protection is not undermined, as required by Article 44. For consent to be specific and informed, the specific transfer and its circumstances must be made known to the data subject.[26] The consent should mention "all countries to which the personal data are being transferred to" and should also provide "information that in the third country there might not be a supervisory authority and/or data processing principles and/or data subject rights might not be provided for in the third country".[27] Another relevant derogation for health research is for transfers "necessary for important reasons of public interest".[28] Note that while the GDPR allows the use of health data for scientific research purposes or public health reasons under Article 9, the transfer derogations are more narrow (Hordern 2022).

---

19 Schrems II, para 134.

20 EDPB. 2020. Recommendations 01/2020 on Measures That Supplement Transfer Tools to Ensure Compliance with the EU Level of Protection of Personal Data | European Data Protection Board. Para. 50.

21 Ibidem, 40.

22 Ibidem, 2.

23 Art. 49 (1) (a), GDPR.

24 Art 4(11) GDPR.

25 Art. 7 GDPR.

26 EDPB. 2018. Guidelines 2/2018 on Derogations of Article 49 under Regulation 2016/679 | European Data Protection Board. 7.

27 Ibidem, 8.

28 Art. 49 (1) (d) GDPR.

21.3 Implications of Cross-Border Transfer Restrictions for Health Research

21.3.1 Problems

The current GDPR regime continues to present challenges for international data flows, in particular for health research which is often pursued through complex international collaborations and interactions between academia, government, and industry actors. The range of Chapter V transfer mechanisms is limited, as are the concrete implementations (e.g., adequacy decisions) provided for by the European Commission. There is persistent legal risk that approved implementations will continue to be invalidated by the courts. Guidance from data protection authorities has also interpreted the avenues for permitted transfers narrowly.[29] Existing transfer mechanisms, such as those that enable commercial data flows, often cannot be used in health research because of inapplicability, conflicts of laws, high costs of implementation, or conversative regulatory interpretations (Bentzen et al. 2021).

The problem is compounded by strict interpretations and persistent uncertainty over generic aspects of the GDPR applicable to health research, such as the broad definition of personal data (including pseudonymized data), and whether or not the GDPR applies to pseudonymized data that is de facto anonymized in the hands of a third party without the re-identification key (Peloquin et al. 2020). The general transparency requirements under the GDPR present an additional, indirect barrier to using mechanisms like the SCCs for transfers of healthcare data. The GDPR transparency obligations generally require controllers to inform data subjects of the intention to transfer data to third countries, and what transfer mechanism will be used.[30] Fully specifying this information in advance can be practically challenging for health research projects, particularly those who seek to eventually share data broadly. Exceptions are possible under the GDPR research exemption where information-giving requirements would be impossible or involve disproportionate effort.[31] In this case controllers may be able to justify providing more general information about international transfers, subsequently allowing a more flexible selection of transfer mechanisms.

Policy maker efforts have prioritized solutions for the private sector which are not fit for purpose for health research contexts.[32] For example, the Canadian adequacy decision applies only to the commercial sector only, and the US-EU Privacy Shield 2.0 only applies to the US private sector. The SCCs are recognized as a viable solution for transfer to private research institutions and private universities, as long as they agree that they do "not intend to share the data with public authorities subject to sovereign immunity, or process the data on a cloud operated by a US cloud provider"(Ursin and Bentzen 2021). Public sector research institutions in third countries often cannot

---

29 EDPB. 2018. Guidelines 2/2018 on Derogations of Article 49 under Regulation 2016/679 | European Data Protection Board.

30 Artt. 13,14 GDPR.

31 Art. 14 (5) (b) GDPR.

32 *Supra* n. 2, ALLEA, EASAC and FEAM joint initiative.

make use of appropriate safeguards due to statutory conflicts.[33] For example, the US National Institutes of Health are not able to accept clauses limiting sovereign immunity. Clauses on judicial redress and indemnification may also be problematic. Academic and non-profit recipients also may face challenges with regards to taking on the obligations and liabilities of the SCCs. For health research, TIAs typically need to be conducted on a project-by-project basis, and to cover many countries participating in a collaboration, which can be practically difficult, cumbersome, and legally risky.

Narrow interpretation of the Article 49 derogations also leaves limited avenues for health research unable to rely on adequacy or SCCs. In the urgent context of the COVID-19 pandemic, the EDPB admitted that the derogations of explicit consent as well as "transfers necessary for important reasons of public interest" may be used, though "mainly as a temporary measure due to the urgency of the medical situation globally" and not for "repetitive transfers of data to third countries part of a long lasting research project".[34] This suggests that even the context of a global pandemic is not a sustainable justification for using the derogations, and even temporary use in other pressing (if not necessary emergency) health research areas may not be justified.

21.3.2 Proposals for Facilitating Cross-Border Transfers for Health Research

The health research community has repeatedly argued for legislators and regulators to adopt a risk-based approach to cross-border transfers, as opposed to a zero-risk approach (Christakis 2024b). A zero-risk approach prohibits transfers even for low-level or theoretical risks of government access. A risk-based approach would consider the nature of the data, severity of the risk of access, and likelihood of the risk materializing.

A risk-based approach may mean adopting health-sector specific rules. Commentators have argued variously for health-sector specific adequacy decisions (Bradford et al. 2020), standard contractual clauses (Bradford et al. 2021),[35] and GDPR codes of conduct including provisions for transfers (Phillips et al. 2020). There have also been calls for existing derogations, namely explicit consent and important reasons of the public interest, to be interpreted more generously by data protection authorities for health research (Hordern 2022). In particular not permitting individuals to explicitly consent to transfers appears to run contrary to ethical principles of respect for human dignity and autonomy, and could also be viewed as legally disproportionate where participation has implications for individual rights to well-being.

---

[33] Ibidem.

[34] EDPB. 2018. "Guidelines 2/2018 on Derogations of Article 49 under Regulation 2016/679 | European Data Protection Board".13. Available at: https://edpb.europa.eu/our-work-tools/our-documents/guidelines/guidelines-22018-derogations-article-49-under-regulation_en.

[35] *Supra* n. 27, ALLEA, EASAC and FEAM joint initiative, 2021, 33.

Health research is done for important societal objectives, that should be balanced against risks to individuals. While health data are sensitive and merit special protection, in the context of health research they are typically pseudonymized, and in some cases can be demonstrated to be effectively anonymized. Data sharing can be subjected to robust security measures. Healthcare and health research organizations in many third countries are already subject to rigorous sector-specific data protection laws and have a demonstrated track record of trustworthiness. Some have also argued that there is a lack of evidence of actual access by intelligence agencies and law enforcement in health research data (Hallinan et al. 2021). Alternatively, health research must hope for general improvements on international transfers. One area of particular importance is efforts by governments to address the scope of trusted government access through international treaties.[36]

21.3.3 Privacy-enhancing Technologies to Circumvent or Facilitate Cross-Border Transfers

There is considerable interest in emerging Privacy-Enhancing Technologies (PETs) to facilitate data sharing and collaborative data analysis while mitigating privacy risks (Centre for Information Policy Leadership, 2023). These technologies could ultimately mean more sensitive data can be put to use by collaborations across institutions and countries for research and innovation, in particular in the area of machine-learning and AI. PETs include a range of different technical safeguards to protect privacy and security during data sharing and use, including differential privacy, synthetic data, federated learning, secure multi-party computing, and homomorphic encryption (The Royal Society 2023, Summary Table).

PETs can also facilitate legal compliance in the context of collaborative analyses by strengthening security safeguards, and supporting realization of the principle of privacy-by-design (The Royal Society 2021). There is an increasing interest from international organizations and policy-makers to promote PET uptake and standardization specifically in support of cross-border data flows and collaborations, albeit as a complement to, not a substitute for, robust legal protections (The Royal Society 2023). Some PETs could permit collaborative cross-border data use while circumventing transfers of personal data altogether (as well as associated restrictions or conditions). The principle of these types of PETs is to demonstrate that any data that are actually transferred are anonymized and thus not subject to personal data protection law. Another strategy is to provide the user with a secure form of remote access to (pseudonymized) data on servers with the source institution and country, i.e., a means of data visiting. The aim here is to enable data to be accessed and used while limiting copying and transfer, in line with evolving requirements on data localization. There are various forms of remote visiting, such as viewing data remotely or executing analytic workflows remotely on data that remain hidden to the user, or hybrids of the two. It is more likely this form of data access could still

[36] OECD. 2022. Declaration on Government Access to Personal Data Held by Private Sector Entities (OECD/LEGAL/0487).

be legally considered an international transfer, but could still facilitate transfer by demonstrably reducing privacy risks.

At least under the EU GDPR, data protection authorities are skeptical that PETs are able to circumvent transfers entirely; they tend to view PETs as "supplementary measures" which can help to ensure compliance with the level of protection required under EU law in a particular third country.[37] This skepticism has also been on display regarding cloud service platforms efforts to satisfy data localization requirements for the EU, including for health research platforms such as the French Health Data Hub. US cloud providers have worked hard to provide data localization on EU servers with strong technical safeguards such as client-side encryption to reduce risks of third-party government access (Christakis 2024a). But EU data protection authorities have still raised concerns about the "risk of extra-territorial access to data stored in Europe" where third-country legislation might have extra-territorial application (Christakis 2024a). Under this view it is not enough that the data is physically stored in the EU, but that the service provider doing so is immune from the application of foreign laws, and so lead to restrictions around headquarters and staff location, and foreign ownership. These demands may prove unworkable for cloud companies, and in any case they only address risks of compelled access and not direct (surreptitious) access by foreign intelligence agencies (Christakis 2024a). At the very least PETs offer a promising strategy to mitigate privacy and security risks in the context of international transfers, including the risk of surreptitious third-country government access where data are made available to a third party researcher or service provider.

Ultimately the particular technical configuration of a PET is a major determinant of its scientific utility, computing, and cost performance, as well as its compliance implications – including if the processing in question legally constitutes an international transfer of personal data (Hallinan et al. 2021). While PETs can facilitate sharing while protecting privacy, many technologies are at an early stage of development and testing, and face knowledge, expertise, standardization, and financial barriers to reaching widespread adoption. PETs should not be viewed as a silver bullet or "complete solution" for international transfers of health research data, but should rather be seen as a valuable component of an overall data governance framework (Bentzen et al. 2021).

21.4 The European Health Data Space: Implications for Cross-border Transfers

The European Health Data Space (EHDS) Regulation reached a political approval by the European Parliament in spring 2024.[38] The EHDS aims on the one hand to promote individuals' access to and control over their electronic health data, including

[37] EDPB. 2020. "Recommendations 01/2020 on Measures That Supplement Transfer Tools to Ensure Compliance with the EU Level of Protection of Personal Data | European Data Protection Board." January 2020. Available at: https://edpb.europa.eu/our-work-tools/our-documents/recommendations/recommendations-012020-measures-supplement-transfer_en.

[38] Council of the European Union,. 2024. "Proposal for a Regulation on the European Health Data Space - Analysis of the Final Compromise Text with a View to Agreement."

the ability to make it available to their healthcare providers at both national and EU levels.[39] It aims on the other hand to facilitate secondary use of electronic health data for research, innovation and policy-making. The discussion here will focus on secondary use within the EHDS and cross-border and international data sharing implications.

Researchers and other prospective data users can make a request to access health data primarily through a national Health Data Access Body (HDAB), who is required to issue a data permit authorizing access if the application meets certain criteria.[40] The EHDS introduces mandates for most health data holders (organizations holding electronic health data) to periodically report their data holdings to their national HDAB for publication in a data catalogue. Health data holders are also required to make data available for approved secondary uses. The HDAB acts as an intermediary, obtaining the data from the health data holder, and providing health data users access to data for an approved purpose in a secure processing environment.

In terms of data localization, the EHDS requires HDABs to process data for secondary use in the Union, or in a third country that provides adequate protection.[41] While this second possibility was added as part of the compromise text, it notably closes off other GDPR Chapter V transfer mechanisms for HDABs, such as standard contractual clauses.

In terms of cross-border data sharing, the EHDS envisages cross-border access requests spanning multiple data holders across multiple Member States. An applicant benefits from a streamlined procedure and can request access from its own HDAB, who will then share information about its decision with the respective HDABs in other jurisdictions. Each HDAB, however, remains responsible for access to data in its remit, though the stated aspiration of the EHDS is for a certain level of cross-border "mutual recognition" between HDABs, given the transparent and harmonized access procedure.[42]

International data users can only obtain access to the EHDS if their country meets the reciprocity principle (equivalent data access and protection).[43] One route for third countries to be recognized is to join the EHDS as authorized participants, demonstrating they provide health data access to EU applicants on the same terms, as approved by implementing act of the European Commission. This is a route for jurisdictions who already have close ties with the EU such as the UK and Switzerland. However, this ability to recognize third countries is deferred for 10 years. Another route is for the third country to demonstrate it provides similar data access, as approved by the EC. Overall there will be substantial delays on international data access to the EHDS, and access is likely to be precluded for the vast majority of countries. The silver lining is that enhanced national and European capacity, maturity, and scale for secondary use can still support the feasibility of international health research collaborations involving segregated analyses and not directly sharing health data.

39 European Commission. 2024. "Q&A on the European Health Data Space." Text. European Commission - European Commission. April 24, 2024.

40 Artt. 45-46 EHDS.

41 Art. 60 EHDS Regulation.

42 Art. 46 (3a) EHDS Regulation.

43 Art. 52 (5) EHDS Regulation.

Adding complexity is that the EHDS Regulation leaves open the discretion of Member States under Article 9(4) GDPR to apply additional restrictions or conditions to international transfers. Perhaps most surprising, the EHDS Regulation leaves it open to the Commission to restrict international transfers of *non-personal* health data held by HDABs, deemed by the EHDS Regulation to be "highly sensitive" in the meaning of the Data Governance Act.[44] These data are defined as health data with a beyond-reasonable likelihood of re-identification. It is unclear why this data merits protection from a risk-based perspective given the already strict anonymization standard under the GDPR. The Commission should generally avoid introducing additional restrictions for these data, which could hamper even those health research collaborations that rely only on exchange of anonymized artifacts.

21.5 Conclusion

This article argues that countries should generally avoid legal restrictions on cross-border transfers, particularly with regards to collaborative, international health research. Even where general restrictions may be justified (balanced against the significant economic costs of limiting data flows), there are several good policy reasons why health research data flows should be exempt. There is a global societal interest in advancing our understanding of human health and disease that needs to be balanced against individual privacy protections. Transfers restrictions alone do not fully protect against the harms they aim to address, namely access by foreign intelligence agencies or bad actors, and can potentially distract from more proximate privacy risks from within an individual's own country. There is also a lack of evidence these data are of interest to intelligence agencies. Any restrictions applied to health research data should be firmly rooted in a risk-based approach. Risks are generally low where data are robustly de-identified (pseudonymized or anonymized), adequate safeguards are applied, and the purpose is to generate generalizable knowledge not to make individual decisions.

It is more important than ever to both support and safeguard international health research data flows. More and more nations are strengthening their data protection regimes and looking to regulate international transfers. The United States, for example, has introduced national security restrictions on the bulk transfer of sensitive personal information (including health and human genomic data) to a subset of nations considered countries of concern.[45] Where transfer restrictions are in place or introduced, generous and clear legal exceptions should be provided to permit transfers for health research collaborations in the public interest subject to appropriate safeguards. While legal restrictions should generally be avoided, this does not preclude efforts and incentives to reduce unnecessary transfers within global health research networks, in line with principles of accountability, privacy-by-design and data minimization. To this end, the health research community should

[44] Art. 61 EHDS Regulation.

[45] Executive Office of the President,. 2024. Preventing Access to Americans' Bulk Sensitive Personal Data and United States Government-Related Data by Countries of Concern. Federal Register.

proactively pursue novel collaborative strategies and the adoption of PETs and socio-technical ecosystems like the EHDS, in line with the spirit of both open science and data protection law.

Acknowledgement

An initial version of this manuscript was developed within the framework of the project "Learning from Clinical Data (LinCDat)", as part of a Mercator Fellowship funded by the Deutsche Forschungsgemeinschaft (DFG, German Research Foundation)– 406103282. The overall project funding was awarded to Eva C. Winkler (University of Heidelberg). The funders had neither involvement in study design; in the collection, analysis and interpretation of data; in the writing of the report, nor in the decision to submit the article for publication. There was no additional external funding received.

I would also like to thank the LinCDat project leads Eva C. Winkler and Christoph Schickhardt for their support with obtaining funding and participating in the fellowship, as well as for conceptual counselling on the fellowship research and feedback on the manuscript.

References

Bentzen, Beate H., R. Castro, R. Fears, G. Griffin, Volker ter Meulen and G. Ursin. 2021. Remove Obstacles to Sharing Health Data with Researchers Outside of the European Union. Nature Medicine 27(8): 1329–33.

Bentzen, Beate H., Kvammen H. Olav and G. Ursin. 2023. Maximizing the GDPR Potential for Data Transfers: First in Europe. The Lancet Regional Health – Europe 27 (April). 100600.

Bradford, L., M. Aboy and K. Liddell. 2020. International Transfers of Health Data between the EU and USA: A Sector-Specific Approach for the USA to Ensure an 'Adequate'Level of Protection. Journal of Law and the Biosciences 7(1): lsaa055.

Bradford, L., M. Aboy and K. Liddell. 2021. Standard Contractual Clauses for Cross-Border Transfers of Health Data after Schrems II. Journal of Law and the Biosciences 8(1): lsab007.

Centre for Information Policy Leadership. 2023. Privacy-Enhancing and Privacy-Preserving Technologies:Understanding the Role of PETs and PPTs in the Digital Age, available at https://www.informationpolicycentre.com/uploads/5/7/1/0/57104281/cipl-understanding-pets-and-ppts-dec2023.pdf.

Christakis, T. 2024. The 'Zero Risk' Fallacy: International Data Transfers, Foreign Governments' Access to Data and the Need for a Risk-Based Approach. SSRN Scholarly Paper. Rochester, NY, 27, 15, 41–42.

Cory, N., D. Castro and E. Dick. 2020. 'Schrems II': What Invalidating the EU-U.S. Privacy Shield Means for Transatlantic Trade and Innovation. Information Technology and Innovation Foundation.

Council of the European Union. 2024. Proposal for a Regulation on the European Health Data Space - Analysis of the Final Compromise Text with a View to Agreement.

Gerke, S. and D. Rezaeikhonakdar. 2023. Privacy Shield 2.0—A New Trans-Atlantic Data Privacy Framework Between the European Union and the United States. Cardozo Law Review 45 (2).

Giske, U. and H.B. Bentzen. 2021. Open Science and Sharing Personal Data Widely – Legally Impossible for Europeans? Acta Oncologica. 60(12): 1555–56.

Hallinan, D., A. Bernier, A. Cambon-Thomsen, F.P. Crawley, D. Dimitrova, B.C. Medeiros, G. Nilsonne, S. Parker, B. Pickering and S. Rennes. 2021. International Transfers of Personal Data for Health

Research Following Schrems II: A Problem in Need of a Solution. European Journal of Human Genetics 29(10): 1502–9, 1503.

Hordern, V. 2022. Why Isn't Transferring Health Data across Borders for Research Purposes Easier?

Knoppers, B.M., A. Bernier, S. Bowers and E. Kirby. 2023. Open Data in the Era of the GDPR: Lessons from the Human Cell Atlas. Annual Review of Genomics and Human Genetics 24(1): 369–91.

Manancourt, V. and Laura Kayali. 2022. US-EU Data Transfers on Life Support after French Google Decision. POLITICO.

Oderkirk, J. 2021. Survey Results: National Health Data Infrastructure and Governance. Paris: OECD.

Peloquin, D., M. DiMaio, B. Bierer and M. Barnes. 2020. Disruptive and Avoidable: GDPR Challenges to Secondary Research Uses of Data. European Journal of Human Genetics 28(6): 697–705.

Phillips, M., F. Molnár-Gábor, J.O. Korbel, A. Thorogood, Y. Joly, D. Chalmers, D. Townend and B.M. Knoppers. 2020. Genomics: Data Sharing Needs an International Code of Conduct. Nature 578 (7793): 31–33.

Stoddart, J., B. Chan and Y. Joly. 2016. The European Union's Adequacy Approach to Privacy and International Data Sharing in Health Research. The Journal of Law, Medicine & Ethics 44(1): 143–55.

The Royal Society. 2021. Protecting Privacy in Practice: The Current Use, Development and Limits of Privacy Enhancing Technologies in Data Analysis.

The Royal Society. 2023. From Privacy to Partnership: The Role of Privacy Enhancing Technologies in Data Governance and Collaborative Analysis, 25.

Théodore Christakis. 2024. Toward a Risk-Based Approach? Challenging the 'zero Risk' Paradigm of EU DPAs in International Data Transfers and Foreign Governments' Data Access.

Tricco, Giovanni. 2024. The New Transatlantic Data Agreement Placed in Context: Decoding the Schrems Saga Within the Digital Economy. Journal of Law, Market & Innovation 3(1): 82–110, 108.

Ursin, G. and H.B. Bentzen. 2021. Open science and sharing personal data widely–legally impossible for Europeans? Acta Oncologica, 60(12): 1555–1556.

Index

S

T

U

V

For Product Safety Concerns and Information please contact our EU representative GPSR@taylorandfrancis.com Taylor & Francis Verlag GmbH, Kaufingerstraße 24, 80331 München, Germany